Conformational Properties
of
Macromolecules

MOLECULAR BIOLOGY

An International Series of Monographs and Textbooks

Editors: BERNARD HORECKER, NATHAN O. KAPLAN, JULIUS MARMUR, AND HAROLD A. SCHERAGA

A complete list of titles in this series appears at the end of this volume.

Conformational Properties
of
Macromolecules

A. J. Hopfinger

Department of Macromolecular Science
Case Western Reserve University
Cleveland, Ohio

1973

ACADEMIC PRESS   *New York and London*
A Subsidiary of Harcourt Brace Jovanovich, Publishers

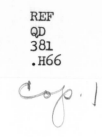

ACADEMIC PRESS, INC.
111 Fifth Avenue, New York, New York 10003

United Kingdom Edition published by
ACADEMIC PRESS, INC. (LONDON) LTD.
24/28 Oval Road, London NW1

Library of Congress Cataloging in Publication Data

Hopfinger, A   J
        Conformational properties of macromolecules.

        (Molecular biology; an international series of
monographs and textbooks)
        Includes bibliographical references.
        1.   Macromolecules.   2.   Conformational analysis.
I.   Title.   II.   Series.   [DNLM: 1.   Macromolecular
systems.   QU 55 H792c]
QD381.H66   539'.6   72–9328
ISBN 0–12–355550–7

To my mother, LaVerne

Contents

Contents

The application of empirical potential functions to determine the spatial properties and, in a few instances, the thermodynamic properties of macromolecules has been the source of considerable controversy among macromolecular scientists. On the positive side, theoretical conformational analysis, as this area of research has come to be known, has provided considerable information about the interrelationship between nonbonded molecular energetics and the spatial organization of macromolecules. From the negative point of view, theoretical conformational analysis lacks a solid foundation in that the essential potential energy functions cannot be ascertained as being reliable. Regardless of one's position on the merits or faults of theoretical conformational analysis, the fact remains that since 1962, when Liquori and co-workers carried out what can be considered the first conformational energy calculation on a polymer-chain molecule, over one hundred and thirty papers have been published involving the use of theoretical conformational energy calculations.

Theoretical conformational analysis has become a popular technique for investigating macromolecular structure. Therefore, it is most appropriate that a comprehensive discussion of this topic be available for reference and review by experienced researchers in the field, and to provide an inclusive guide to those workers just entering the field. This particular book provides a wealth of molecular parameters very often needed by structural chemists, X-ray crystallographers, biophysicists, physical chemists, and macromolecular scientists. Thus, while the central topic of the book, conformational analysis of macromolecules, is of explicit interest to only a select group of researchers, the data contained in the text is of general interest to most physical and biological scientists who are concerned with molecular structure.

The application of theoretical conformational analysis to the study of the structural properties of polypeptides is given considerable emphasis in the text. There are two reasons for this. First, the structural properties of polypeptides are of special interest to the author. Second, a large portion of the research which has been performed using conformational energy calculations has been concerned with the spatial organization of polypeptide chains. Notable are the studies of Scheraga and Ramachandran. Those readers who do not work in protein chemistry should consider the structural studies of polypeptide chains as examples of how conformational energy calculations can be applied. The conformational potential energy functions, geometric algorithms, transition theories, and fluctuation models presented in the text are not restricted to polypeptides, but, in general, are applicable to most macromolecules. At the same time, however, some of the discussions dealing with polypeptide chain structure may, in fact, be of special interest to protein chemists.

The book is divided into five chapters. Chapter 1 deals with the classification and characterization of macromolecular structure. Application of linear algebra to generate macromolecular conformations is discussed extensively. Chapter 2 deals with molecular energetics. Special consideration is given to the various types of semiempirical potential functions used in conformational energy calculations. All significant contributions to the total conformational free energy are discussed in detail. Chapter 3 presents the results of a number of significant studies involving the use of theoretical conformational analysis. It is left to the reader to judge, from the reported investigations, how reliable, and useful, conformational energy calculations might be for a specific type of structural investigation. Chapter 4 discusses structural transitions in macromolecules. Specific emphasis is given to the description of how conformational energy calculations might be useful to help describe the nature and properties of such transitions. Chapter 5 presents a number of research topics where the application of conformational energy calculations might be beneficial.

In writing this book I have endeavored to assemble a comprehensive ensemble of information about conformational energy calculations. I have tried to point out all of the tenuous assumptions which have been made in this area of research. Keeping this in mind, I believe this area of study has its greatest days yet ahead. There are several reasons for this optimism. The two most significant factors are (a) advances in computer technology and applied mathematical methods, and (b) the high quality of the researchers interested in structural calculations of macromolecules.

Last, I would like to thank my wife, Kathleen, for her long hours of patient proofreading, and my secretary, Ina Cael, for her perseverance in typing and proofreading the manuscript.

Chapter 1 | Macromolecular Geometry

I. Introduction

The architecture of macromolecules is as fascinating as it is complex. While the basic building blocks, the atoms, have rather simple bonding geometries, the versatility in the number of ways they may be combined coupled with the large numbers of atoms in macromolecules results in virtually an unlimited number of different spatial macromolecular assemblies. Some linear polymeric molecules adopt highly symmetric spatial structures such as helices. Other linear polymeric molecules, most notably the globular proteins, adopt spatial structures reminiscent of a discarded length of used baling wire. Perhaps even more incredible than the variations in known macromolecular structures is the fact that each globular protein has its own unique, yet seemingly random, gross structure. This would suggest that the spatial structure of macromolecules can be extremely sensitive to the functional needs of the molecule.

The flexibility of many macromolecular chains give rise to an ensemble of spatial geometries not having analogs in the world of "micromolecules." Among these phenomena are (i) chain folding in which a long polymeric molecule folds back on itself to form a minicrystal, (ii) interchain supercoiling in which two or more macromolecules wrap around one another to generate a braided-rope structure, and (iii) extended aggregation in which polymeric chains are cross-linked to form two-dimensional networks which act as unique molecular entities.

Much of the commercial value of polymers derives from the geometric versatility of these molecules. The question to which we will address ourselves in this chapter is, "How can we classify and generate macromolecular structures?" We will say nothing about which geometric structure is most probable for a given macromolecule, but rather characterize the nature of those structures which are possible.

II. Classification of Macromolecular Structure

A. General Comments

It is very difficult to describe precisely macromolecular geometries because of the immense size of these molecules and the large number of possible degrees of structural freedom. In order to have the capability of describing the spatial properties of any linear macromolecule four levels of structural specification have been defined. In the next section these four levels of structural identity are discussed in detail.

B. Levels of Structural Identity

1. Primary Structure

This is the structural formula of the monomer units, including the bond lengths and bond angles of the chemical bonds, and a list as to how the monomer units are joined to one another. Ultimately the primary structure dictates the resultant spatial geometry of the macromolecule. That this is the case is conceptually clear when one considers the possible change in total macromolecular geometry that can accompany a change from, say, tetrahedral to trigonal symmetry in one or more backbone bond angles. An even more obvious factor which can modify the total macromolecular geometry is interchanging different types of monomer units along the macromolecular chain.

Of fundamental importance in being able to predict macromolecular structure, or even rationalize experimental findings, is the availability of a reliable set of values for bond lengths and bond angles as well as the deviation values. The sensitivity of total macromolecular geometry to small changes in the values of bond lengths and bond angles (of the magnitudes of the observed deviations) is, as one would expect, specific to the particular macromolecule. A limited number of studies indicate that the resultant macromolecular geometry is rather insensitive to small changes in the values of bond lengths and bond angles. Table 1-1 contains a list of bond lengths and bond angles.

2. Secondary Structure

This is the manner in which monomer units are folded to generate simple geometric structures. For linear nonbranching macromolecules, which are the only type considered in this book, virtually all secondary structures possess cylindrical symmetry. In fact, nearly every secondary structure is helical. There is some debate as to whether or not β and zigzag structures are really helical. Mathematically, these structures can be considered helices having elliptical cross sections. Thus a discussion of secondary structure must include a description of the properties of helices.

TABLE 1-1

A General List of Acceptable Bond Lengths and Bond Angles in Some Specific
Classes of Molecular Structure

I. Bond lengths

Group	Bond length (Å)	Group	Bond length (Å)
C—C		C==O	
(a) Any molecule	1.54	(a) Carbonyl	1.27
C==C		(b) Carboxyl	1.25
(a) Aromatic rings	1.38	C=O	
(b) Olefinic linear chains	1.40	(a) Carbon near sp²	1.23
C=C		(b) Carbon near sp	1.16
(a) Any molecule	1.34	C—S	
C≡C		(a) *n*-Paraffins	1.82
(a) Any molecule	1.21	(b) Cyclic molecules	1.75
C—N		C—F	
(a) Polyamides	1.49	(a) Any molecule	1.39
(b) All other molecules	1.47	C—Cl	
C==N		(a) Any molecule	1.77
(a) Polypeptides	1.33	N=O	
(b) Five- and six-membered rings	1.36	(a) Any molecule	1.20
C=N		N—H	
(a) Any molecule	1.27	(a) Any molecule	1.01
C≡N		O—H	
(a) Any molecule	1.16	(a) Any molecule	0.96
C—H		O—F	
(a) *n*-Paraffin chains	1.09	(a) Any molecule	1.40
(b) Ethylenic chains	1.06	O—Cl	
(c) Aromatic rings	1.08	(a) Any molecule	1.68
C—OX		H—S	
(a) X = hydrogen	1.33	(a) Any molecule	1.33
(b) X ≠ hydrogen	1.42	S–S	
		(a) Any molecule	2.05

TABLE 1-1 (continued)

II. Bond angles

Group	Bond angle (deg)	Group	Bond angle (deg)
C—C—C		C—N=O	112.0–116.0
(a) Short linear chains	109.5–110.5	O—N—O	122.5–135.0
(b) *n*-paraffin chain	111.0–115.0	Cl—N=O	113.2
polymers (backbone)		O—N=O	113.5–119.6
H—C—H	109.5–112.0	C—O—C	110.0–114.0
H—C—F	111.6	Cl—O—O	105.6–109.3
H—C—Cl	111.8	H—O—C	106.7–109.0
F—C—F	108.0–111.0	Cl—O—Cl	110.8
Cl—C—Cl	109.5–113.0	F—O—F	103.2
H—C—C	110.2		
F—C—C	109.4	O—P—O	101.0–103.0
Cl—C—F	100.0	O—P—O′	114.7–117.3
		C—P—C	100.4
F—C(sp²)=C	125.2	O—P—S	116.5
N—C(sp²)=S	124.8		
C—C(sp²)—S	125.0	H—S—C	100.3
N—C(sp²)=O	121.5	O—S—O	120.0
O—C(sp²)=O′	124.3	Cl—S—Cl	101.0
C—C(sp²)=O	122.0–124.5	Cl—S—S	105.0
		F—S—F	93.5

In constructing Table 1-1 typical representative values for bond lengths and angles are reported. Obviously there can be deviations in these "typical" values. This is especially true for the bond angles where, in some cases, a range of values had to be given.

The simplest helix can be generated by uniformly winding a string around a cylinder of constant radius as shown in Fig. 1-1. As indicated in Fig. 1-1 there are three fundamental parameters which uniquely describe such a simple helix: r, the radius of the helix, l, the axial advancement of the helix for one complete turn, and ω, the frequency constant of rotation. Using these three parameters we can construct the parametric equations, in t, for such a simple helix.

$$x = r \sin \omega t, \qquad y = r \cos \omega t, \qquad z = lt \qquad (1\text{-}1)$$

The parameters ω and l can be defined, for a polymer composed of a sequence of identical monomer units, in terms of n, the number of monomer units per turn, $\hat{\omega}$, the rotational advancement per monomer unit, and $\hat{l}$, the axial advancement per monomer unit.

$$\hat{\omega} = n\omega, \qquad \hat{l} = nl$$

so that

$$x = r \sin n\hat{\omega}t, \qquad y = r \cos n\hat{\omega}t, \qquad z = n\hat{l}t \qquad (1\text{-}2)$$

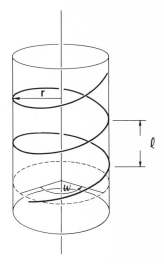

Fig. 1-1. A simple helix generated by wrapping a string about a cylinder. The helical parameters, r, l, and ω are also defined.

It is possible to establish a relationship between the helical parameters (r, ω, l) and the internal geometry of the helix including internal bond rotations (1). Using Fig. 1-2 as a reference, we can establish a relationship between (r, ω, l) and (α, d, χ). To accomplish this, first let us choose sets of right-handed rectangular coordinate systems in the following way (Fig. 1-2c).

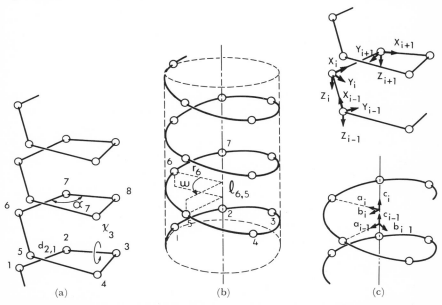

Fig. 1-2. The helical geometries needed to relate (r, ω, l) to (α, d, χ). (a) Internal co-ordinates, (b) cylindrical coordinates, (c) the two sets of rectangular coordinates.

The origin of the ith coordinate system coincides with the location of the ith atom such that the X_i axis coincides with the real or virtual directed bond, $d_{i+1,i}$, defined by atoms i and $i + 1$. The Y_i axis lies in the plane determined by the two bonds $d_{i,i-1}$ and $d_{i+1,i}$ in such a way that the angle between r_i and the positive direction of Y_i is acute. The positive direction of Z_i is taken so as to make the coordinate system right handed. Then coordinate system i can be transformed into coordinate system $i - 1$ by a linear operation

$$\mathbf{U}_{i-1} = \mathbf{A}_i \mathbf{U}_i + \mathbf{b}_{i,\,i-1} \tag{1-3}$$

where $\mathbf{U}_i$ is a column vector containing elements X_i, Y_i, Z_i;

$$\mathbf{b}_{i,\,i-1} = \begin{pmatrix} d_{i,\,i-1} \\ 0 \\ 0 \end{pmatrix}$$

and

$$\mathbf{A}_i = \begin{pmatrix} 1 & 0 & 0 \\ 0 & \cos\chi_i & \sin\chi_i \\ 0 & -\sin\chi_i & \cos\chi_i \end{pmatrix} \begin{pmatrix} \cos[-(\pi - \alpha_i)] & \sin[-(\pi - \alpha_i)] & 0 \\ -\sin[-(\pi - \alpha_i)] & \cos[-(\pi - \alpha_i)] & 0 \\ 0 & 0 & 1 \end{pmatrix}$$

or for $\alpha_i > \pi/2$ as is normally the case

$$\mathbf{A}_i = \begin{pmatrix} -\cos\alpha_i & -\sin\alpha_i & 0 \\ \sin\alpha_i\cos\chi_i & -\cos\alpha_i\cos\chi_i & -\sin\chi_i \\ \sin\alpha_i\sin\chi_i & -\cos\alpha_i\sin\chi_i & \cos\chi_i \end{pmatrix}$$

An alternate set of right-handed rectangular coordinate systems (a_i, b_i, c_i) may also be defined. For all these frames the origins of the ith coordinates coincide with the foot of the perpendicular to the axis of the helix drawn from the ith atom, the a_i axis being on this perpendicular with its positive direction pointing toward the ith atom. The c_i axis lies on the axis of the helix and the b_i axis is perpendicular to both a_i and c_i. The geometry is shown in Fig. 1-2c. The transformations of (a_i, b_i, c_i) to $(a_{i-1}, b_{i-1}, c_{i-1})$ can be expressed as

$$\mathbf{V}_{i-1} = \mathbf{C}_{i-1} \mathbf{V}_i + \mathbf{D}_{i,\,i-1} \tag{1-4}$$

where $\mathbf{V}_i$ is the column vector containing the a_i, b_i, c_i;

$$\mathbf{D}_{i,\,i-1} = \begin{pmatrix} 0 \\ 0 \\ l_{i,\,i-1} \end{pmatrix} \tag{1-5}$$

and

$$\mathbf{C}_{i-1} = \begin{pmatrix} \cos\omega_{i-1} & -\sin\omega_{i-1} & 0 \\ \sin\omega_{i-1} & \cos\omega_{i-1} & 0 \\ 0 & 0 & 1 \end{pmatrix} \tag{1-6}$$

Let us now establish a relationship between the two types of coordinate systems referred to above. The transformation of $\mathbf{V}_j$ to $\mathbf{U}_j$ has the general form

$$\mathbf{U}_j = \mathbf{T}_j(\mathbf{V}_j + \mathbf{S}_j) \tag{1-7}$$

where $\mathbf{S}_j$ can be immediately identified to be

$$\mathbf{S}_j = \begin{pmatrix} -r_j \\ 0 \\ 0 \end{pmatrix} \tag{1-8}$$

and $\mathbf{T}_j$ remains completely unspecified for the moment with the exception that we know it must be orthogonal:

$$\mathbf{T}_j = \begin{pmatrix} t_{11} & t_{12} & t_{13} \\ t_{21} & t_{22} & t_{23} \\ t_{31} & t_{32} & t_{33} \end{pmatrix} \tag{1-9}$$

However, from the way the two sets of coordinate frames have been defined we can learn much about $\mathbf{T}_j$. From Eqs. (1-3) and (1-7) we have

$$\mathbf{T}_{j-1}(\mathbf{V}_{j-1} + \mathbf{S}_{j-1}) = \mathbf{A}_j \, \mathbf{T}_j(\mathbf{V}_j + \mathbf{S}_j) + \mathbf{b}_{j,\,j-1} \tag{1-10}$$

or for

$$\mathbf{T}_{j-1} = \mathbf{T}_j = \mathbf{T} \quad \text{and} \quad \mathbf{S}_{j-1} = \mathbf{S}_j = \mathbf{S}$$
$$\mathbf{V}_{j-1} = \mathbf{T}^{-1}\mathbf{A}_j \, \mathbf{T}\mathbf{V}_j + (\mathbf{T}^{-1}\mathbf{A}_j \mathbf{T} - I)\,\mathbf{S} + \mathbf{T}^{-1}\mathbf{b}_{j,\,j-1} \tag{1-11}$$

where I is the identity matrix. This then implies all the $\mathbf{A}_i$, etc. are identical. Hence, there is no further need for subscripts due to the uniform geometry. We will see shortly that this restriction only limits our discussion to helices for which identical *virtual* bonds can be constructed for a uniform set of backbone atoms. Equation (1-11) must now be equivalent to Eq. (1-4) and as a consequence,

$$\mathbf{C} = \mathbf{T}^{-1}\mathbf{AT} \tag{1-12}$$

and

$$\mathbf{D} = (\mathbf{T}^{-1}\mathbf{AT} - I)\,\mathbf{S} + \mathbf{T}^{-1}\mathbf{b} \tag{1-13}$$

The first of these two equations implies that the orthogonal transformation applied to $\mathbf{A}$ gives $\mathbf{C}$ and, therefore, the sum of the diagonal elements of these two matrices must be the same. Therefore,

$$1 + 2\cos\omega = -\cos\alpha - \cos\alpha\cos\chi + \cos\chi \tag{1-14}$$

or

$$\cos\omega = \tfrac{1}{2}[-\cos\alpha + \cos\chi - \cos\alpha\cos\chi - 1] \tag{1-15}$$

From the second equation coming from the joint expressions for $\mathbf{V}_{j-1}$ along with Eq. (1-8) we obtain the following expressions for r and l:

$$r^2 = 2d^2(1 + \cos\alpha)/(3 + \cos\alpha - \cos\chi + \cos\alpha\cos\chi)^2 \tag{1-16}$$

$$l^2 = d^2(1 - \cos\chi)\,(1 - \cos\alpha)/(3 + \cos\alpha - \cos\chi + \cos\alpha\cos\chi) \tag{1-17}$$

These last three equations define (r, ω, l) as a function of (α, d, χ) for a helical chain having atoms uniformly distributed.

Since the bonds may be virtual, as briefly discussed earlier, any macromolecule possessing a uniformly repetitious backbone monomer unit may be treated by the above formulation. All that is necessary is to redefine $\mathbf{A}$ and $\mathbf{b}$ in terms of the specific monomer geometry. This is perhaps best demonstrated by an example, and, to be consistent, we will again refer to polypeptides. We may use the set of C^α atoms of the polypeptide chain as the set of uniformly distributed points so that the calculated (r, ω, l) will refer to a planar peptide unit. Figure 1-3 contains a planar peptide unit and defines the monomer unit bond angles and bond lengths needed to compute $\mathbf{A}$ and $\mathbf{b}$. In order to transform the coordinate system located at C_i^α to C_{i-1}^α along a path defined by the monomer unit geometry $\mathbf{A}$ must have the form

$$\mathbf{A} = \mathbf{B}_3\,\mathbf{B}_2\,\mathbf{B}_1 \tag{1-18}$$

where

$$\mathbf{B}_i = \begin{pmatrix} -\cos\beta_i & -\sin\beta_i & 0 \\ \sin\beta_i\cos\theta_i & -\cos\beta_i\cos\theta_i & -\sin\theta_i \\ \sin\beta_i\sin\theta_i & -\cos\beta_i\sin\theta_i & \cos\theta_i \end{pmatrix} \tag{1-19}$$

provided $\theta_2 = 0$. The translation $\mathbf{b}$ turns out to be given by

$$\mathbf{b} = \mathbf{B}_3\,\mathbf{B}_2\,\boldsymbol{\rho}_1 + \mathbf{B}_3\,\boldsymbol{\rho}_2 + \boldsymbol{\rho}_3 \tag{1-20}$$

The form of **C**, **D**, and **S** remains the same as first given so Eqs. (1-12) and (1-13) may be used at this point to calculate (r, ω, l) as a function of $(\rho_1, \rho_2, \rho_3, \beta_1, \beta_2, \beta_3, \theta_1, \theta_2, \theta_3)$.

An alternate means of estimating the parameters (r, ω, l) of a helix involves a linear least square curve fit for a set of uniformly distributed points (atoms) on the helix. If the number of points corresponds to an integral number of turns the calculation is exact. Under any other conditions the calculation is only approximate with the approximation becoming better as the number of

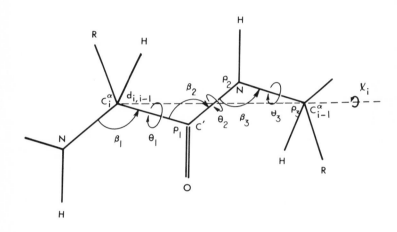

Fig. 1-3. The geometry of a planar peptide unit used to determine the helical parameters (r, ω, l).

points becomes larger. This procedure also makes it possible to transform a helix from any position and orientation in space to a position and orientation about some prechosen axis. The method is based upon the fact that the linear least square curve fit of a set of uniformly distributed points on a helix is the helical axis.

If X_i, Y_i, and Z_i are the coordinates of the m_ith member of the set of uniformly distributed points on the helix and there are n points, then the following sequence of calculations results in the desired transformation so that the helix axis and the x axis of the coordinate system coincide,

$$\Phi_{xy} = \tan^{-1}\left[\frac{n \sum_{i=1}^{n} X_i Y_i - (\sum_{i=1}^{n} X_i)\cdot(\sum_{i=1}^{n} Y_i)}{n \sum_{i=1}^{n} X_i^2 - (\sum_{i=1}^{n} X_i)\cdot(\sum_{i=1}^{n} X_i)}\right] \tag{1-21}$$

$$\left.\begin{array}{l} X_i' = X_i \cos \Phi_{xy} + Y_i \sin \Phi_{xy} \\ Y_i' = -X_i \sin \Phi_{xy} + Y_i \cos \Phi_{xy} \\ Z_i' = Z_i \end{array}\right\} \quad \text{for} \quad i = 1, n \qquad (1\text{-}22)$$

$$\Phi_{x'z} = \tan^{-1} \left[\frac{n \sum_{i=1}^{n} X_i' Z_i - (\sum_{i=1}^{n} X_i') \cdot (\sum_{i=1}^{n} Z_i)}{n \sum_{i=1}^{n} X_i'^2 - (\sum_{i=1}^{n} X_i') \cdot (\sum_{i=1}^{n} X_i')} \right] \qquad (1\text{-}23)$$

$$\left.\begin{array}{l} X_i'' = X_i' \cos \Phi_{x'z} + Z_i \sin \Phi_{x'z} \\ Y_i'' = Y_i' \\ Z_i'' = -X_i' \sin \Phi_{x'z} + Z_i' \cos \Phi_{x'z} \end{array}\right\} \quad \text{for} \quad i = 1, n \qquad (1\text{-}24)$$

$$\bar{Y} = \frac{\sum_{i=1}^{n} Y_i''}{n} \qquad (1\text{-}25)$$

$$\bar{Z} = \frac{\sum_{i=1}^{n} Z_i''}{n} \qquad (1\text{-}26)$$

Now, if (x, y, z) is the coordinate vector of any atom in the macromolecule before the transformation, then the coordinate vector (x_h, y_h, z_h) of the same atom in the cylindrical frame is given by

$$\begin{array}{l} x' = \cos \Phi_{xy} + y \sin \Phi_{xy} \\ y' = -x \sin \Phi_{xy} + y \cos \Phi_{xy} \\ z' = z \end{array} \qquad (1\text{-}27)$$

$$\begin{array}{l} x'' = x' \cos \Phi_{x'z} + z \sin \Phi_{x'z} \\ y'' = y' \\ z'' = -x' \sin \Phi_{x'z} + z \cos \Phi_{x'z} \end{array} \qquad (1\text{-}28)$$

$$\begin{array}{l} x_h = x'' \\ y_h = y'' - \bar{Y} \\ z_h = z'' - \bar{Z} \end{array} \qquad (1\text{-}29)$$

The helical parameters (r, ω, l) may be estimated by computing their values using the complete set of uniformly distributed points $(X_{h(i)}, Y_{h(i)}, Z_{h(i)})$:

$$l \equiv \langle l \rangle = \sum_{i=2}^{n} \frac{1}{n-1} [X_{h(i)} - X_{h(i-1)}] \qquad (1\text{-}30)$$

$$r \equiv \langle r \rangle = \sum_{i=1}^{n} \frac{1}{n} (Y_{h(i)}^2 + Z_{h(i)}^2)^{1/2} \tag{1-31}$$

$$\omega = \langle \omega \rangle = \sum_{i=2}^{n} \frac{1}{n-1} \left(\cos^{-1} \left[\frac{Y_{h(i)} Y_{h(i-1)} + Z_{h(i)} Z_{h(i-1)}}{(Y_{h(i)}^2 + Y_{h(i-1)}^2)^{1/2} + (Z_{h(i)}^2 + Z_{h(i-1)}^2)^{1/2}} \right] \right) \tag{1-32}$$

Now that a relationship has been established between (r, ω, l) and the geometry of the polymer backbone it is possible to construct plots of r, ω, and l versus the internal bond rotations. Normally l and $n = 2\pi/\omega$, the number of monomer units per helical turn are plotted as a function of the internal bond rotations. Figure 1-4a is a plot of n and l versus χ for a polyethylene backbone and Fig. 1-4b is a plot of n and l versus ϕ and ψ for a polypeptide backbone

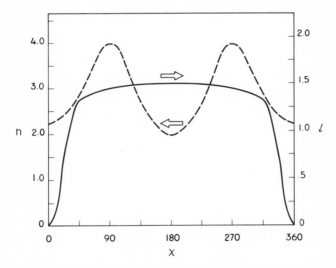

Fig. 1-4a. The helical parameter plot of a polyethylene backbone. The dashed curve and left ordinate refer to the number of monomer units per turn (n) and the solid curve and right ordinate refer to monomer repeat length along the helical axis (l). $\chi = 180°$ defines the planar zigzag conformation.

Polyethylene backbone

(2a). When these plots are overlayed upon the appropriate potential energy maps (to be described later) it is possible to rapidly identify the helical geometry of stable structures.

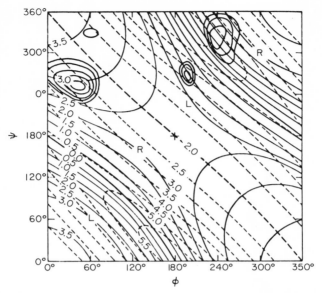

Fig. 1-4b. A helical parameter plot of a polypeptide backbone calculated as a function of nonrigid skeleton backbone. Here dashed and solid curves are defined the same as in part (a), and L and R refer to left and right helices. The convention proposed by Edsall *et al.* (2b) has been used to indicate the angles of rotation. Throughout the book, we will use the Edsall *et al.* convention. As a point of reference however, this convention has been superseded. The new convention is described in (2c). The new convention defines $\phi = 0°$, $\psi = 0°$ to be at the old $\phi = 180°$, $\psi = 180°$ values. The sense of rotation remains the same.

Polypeptide backbone

There are several notations to define the spatial properties of a helix. The one adopted throughout this book is of the form $\pm M_T$, where M is the number of monomer units in T turns of the helix. The choice of plus (+) or minus (−) determines the sense of the helix. By convention the plus (+) refers to a right-handed helix and the minus (−) to a left-handed helix.

3. Tertiary Structure

This is the relative arrangement of the secondary structure(s) in a macromolecule. For the vast majority of macromolecules the tertiary structure is either extended or folded.

(*a*) *Extended* That is, the macromolecule assumes an overall rodlike shape. This structure can result from introducing small periodic variations in

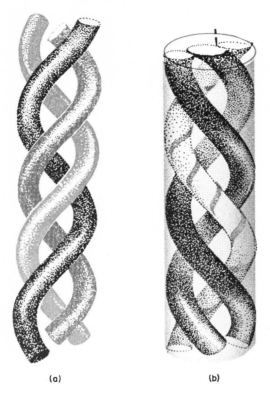

Fig. 1-5. Schematic illustrations of the quaternary structures of some macromolecules which exist as coiled-coils: (a) collagen, (b) α-keratin.

a helical secondary structure. These small variations give rise to a second helix so that the resulting tertiary structure is equivalent to a spring, the helix of which corresponds to the secondary structure, wrapped around a cylinder so as to generate a second helix. Examples of macromolecules which can assume this structure, called a *coiled-coil*, are poly-L-alanine in the α form, collagen, α-keratin, and numerous polytripeptides such as $(Gly-Pro-Pro)_n$. Two of these biopolymers, in their quaternary structures, are shown in Fig. 1-5.

The coiled-coil structure is normally stabilized by intermolecular interactions resulting from one coiled-coil "wrapping around" a second coiled-coil. Each of the aforementioned macromolecules which are coiled-coils are stabilized in this manner. More will be said about this when we discuss quaternary structure. Figure 1-6 illustrates a coiled-coil and defines the

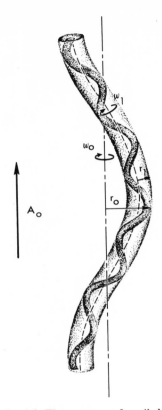

Fig. 1-6. The geometry of a coiled-coil.

geometric parameters of such a structure. The corresponding parametric equations in t, for a coiled-coil are (3)

$$x = r_0 \cos \omega_0 t + \Delta \cos[(\omega_0 + \omega_1) t] + \delta \cos[(\omega_0 - \omega_1) t] \qquad (1\text{-}33)$$

$$y = r_0 \sin \omega_0 t + \Delta \sin[(\omega_0 + \omega_1) t] + \delta \sin[(\omega_0 - \omega_1) t] \qquad (1\text{-}34)$$

$$z = A_0 t - r_1 \sin \alpha \sin \omega_1 y \qquad (1\text{-}35)$$

where

$$\Delta = r_1(1 + \cos \alpha)/2, \qquad \delta = r_1(1 - \cos \alpha)/2, \qquad \alpha = \tan^{-1}(\omega_0 r_0/A_0)$$

A secondary helical structure can be identical to the tertiary structure of the macromolecule. When this is the situation the tertiary structure may be extended. Equivalence of the secondary and tertiary structures of a macro-molecule may result from: (i) The rigid intramolecular stereochemical restrictions of the macromolecule. Polyisobutylene, the primary structure of which is shown in Fig. 1-7a, is an example of a polymer possessing considerable stereochemical restrictions; (ii) intermolecular packing considerations such as those which occur in a crystal.

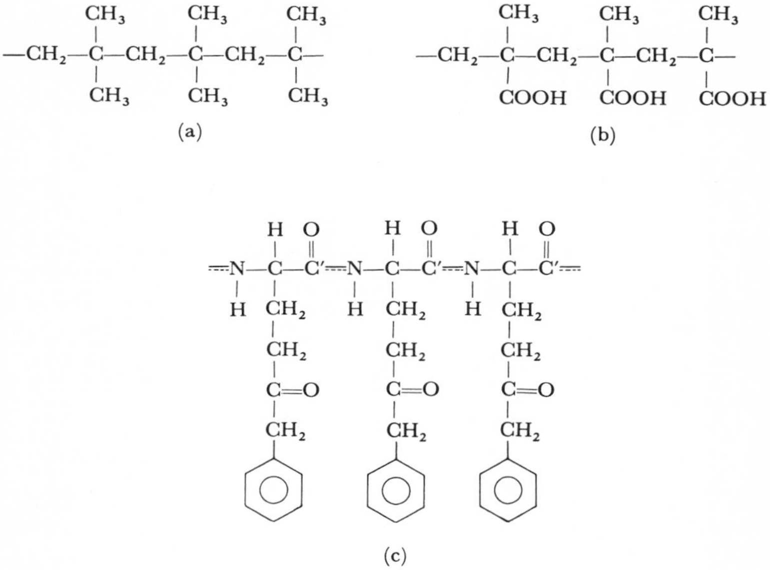

(a)

(b)

(c)

Fig. 1-7. Primary structure of (a) polyisobutylene, (b) poly(methacrylic acid), and (c) polybenzyl-L-glutamate.

(*b*) *Folded*   That is, segments of the chain fold back to form tertiary structures which can be classed into three groups depending upon the extent of local organization associated with the folded chain segments.

(*i*) *Ordered-folded tertiary structures*   In such structures there is a charac-teristic fold length common to the segments of the macromolecule. Each of the segments act as virtually independent chains to pack into an ordered array which can be considered to be a minicrystal. The lattice packing varies with the specific macromolecule and the fold length depends upon the thermo-dynamic and chemical environments. Obviously, macromolecules which possess heterogeneous primary sequences and/or *randomly* distributed struc-tural defects will not be as likely to assume the ordered-folded tertiary structures because close packing will not be possible. Hence a necessary, though not sufficient, requirement in order that a macromolecule have an ordered-folded tertiary structure is an ordered primary structure and/or ordered sequence of structural defects. The repetitious primary structure insures the possibility of a mode of close packing. The same is true for ordered structural defects. The classical example of a macromolecule which can take up an ordered-folded tertiary structure is polyethylene (PE). Figure 1-8

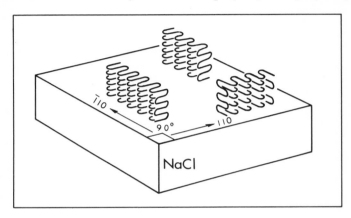

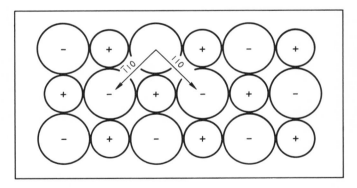

Fig. 1-8. The lattice packing of chain-folded polyethylene chains.

shows the lattice packing of chain-folding polyethylene chains. Other examples of macromolecules which may have ordered-folded tertiary structures are listed in Table 1-2 along with their respective packing lengths. The macroscopic shapes of ordered-folded tertiary structures are those of "minicrystal" polyhedrons.

(*ii*) *Globular-folded tertiary structures* Some macromolecules are heterogeneous in primary structure which prohibits the formation of ordered folds in the macromolecule and the gross shape of the macromolecule is globular. However, a specific tertiary structure forms each time the macromolecule is subjected to a given set of thermodynamic and chemical conditions. Proteins

TABLE 1-2

Chain-Folding Characteristics of Several Polymers under Varying Conditions

Polymer	Solvent	Temperature (°C)	Long period (Å)
Polyacrylonitrile	Propylene carbonate	90–100	~100
Polyethylene oxide	Cyclohexene–paraxylene		<100
Isotactic poly-4-methyl pentene-1	Xylene		
Nylon 66	Glycerin		40–60
Polyproplene (isotactic)			
Polybenzylglutamate[a]	Mesitylene	100	~600
Polytetrafluoroethylene	Perfluorinated olefin solutions		200–500
Cellulose esters	Nitromethane–butanol		180–200
1,10-Decanediol and an unreported ester	Ethanol		15–35
1,20-Eicosanediol and an unreported acid	Amyl acetate		40–70
Xylan	Water		~50
trans-Polyisoprene	Chloroform		~200
Nylon 6	Glycerin		~55
Polyethylene[a]	Xylene	72	110
	Xylene	78	140
	Tetrachloroethylene	50	112
Polyhexamethylene Terepthalate	Tetrachloroethylene		275
Polyoxymethylene			~80
Polyisopropylsiloxane	*n*-Butyl acetate	5	~80
Polytryrosine			
DNA			
Polyethylene terepthalate			
Polybutene			
Isotactic polystyrene			
Polypropylene oxide			

[a] Indicates epitaxial crystallization upon alkali halides.

are the most common example of macromolecules having globular-folded tertiary structures. Figure 1-9 illustrates part of the molecular structure of the protein lysozyme. It is depicted in such a way that gross shape of the macromolecule is obvious. However, equally obvious are the regions of

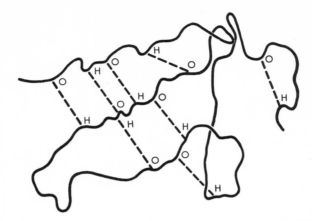

Fig. 1-9. A portion of the backbone tertiary structure, denoted by the solid line of lysozyme. The dashed lines indicate the formation of hydrogen bonds resulting from chain folding. This precise mode of hydrogen bonding is a specific property of the structure of this molecule.

ordered-folded structures which distinguishes this macromolecule from all others.

(*iii*) *Random tertiary structures*   Some macromolecules possess locally ordered regions which are not reproducible for a given set of thermodynamic and/or chemical conditions. Often the total content of locally ordered structure is constant, but the particular chain segments, and their lengths, involved in the local order are randomly distributed over the entire macromolecular chain. The resulting tertiary structure may be globular, as is the case of poly(methacrylic acid) (see Fig. 1-7b for the primary structure) in aqueous solution. Other times the tertiary structure may be partially "extended" in that the macromolecule is part of a bulk amorphous ultrastructure. Polybenzyl-L-glutamate whose primary structure is given in Fig. 1-7c, is an example of a macromolecule which may exist in a random tertiary structure in an amorphous ultrastructure. Polymer chains which have degrees of freedom that can be described by random statistics can often be mathematically modeled. From such models certain basic parameters can be derived which reflect the characteristic properties of the polymer chain. By determining the experimental values of these parameters for polymers which have apparent randomly ordered tertiary structures it is sometimes possible to deduce the actual extent of random structure.

In the following few paragraphs we will derive expressions for the mean-square end-to-end distance $\langle h^2 \rangle$ for linear polymers having different types of random structuring. $\langle h^2 \rangle$ is perhaps the most often employed parameter used to describe the overall shape of a polymer chain.

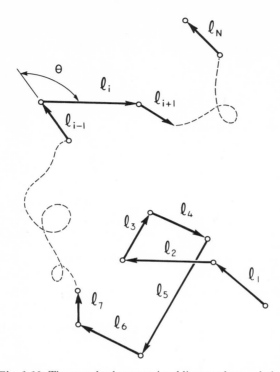

Fig. 1-10. The completely unrestricted linear polymer chain.

a. *The completely unrestricted polymer chain* Consider a linear sequence of $N + 1$ equivalent elements joined by N bonds each of different fixed lengths. All bond angles will be equally probable. The length and direction of each bond can be characterized by a vector $\boldsymbol{l}_i$. Such a chain is shown in Fig. 1-10. The end-to-end vector for any configuration is

$$\mathbf{h} = \sum_{i=1}^{N} \boldsymbol{l}_i \tag{1-36}$$

The square of the average end-to-end distance, a characteristic property of the chain, is defined as

$$\langle h^2 \rangle = \overline{h^2} \equiv \overline{\mathbf{h} \cdot \mathbf{h}} = \overline{\left(\sum_{i=1}^{N} \boldsymbol{l}_i \right) \cdot \left(\sum_{j=1}^{N} \boldsymbol{l}_j \right)} \tag{1-37}$$

Carrying out the indicated dot product

$$\langle h^2 \rangle = \sum_{i=1}^{N} \sum_{j=1}^{N} \overline{(\boldsymbol{l}_i \cdot \boldsymbol{l}_j)} \tag{1-38}$$

If θ is the angle between $\boldsymbol{l}_{i-1}$ and $\boldsymbol{l}_i$,

$$\boldsymbol{l}_{i-1} \cdot \boldsymbol{l}_i = l_{i-1}\, l_i \cos \theta \tag{1-39}$$

If all possible values of θ are equally probable, then

$$\overline{\boldsymbol{l}_{i-1}\boldsymbol{l}_i} = l_{i-1}\, l_i \overline{\cos \theta} = 0 \tag{1-40}$$

since positive and negative values of θ are equally likely. In the same manner all terms in which i is not equal to j also vanish. What is left is

$$\langle h^2 \rangle = \sum_{i=1}^{N} \overline{l_i^2} = N\overline{l_i^2} = Nl_{av}^2 \tag{1-41}$$

where $(l_i^2)^{1/2} \equiv l_{av}$ is the root mean square bond length in the molecule.

  b. *Fixed bond length and bond angle polymer chains*   Here we consider a linear polymer containing identical bond lengths l and bond angles θ, for which there is free rotation about the bonds χ. An illustration of such a polymer chain is shown in Fig. 1-11. The derivation of an expression for $\langle h^2 \rangle$ is the same as before up to and including

$$\langle h^2 \rangle = \overline{h^2} = \sum_{i=1}^{N} \sum_{j=1}^{N} \overline{(\boldsymbol{l}_i \cdot \boldsymbol{l}_j)} \tag{1-42}$$

where N is again the number of atoms.

  The manner in which we sum up the terms on the right in the above equation now differs from the completely unrestricted chain. There will be N terms of the form $\overline{\boldsymbol{l}_i \cdot \boldsymbol{l}_i} = l^2$; there will be $2(N-1)$ terms of the form $\overline{\boldsymbol{l}_i \cdot \boldsymbol{l}_{i+1}} = l^2 \cos \theta$, next there will be $2(N-2)$ terms of the form $\overline{\boldsymbol{l}_i \cdot \boldsymbol{l}_{i+2}}$. To evaluate these terms let us split $\boldsymbol{l}_{i+2}$ into two components, $\boldsymbol{l}_{\parallel}$ parallel to $\boldsymbol{l}_{i+1}$ and $\boldsymbol{l}_{\perp}$ perpendicular to $\boldsymbol{l}_{i+1}$ so that

$$\overline{\boldsymbol{l}_i \cdot \boldsymbol{l}_{i+2}} = \overline{\boldsymbol{l}_i \cdot \boldsymbol{l}_{\parallel}} + \overline{\boldsymbol{l}_i \cdot \boldsymbol{l}_{\perp}} \tag{1-43}$$

$\boldsymbol{l}_{\parallel}$ is fixed and independent of ϕ while $\boldsymbol{l}_{\perp}$ can terminate anywhere on the indicated circle depending upon the value of ϕ. Since we have stipulated free rotation about bonds, all terminal points on the circle are equally probable. Thus,

$$\overline{\boldsymbol{l}_i \cdot \boldsymbol{l}_{\perp}} = 0 \tag{1-44}$$

Further, since $\boldsymbol{l}_i \cdot \boldsymbol{l}_{\parallel} = l_i l_{\parallel} \cos \theta$ and $l_{\parallel}$ itself is equal to $l \cos \theta$, we finally find that

$$\boldsymbol{l}_i \cdot \boldsymbol{l}_{i+2} = l^2 \cos^2 \theta \tag{1-45}$$

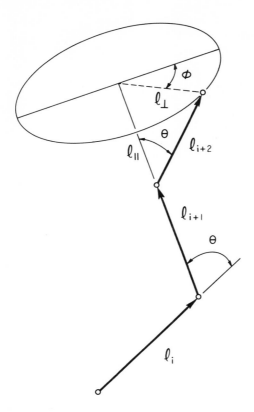

Fig. 1-11. The general geometry of a linear polymer chain having fixed bond lengths and bond angles, but unrestricted bond rotations.

The $l_i \cdot l_{i+3}$ terms, of which there are $2(N-3)$, may also be calculated by splitting the l_{i+3} vector into parallel and perpendicular components relative to vector l_{i+2}. Upon carrying out a goemetric analysis of this situation one can easily reach the equation

$$\overline{l_i \cdot l_{i+3}} = l^2 \cos^3 \theta \qquad (1\text{-}46)$$

In general we find

$$\overline{l_i \cdot l_{i+k}} = l^2 \cos_k \theta \qquad (1\text{-}47)$$

and there are $2(N-k)$ such terms. Substituting these terms into the general expression for $\langle h^2 \rangle$ yields the series expression

$$\langle h^2 \rangle = l^2[N + 2(N-1)\cos\theta + 2(N-2)\cos^2\theta + \cdots + 2(N-k)\cos^k\theta$$
$$+ \cdots + 2\cos^{N-1}\theta] \qquad (1\text{-}48)$$

For very large N (e.g., normally a few hundred to a few thousand in most polymer chains), $\langle h^2 \rangle$ can be approximated by

$$\langle h^2 \rangle \cong l^2 N(1 + 2\cos\theta + 2\cos^2\theta + \cdots) \qquad (1\text{-}49)$$

$$\langle h^2 \rangle \cong l^2 N(1 + 2 \sum_{j=1}^{N-1}\cos^j\theta) \qquad (1\text{-}50)$$

This expression, in turn, for $\cos\theta < 1$ converges to an analytic-finite sum,

$$\langle h^2 \rangle \cong Nl^2 \left[\frac{1 + \cos\theta}{1 - \cos\theta}\right] \qquad (1\text{-}51)$$

In general, $\langle h^2 \rangle$ can always be given by

$$\langle h^2 \rangle = \Phi^2 N \qquad (1\text{-}52)$$

where Φ is a constant characteristic of the polymer.

For restrictions to rotation about ϕ which are not too severe Benoit and Sadron (4) were able to show that

$$\Phi^2 = l^2 \left(\frac{1 + \cos\theta}{1 - \cos\theta}\right)\left(\frac{1 + \overline{\cos}\,\phi}{1 - \overline{\cos}\,\phi}\right) \qquad (1\text{-}53)$$

where $\overline{\cos}\,\phi$ is the average value of the $\cos\phi$. Expressions for Φ have also been determined for cellulose (5), silicone, and other linear chains (6). Flory and co-workers (7–9) have made use of conformational potential energy calculations to compute, essentially, empirical values of Φ for various ordered biopolymers.

c. *A second basic parameter* often used to describe the bulk shape of a linear polymer chain is the average radius of gyration. Consider a polymer chain as a collection of mass elements m_i each located r_i from the center of mass when the polymer chain is in some specific tertiary structure. The radius of gyration R for this configuration is defined as

$$R^2 \equiv \left(\sum_{i=1}^{N} m_i r_i^2\right) / \left(\sum_{i=1}^{N} m_i\right) \qquad (1\text{-}54)$$

where N is the number of atoms in the chain. Since we presuppose that the macromolecular chain can take on a variety of different configurations, that is, tertiary structures, we must average over the entire set of R resulting from the possible configurations. We define the average radius of gyration $\langle R \rangle$ as

$$\langle R \rangle = (\overline{R^2})^{1/2} \equiv [(\overline{\textstyle\sum_{i=1}^{N} m_i r_i^2})/(\textstyle\sum_{i=1}^{N} m_i)]^{1/2} \tag{1-55}$$

since, however,

$$\overline{\textstyle\sum_i^N m_i r_i^2} = \textstyle\sum_i^N \overline{m_i r_i^2} \tag{1-56}$$

we obtain

$$\langle R \rangle = [(\textstyle\sum_{i=1}^{N} \overline{m_i r_i^2})/(\textstyle\sum_{i=1}^{N} m_i)]^{1/2} \tag{1-57}$$

If all the masses are equal, $m_i = m$, then

$$\langle R \rangle = (\textstyle\sum_{i=1}^{N} \overline{r_i^2}/N)^{1/2} \tag{1-58}$$

4. Quaternary Structure

This is the specific arrangement of two or more macromolecules to form a unique macromolecular organization. This type of structure is not to be confused with crystal or amorphous structures composed of macromolecules. Quaternary structures involve only a few macromolecules, at least when compared to crystal or amorphous structures, and the resulting quaternary structure usually has all the chemical properties of a distinct entity.

When we discuss the quaternary structure of a macromolecular species we have to concern ourselves with not only the conformation of the constituent macromolecules, but also with the way in which the constituent macromolecules "fit" together. The classic example of the importance of correctly fitting together macromolecular chains is the DNA double helix. Once Watson and Crick (10) were able to establish the quaternary structure of this molecule, our understanding of cell reproduction and the transfer of genetic information had a molecular basis. Not to understate the importance of elucidating the quaternary structure of DNA as well as coiled-coils, these are, nonetheless, the simplest type of quaternary structures. In general, the precise quaternary structure of a coiled-coil results from the wrapping of macromolecules about one another followed by variations in the helical geometry so that the total intra- and interchain interaction free energy is minimized.

Other macromolecular systems which have unique quaternary structures have features which make their assemblage more complex. The majority of such macromolecular systems are biological. Assembly problems are introduced not only by the need for stereochemical compatibility, but also through the requirement for chemical activity, e.g., biological functioning. Some quaternary structures involve only a few subunits; DNA has two polynucleotide chains, hemoglobin is composed of four polypeptide chains, and collagen three polypeptide chains. Other molecules may be composed of many subunits. Among the proteins containing many subunits are lipoic reductase-transacetylase with 60 subunits, chlorocruorin with 12 subunits, and hemerythrin with 8 subunits. [See Klotz (11) for further references to these proteins.]

5. *Discussion*

In this section we have discussed the first four levels of structural identity in macromolecular systems. It is tempting to allot some space to the fifth level of structural identity; pentiary structure. Many macromolecular scientists believe that the next fifteen years of research will become increasingly directed toward the elucidation of pentiary structure. Now just what is meant by pentiary structure? This term is meant to include the structural organization of a macromolecule with some second molecular component. This second component might or might not be some *different* type of macromolecule. Thus composites are examples of pentiary structures. A macromolecule in solution would qualify as a pentiary structure. The most abundant type of pentiary structural studies probably involve the mutual organization of two different macromolecules. Common examples are, DNA with histones, collagen with mucopolysaccharides and virus–antibody complexes.

III. Generation of Macromolecular Conformations and Configurations

A. General Comments

Now that we have classified macromolecular structure it is appropriate to derive some equations which make it possible to generate secondary, tertiary, quaternary and intermolecular structures from a knowledge of primary structure. In establishing expressions between the internal bond rotations and the parameters of a helix, as done in the previous section, we carried out rather special derivations using two different approaches to generate intramolecular structure. In this section we develop two general means of generating molecular structure as a function of internal bond rotations. Lastly, in this chapter we will present a brief discussion of intermolecular geometry as it relates to helical chains of polymer molecules. These techniques to generate conformations of macromolecules, and assemblies of macromolecules, are needed in order to be able to compute the conformational and/or configurational energies of such systems.

B. Intrachain Geometry

There are two major ways to generate macromolecular chain conformations. The first is the general linear chain model which is applicable to virtually all macromolecules. The second chain-generating technique is the residue model, which in terms of computation time is much more efficient, in most cases, than the linear chain model. However, the residue model is restricted to those macromolecules having characteristic repeat units (the residues). Still, this technique can often be employed since many macromolecules can be thought of as a sequence of residues.

1. The General Linear Chain Model

A *general linear chain* is defined as any set of points which contains a subset of points called the *backbone* such that all points in the backbone can be

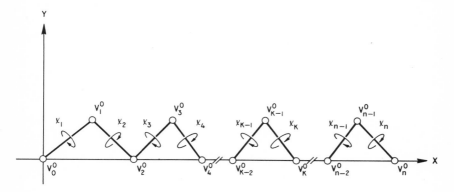

Fig. 1-12. The general linear chain (skeletal atoms only) in the standard chain conformation.

connected by one unique broken line. When the sequence involves atoms, the backbone atoms are called the *skeletal atoms* of the chain and the remaining sets of atoms which can be joined to the various skeletal atoms are called *branch atoms*. A general linear chain (skeletal atoms only) is shown in Fig. 1-12. It should be clear that most macromolecules can be considered as a set of linear chains in which one chain serves as the molecular backbone and the remaining chains compose the set of side-chain groups.

The *standard chain conformation* of a linear chain is defined as that conformation for which all the skeletal atoms are in the XY plane of the frame associated with the chain *and* each successive skeletal atom has an associated x coordinate which has greater positive value than the x coordinate of any preceding atom (any atom whose label number is a smaller integer). If more than one conformation is possible using the above definition, then, to make the definition unique, the conformation with the maximum end-to-end distance is chosen.

The notation associated with a linear chain is as follows:

χ_l lth rotation,

$\mathbf{v}_j{}^0$ vector position of the jth skeletal atom in the standard conformation,

$\mathbf{V}_{j,k}^0$ vector position of the kth branch atom on the jth skeletal atom in the standard conformation,

$\mathbf{v}_j^{[l]}$ vector position of the jth skeletal atom with χ_l being the last rotation performed,

$\mathbf{V}_{j,k}^{[l]}$ vector position of the kth branch atom on the jth skeletal atom with χ_l being the last rotation performed.

Using Fig. 1-12 we can write down, although trivial, the coordinate vector of the nth atom after rotation χ_n has been carried out:

$$\mathbf{v}_n^{[n]} = \mathbf{T}_{\chi_n}(\mathbf{v}_n{}^0 - \mathbf{v}_{n-1}^0) + \mathbf{v}_{n-1}^0 \tag{1-59}$$

We can also write down $\mathbf{v}_n^{[n-1]}$ in the same way:

$$\mathbf{v}_n^{[n-1]} = \mathbf{T}_{\chi_{n-1}}(\mathbf{v}_n^{[n]} - \mathbf{v}_{n-2}^0) + \mathbf{v}_{n-2}^0 \tag{1-60}$$

$\mathbf{v}_n^{[n-1]}$ can be expressed in terms of standard conformation vectors by substituting (1-59) into (1-60):

$$\mathbf{v}_n^{[n-1]} = \mathbf{T}_{\chi_{n-1}}\mathbf{T}_{\chi_n}(\mathbf{v}_n{}^0 - \mathbf{v}_{n-1}^0) + \mathbf{T}_{\chi_{n-1}}(\mathbf{v}_{n-1}^0 - \mathbf{v}_{n-2}^0) + \mathbf{v}_{n-2}^0 \tag{1-61}$$

Inducting on the above scheme, the jth skeletal coordinate vector, $\mathbf{v}_j$, relative to the chain coordinate system for the linear chain in any conformation is given by

$$\mathbf{v}_j = \mathbf{v}_0{}^0 + \sum_{m=1}^{j} \left\{ \prod_{l=1}^{m} \mathbf{T}_{\chi_l}(\mathbf{v}_m{}^0 - \mathbf{v}_{m-1}^0) \right\} \tag{1-62}$$

In the same way the kth branch atom on the jth skeletal atom can be expressed by

$$\mathbf{V}_{j,k} = \mathbf{v}_0{}^0 + \sum_{m=1}^{j-1} \left\{ \prod_{l=1}^{m} \mathbf{T}_{\chi_l}(\mathbf{v}_m{}^0 - \mathbf{v}_{m-1}^0) \right\} + \prod_{l=1}^{j} \mathbf{T}_{\chi_l}(\mathbf{V}_{j,k}^0 - \mathbf{v}_{j-1}) \tag{1-63}$$

where

$$T_{\chi_l} = T_{\alpha_l}^{-1} \, \bar{T}_{\chi_l} \, T_{\alpha_l} \tag{1-64}$$

Each of the matrices on the right in (1-64) are unitary with the forms

$$T_{\alpha_l} = \begin{pmatrix} \cos \alpha_l & \sin \alpha_l & 0 \\ -\sin \alpha_l & \cos \alpha_l & 0 \\ 0 & 0 & 1 \end{pmatrix} \tag{1-65}$$

where

$$\alpha_l = \tan^{-1} \left[\frac{y_l{}^0 - y_{l-1}^0}{x_l{}^0 - x_{l-1}^0} \right] \tag{1-66}$$

with the $x_j{}^0$'s and $y_j{}^0$'s the x and y components of $\mathbf{v}_j{}^0$:

$$\bar{T}_\chi = \begin{pmatrix} 1 & 0 & 0 \\ 0 & \cos \chi_l & \sin \chi_l \\ 0 & -\sin \chi_l & \cos \chi_l \end{pmatrix} \tag{1-67}$$

Since matrix multiplication is a time-consuming operation, it is best to express T_{χ_l} explicitly:

$$T_\chi = \begin{pmatrix} [\cos^2 \alpha_l + \cos \chi_l \sin^2 \alpha_l] & [\cos \alpha_l \sin \alpha_l (1 - \cos \chi_l)] & [-\sin \alpha_l \sin \chi_l] \\ [\cos \alpha_l \sin \alpha_l (1 - \cos \chi_l)] & [\sin^2 \alpha_l + \cos^2 \alpha_l \cos \chi_l] & [\cos \alpha_l \sin \chi_l] \\ [\sin \chi_l \sin \alpha_l] & [-\sin \chi_l \cos \alpha_l] & [\cos \chi_l] \end{pmatrix} \tag{1-68}$$

In principle, a set of linear chains can be used to simulate any macromolecule. Macromolecules which contain nonlinear chains can still be approximated by linear chains by decomposing the nonlinear chains into the linear components. It should also be noted that to use the general linear-chain theory, all one need know is the set of skeletal coordinate vectors and the set of branch coordinate vectors for the standard conformation.

This method of generating conformations is not well suited for situations in which bond lengths and bond angles are distorted from their initial values. If the bond defined by coordinate vectors $\mathbf{v}_{j-1}^0$ and $\mathbf{v}_j{}^0$ is perturbed, then all standard coordinate vectors $\mathbf{v}_j{}^0$, $\mathbf{v}_{j+1}^0$, ..., $\mathbf{v}_n{}^0$ need be corrected. The same holds for bond angle distortions. Each time a bond angle or distance distortion is allowed in the macromolecule the standard coordinate vectors must be redefined. Using Fig. 1-13, we can arrive at expressions which define the ith standard coordinate vector after the associated bond angle and bond distance distortions have occurred.

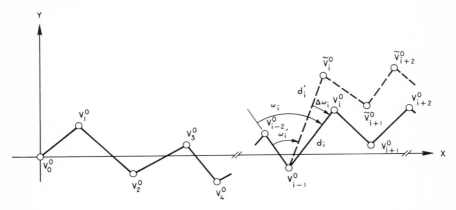

Fig. 1-13. The change in the geometry of the coordinate position vectors of the skeletal atoms i, $i+1$, ..., n due to the bond angle distortion, $\Delta\omega_i$, and the bond length distortion, Δd_i. $\mathbf{v}_k{}^0 = (x_k{}^0, y_k{}^0, z_k{}^0)$ refers to a nondistorted coordinate vector and $\tilde{\mathbf{v}}_k{}^0 = (\tilde{x}_k{}^0, \tilde{y}_k{}^0, \tilde{z}_k{}^0)$ refers to a distorted coordinate position vector.

For the x coordinates,

$$\tilde{x}_i{}^0 = x_{i-1}^0 + d_i{}' \cos(\alpha_i - \Delta\omega_i) \tag{1-69}$$

where

$$\alpha_i = \tan^{-1}\left[\frac{y_i{}^0 - y_{i-1}^0}{x_i{}^0 - x_{i-1}^0}\right] \quad \text{or} \quad \alpha_i = 180° - (\tilde{\alpha}_{i-1} + \omega_i) \tag{1-70}$$

and

$$d_i{}' = d_i + \Delta d_i \tag{1-71}$$

The tilde ($\sim$) above the x's denotes the coordinate positions result from bond length and/or bond angle deformations. The appropriate y coordinates can also be derived from inspection of Fig. 1-13

$$\tilde{y}_i{}^0 = y_{i-1}^0 + d_i{}' \sin(\alpha_i - \Delta\omega_i) \tag{1-72}$$

The z coordinates are independent of bond length and angle deformations when the chain is in the standard chain conformation.

Suppose now that every bond and bond angle, in the skeletal backbone, can be deformed. What is the net effect on the coordinate position of skeletal atom i in the backbone? We can arrive at the expressions for the x and y coordinates of skeletal atom i as follows:

for atom i: $\qquad\qquad \tilde{x}_i^{\,0} = \tilde{x}_{i-1}^{0} + d_i' \cos(\tilde{\alpha}_i - \Delta\omega_i)$ $\qquad$ (1-73)

for atom $i-1$: $\qquad \tilde{x}_{i-1}^{0} = \tilde{x}_{i-2}^{0} + d_{i-1}' \cos(\tilde{\alpha}_{i-1} - \Delta\omega_{i-1})$ $\qquad$ (1-74)

Substituting (1-74) into (1-73), we obtain

$$\tilde{x}_i^{\,0} = \tilde{x}_{i-2}^{0} + d_{i-1}' \cos(\tilde{\alpha}_{i-1} - \Delta\omega_{i-1}) + d_i' \cos(\tilde{\alpha}_i - \Delta\omega_i) \qquad (1\text{-}75)$$

Inducting upon the above result expressed in Eq. (1-75), we obtain the general expression

$$\tilde{x}_i^{\,0} = x_0^{\,0} + \sum_{j=1}^{i} d_j' \cos(\tilde{\alpha}_j - \Delta\omega_j) \qquad (1\text{-}76)$$

In the same way we obtain the following general expression for the y coordinate:

$$\tilde{y}_i^{\,0} = y_0^{\,0} + \sum_{j=1}^{i} d_j' \sin(\tilde{\alpha}_j - \Delta\omega_j) \qquad (1\text{-}77)$$

The tilde ($\sim$) above the α_k indicates that this angle is dependent upon the bond length and bond angle deformations. Hence it is necessary to compute the coordinate positions $(\tilde{x}_k^{\,0}, \tilde{y}_k^{\,0})$ sequentially from "left" to "right." Thus, for example, $(\tilde{x}_4^{\,0}, \tilde{y}_4^{\,0})$ must be computed in order to determine $\tilde{\alpha}_5$ which, in turn, is needed before $(\tilde{x}_5^{\,0}, \tilde{y}_5^{\,0})$ can be found. By substituting Eqs. (1-76) and (1-77) into Eq. (1-62) it is possible to arrive at an expression which implicitly relates both bond rotations and bond length and bond angle distortions to the coordinate positions of the backbone skeletal atoms:

$$\tilde{\mathbf{v}}_j = \tilde{\mathbf{v}}_0^{\,0} + \sum_{m=1}^{j} \left\{ \prod_{l=1}^{m} \tilde{\mathsf{T}}_{x_l} \left[\begin{pmatrix} d_m' \cos(\tilde{\alpha}_m - \Delta\omega_m) \\ d_m' \sin(\tilde{\alpha}_m - \Delta\omega_m) \\ 0 \end{pmatrix} \right] \right\} \qquad (1\text{-}78)$$

where

$$\tilde{\mathsf{T}}_{x_l} = \mathsf{T}_{\tilde{\alpha}_l}^{-1} \mathbf{F}_{x_l} \mathsf{T}_{\tilde{\alpha}_l}$$

This expression, Eq. (1-78), cannot be used in its given form unless the $\tilde{\alpha}_l$ are known, which requires that the $\tilde{\mathbf{v}}_k^{\,0}$ for k less than j have already been determined.

An analogous expression for $\tilde{\mathbf{V}}_{j,k}$, the branch atoms of the molecule, can be written by inspection and it is left to the reader to construct the appropriate equation.

2. The Residue Chain Model

Whenever a macromolecule can be considered as a sequence of groups of atoms the residue chain model can be used to generate the conformations. The groups of atoms are called residues. The real advantage of this technique is evident when the residues are composed of large numbers of atoms which have fixed geometries. Then one unique transformation relates all the atoms of a given residue to the atoms of an adjacent residue. In the linear chain model there is a near one-to-one correspondence between atoms and transformations in the generation of chain conformations. Thus, provided the interresidue transformation matrices are easily computed, the residue chain model requires less computation time than the linear chain model.

Suppose we are given a residue r_i. We can place this residue in a cartesian frame in some specific manner and define the positions of all atoms of the residue relative to the coordinate system. We call this frame the ith residue coordinate system and denote it by $(\text{rcs})^i$. The position of the jth atom in this the ith frame is denoted by $\mathbf{q}_{ij} = (x_j, y_j, z_j)^i$. It should be clear that the general linear chain model could be used to define the positions of the atoms in the ith residue. In actual practice this is the procedure often followed. The residues are defined via the general linear chain model and the macromolecular conformation is defined via the residue chain model. Now the question arises: What are the positions of the atoms of residue i with respect to reside $i - 1$? Quite clearly, we can answer this question once we have specified what operations must be performed in order that $(\text{rcs})^i$ superimposes $(\text{rcs})^{i-1}$. From elementary analysis we know that any two orthogonal frames in space are related by one rotation and one translation. This is illustrated in Fig. 1-14 and expressed by

$$\mathbf{q}_{i,j}^{[i-1]} = \mathsf{T}_{i(i-1)} \mathbf{q}_{i,j}^{[i]} + \mathbf{R}_{i(i-1)} \tag{1-79}$$

where $\mathsf{T}_{i(i-1)}$ is the compound rotation between frames and $\mathbf{R}_{i(i-1)}$ is the translation between frames. The superscripts above the $\mathbf{q}_{i,j}$ denote the frame of reference.

Perhaps the best-known example of application of the residue chain model is in generating the backbone conformations of polypeptide chains. Polypeptides will be discussed in much greater length later in this book and for now it will suffice to say that polypeptides are composed of residues of the form

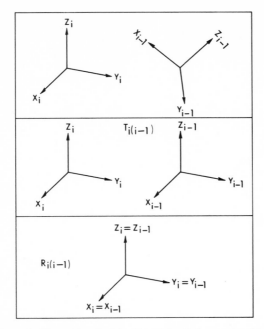

Fig. 1-14. A schematic illustration that $(rcs)^i$ and $(rcs)^{i-1}$ are related by, at most, a compound rotation matrix, $T_{i(i-1)}$, and a translation, $R_{i(i-1)}$.

which are always planar units (except for R_i and H_i') because of the partial double bond character of the $C'\!\!=\!\!=\!\!N$ bond (the peptide bond). The R_i group represents the side chain of the ith residue and does not enter into backbone calculations. The various backbone conformations result from internal bond rotations about C^α—C' bonds (denoted by ψ) and bond rotations about N—C^α bonds (denoted by ϕ). The complete geometry of two residues ($i-1$ and i) is shown in Fig. 1-15.

In this case the translation vector $\mathbf{R}_{i(i-1)}$ is simply given as

$$\mathbf{R}_{i(i-1)} = \begin{pmatrix} |C^\alpha_{i-2} - C_{i-1}| \\ 0 \\ 0 \end{pmatrix} \qquad (1\text{-}80)$$

Since any compound rotation matrix can be considered as a product of unitary rotation matrices, $T_{i(i-1)}$ can be given by

$$T_{i(i-1)} = T_{\alpha_{i-1}} T_{\phi_{i-1}} T_{\beta_i} T_{\psi_i} T_{\gamma_i} \qquad (1\text{-}81)$$

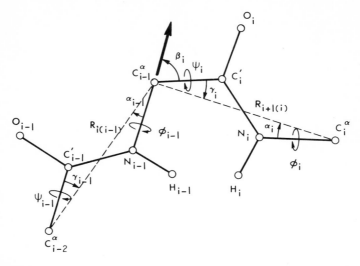

Fig. 1-15. The geometry of the $(i-1)$th and ith planar peptide units (residues) for $\phi_{i-1} = 0$ and $\psi_i = \pi$.

The respective unitary matrices have component form

$$T_{\gamma_i} = \begin{pmatrix} \cos\gamma_i & \sin\gamma_i & 0 \\ -\sin\gamma_i & \cos\gamma_i & 0 \\ 0 & 0 & 1 \end{pmatrix} \qquad (1\text{-}82)$$

T_{β_i} and $T_{\alpha_{i-1}}$ are similar matrices with β_i and α_{i-1} replacing γ_i, respectively:

$$T_{\psi_i} = \begin{pmatrix} 1 & 0 & 0 \\ 0 & \cos\psi_i & \sin\psi_i \\ 0 & -\sin\psi_i & \cos\psi_i \end{pmatrix} \qquad (1\text{-}83)$$

and $T_{\phi_{i-1}}$ is of the same form as T_{ψ_i} with ϕ_{i-1} replacing ψ_i. To enhance the meaning of these rotation parameters, it should be pointed out that α_{i-1} is a function of the residue coordinates of C_{i-1}^{α} and N_{i-1} while γ_i is determined by the residue coordinates of $C_i{}'$ and the origin of $(\text{rcs})^i$. However, β_i is completely a function of the bonding geometry of C_{i-1}. If the bonding geometry of C_{i-1}^{α} is perfectly tetrahedral, then $\beta_i = 70.5°$.

We can write down an equation of identical form to (1-79) relating the jth coordinate vector of an atom in $(rcs)^{i-1}$ to $(rcs)^{i-2}$,

$$\mathbf{q}_{i,j}^{[i-2]} = \mathsf{T}_{i-1(i-2)}\,\mathbf{q}_{i,j}^{[i-1]} + \mathbf{R}_{i-1(i-2)} \tag{1-84}$$

Now if we were to substitute (1-79) into (1-84) for $\mathbf{q}_{i,j}^{[i-1]}$ we would have an expression for the jth coordinate vector of an atom in $(rcs)^i$ relative to $(rcs)^{i-2}$. This equation is

$$\mathbf{q}_{i,j}^{[i-2]} = \mathsf{T}_{i-1(i-2)}\,\mathsf{T}_{i(i-1)}\,\mathbf{q}_{i,j}^{[i]} + \mathsf{T}_{i-1(i-2)}\,\mathbf{R}_{i(i-1)} + \mathbf{R}_{i-1(i-2)} \tag{1-85}$$

We can continue this sequential backward substitution until we have expressed the jth coordinate vector of $(rcs)^i$ relative to $(rcs)^1$. The general transform equation from $(rcs)^i$ to $(rcs)^1$ is given by

$$q_{i,j}^{[1]} = \mathbf{R}_{2(1)} + \sum_{\mu=2}^{i-1}\left[\left(\prod_{\lambda=2}^{\mu}\mathsf{T}_{\lambda(\lambda-1)}\right)\mathbf{R}_{\mu+1(\mu)}\right] + \left(\prod_{k=2}^{i}\mathsf{T}_{k(k-1)}\right)\mathbf{q}_{i,j}^{[i]} \tag{1-86}$$

Then, if $(rcs)^1$ is defined to be the fixed coordinate system, (fcs), Eq. (1-86) allows us to express the coordinate vector of all atoms in any polypeptide backbone in any conformation relative to one common coordinate system, namely (fcs).

From the derivation of Eq. (1-86) it is quite obvious that once we were able to express $\mathbf{q}_{i,j}^{[i]}$ as $\mathbf{q}_{i,j}^{[i-1]}$ we had solved the problem of chain generation. The rest of the derivation amounted to "mathematical bookkeeping." The application of Eq. (1-86) when bond distance and bond angle distortions are allowed is simple compared to use of the general linear chain model. Corrections due to the distortions in the bonding topology manifest themselves in the set of $\mathsf{T}_{k(k-1)}$ and the set of $\mathbf{R}_{k(k-1)}$. Once the elements of these operators are computed the interresidue geometry is completely specified. For polypeptides the distortions in the interresidue geometry are completely defined by the values of $(C_i^\alpha - C_{i-1}^\alpha)$ and β_i. However, computing the set of $\mathbf{q}_{i,j}^{[i]}$ is not easily achieved and the general linear chain model, even though cumbersome to use, is generally employed to achieve this end.

The changes in the coordinate positions of the atoms of the polypeptide due to a rotation ω_i about the ith amide bond $C'\!\!=\!\!N$ may be formulated by first describing the coordinate positions of the atoms of residue i relative to the $(rcs)^i$. Let us first translate the origin of $(rcs)^i$ to the coordinate position of the carbonyl carbon, atom $j = 1$. The atomic coordinates in this shifted frame are

$$^1\mathbf{q}_{i,j}^{[i]} = {}^0\mathbf{q}_{i,j}^{[i]} - {}^0\mathbf{q}_{i,j}^{[i]} \tag{1-87}$$

where the 0 and 1 superscripts preceding the q refer to the atomic coordinates before and after, respectively, carrying out the coordinate shift. Next we carry out a rotation through an angle $-\delta_i$, i.e. the angle between the x axis and C'===N, followed, in turn, by the desired ω_i rotation about the C'===N bond. Lastly, we rotate through δ_i and shift the origin of (rcs)^i back to its original location. The corresponding rotation matrix is

$$\mathbf{T}_{\omega_i} \equiv \mathbf{T}_{\delta_i}\, \mathbf{T}_{\omega_i}\, \mathbf{T}_{\delta_i}^{-1} \tag{1-88}$$

and the new coordinate positions of atoms three through five of the backbone due to rotation ω_i are

$$\mathbf{q}_{i,j}^{[i]} = \mathbf{T}_{\omega_i}\,{}^{0}\mathbf{q}_{i,j}^{[i]} - \mathbf{T}_{\omega_i}\,{}^{0}\mathbf{q}_{i,1}^{[i]} + {}^{0}\mathbf{q}_{i,1}^{[i]} \tag{1-89}$$

for $j = 3, 5$ and all sidechain atoms. By making the appropriate substitutions into Eqs. (1-79)–(1-86), to account for the contributions of the ω rotations, we can derive an expression which is analogous to Eq. (1-86) and includes the ω bond rotations.

$$
\begin{aligned}
\mathbf{q}_{i,j}^{[1]} = \mathbf{R}_{2,1} &+ \sum_{\mu=3}^{i-1}\left\{\left[\prod_{\lambda=2}^{\mu-1}\mathbf{T}_{\lambda(\lambda-1)}\,\mathbf{T}_{\omega_\lambda}\right]\mathbf{T}_{\mu(\mu+1)}\,\mathbf{q}_{\mu,1}^{[\mu]}\right\} \\
&+ \sum_{\mu=2}^{i-1}\left\{\left[\prod_{\lambda=2}^{\mu}\mathbf{T}_{\lambda(\lambda-1)}\,\mathbf{T}_{\omega_\lambda}\right](\mathbf{R}_{\mu+1(\mu)} - \mathbf{q}_{\mu,1}^{[\mu]})\right\} \\
&+ \prod_{k=2}^{i-1}(\mathbf{T}_{k(k-1)}\,\mathbf{T}_{\omega_k})\,\mathbf{T}_{i(i-1)}\,\mathbf{q}_{i,1}^{[i]} + \mathbf{T}_{2(1)}\,\mathbf{q}_{2,1}^{[2]} \\
&+ \left[\prod_{k=2}^{i}\mathbf{T}_{k(k-1)}\,\mathbf{T}_{\omega_k}\right]\mathbf{T}_{i(i-1)}\,\mathbf{T}_{\omega_i}(\mathbf{q}_{ij}^{[i]} - \mathbf{q}_{i1}^{[i]}) \tag{1-90}
\end{aligned}
$$

(include only for $j \geqslant 4$, else use identity matrix)

The j index is assigned to the peptide atoms as follows:

$$
\begin{array}{ccccccc}
j: & 1 & 2 & 3 & 4 & 5 & >6 \\
\text{Atom:} & \text{C}' & \text{O} & \text{N} & \text{H} & \text{C}^\alpha & \text{side chain atoms}
\end{array}
$$

C. Interchain Geometry

There is no single formulation to describe interchain geometry in general. Rather each intermolecular geometry must be treated in a specific manner. In this section we will discuss perhaps the most often reoccurring type of interchain geometry, namely the packing of helical polymer chains.

Suppose one places a helix in a cartesian coordinate system such that the helix axis is parallel to the Z axis. There are four spatial variables which define the position and orientation of the helix in the cartesian frame relative to its initial position. With the aid of Fig. 1-16 we can define these four variables: $\mathbf{R}$ is a vector defined by the origin of the coordinate frame and the point of intersection of the helix axis with the XY plane. $\boldsymbol{\theta}$ is the angle measured from the positive X axis to the $\mathbf{R}$ vector. The angle is measured counter-

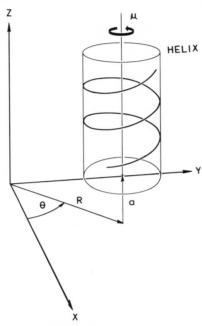

Fig. 1-16. The spatial variables $\mathbf{R}$, $\boldsymbol{\theta}$, $\mathbf{a}$, $\boldsymbol{\mu}$ indicating the position and orientation of a helix in a general cartesian frame.

clockwise. $\mathbf{a}$ is a vector parallel to the Z axis which originates at the point of intersection of the helix axis with the XY plane. $\boldsymbol{\mu}$ is a rotation about the axis of the helix. The sense of the rotation is counterclockwise when looking toward the XY plane from a position corresponding to a positive value of z.

Consider a collection of helices in the cartesian frame. The ith helix would have a position and orientation defined by $\mathbf{R}_i$, $\boldsymbol{\theta}_i$, $\mathbf{a}_i$, and $\boldsymbol{\mu}_i$. In Fig. 1-17 is shown the cartesian frame with a few helices included. The relative position and orientation of one helix, i, with respect to a second helix, j, can be expressed by

$$\mathbf{R}_i^{(j)} = \mathbf{R}_j - \mathbf{R}_i, \qquad \boldsymbol{\theta}_i^{(j)} = \cos^{-1}\{(R_i^{(j)})_x/R_i^{(j)}\} \qquad (1\text{-}91)$$

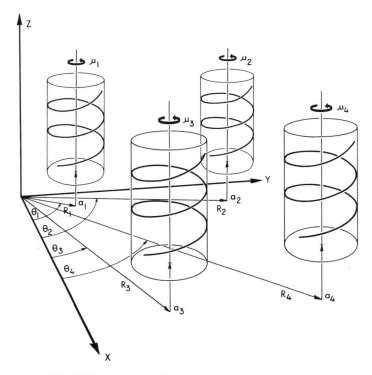

Fig. 1.17. A number of helices located in the cartesian frame.

where x denotes the x component of $R_i^{(j)}$. Unless the helices are identical the relationship between the $\mathbf{a}$ and the $\boldsymbol{\mu}$ has no meaning. In the event the helices are identical

$$\mathbf{a}_i^{(j)} = \mathbf{a}_j - \mathbf{a}_i \qquad \text{and} \qquad \boldsymbol{\mu}_i^{(j)} = \boldsymbol{\mu}_j - \boldsymbol{\mu}_i \qquad (1\text{-}92)$$

Of course the spatial variables $\mathbf{R}_i^{(j)}$, $\boldsymbol{\theta}_i^{(j)}$, $\mathbf{a}_i^{(j)}$, and $\boldsymbol{\mu}_i^{(j)}$ are measured relative to a new frame centered at the intersection of the helix axis j with the XY plane and parallel to the old frame.

The parameters $\mathbf{R}$ and $\boldsymbol{\theta}$ could be reexpressed as vector translations $\mathbf{b}$ and $\mathbf{c}$ along the X and Y axes, respectively, by the transformation

$$b = R\cos\theta, \qquad c = R\sin\theta \qquad (1\text{-}93)$$

Throughout the remainder of this book we will present other types of intermolecular geometries. Those geometries involving small molecules interacting with one another or with large substrates may be of particular interest.

Much of the present work in polymer science is concerned with gaining an understanding of how small molecules interact with macromolecules and the effect of such interactions on the entire complex.

References

1. H. Sugeta and T. Miyazawa, *Biopolymers* **5**, 673 (1967).
2a. A. M. Liquori and P. De Santis, *Biopolymers* **5**, 815 (1967).
2b. J. T. Edsall *et al.*, *Biopolymers* **4**, 121 (1966).
2c. IUPAC–IUB Commission on Biochemical Nomenclature, *Biochemistry* **9**, 3471 (1970).
3. F. H. C. Crick, *Acta Crystallogr.* **6**, 685 (1953).
4. H. Benoit and C. Sadron, *J. Polym. Sci.* **4**, 473 (1949).
5. H. Benoit, *J. Polym. Sci.* **3**, 376 (1948).
6. P. J. Flory, "Principles of Polymer Chemistry." Cornell Univ. Press, Ithaca, New York, 1953.
7. D. A. Brant, W. G. Miller, and P. J. Flory, *J. Mol. Biol.* **23**, 47 (1967).
8. D. A. Brant and P. J. Flory, *J. Mol. Biol.* **23**, 67 (1967).
9. P. R. Schimmel and P. J. Flory, *Proc. Nat. Acad. Sci. U.S.* **58**, 52 (1967).
10. J. D. Watson and F. H. C. Crick, *Nature (London)* **171**, 737, 964 (1953); F. H. C. Crick and J. D. Watson, *Proc. Roy. Soc. London* **A223**, 80 (1954).
11. I. M. Klotz, "Handbook of Biochemistry" (H. A. Soker, ed.). Chem. Rubber Co., Cleveland, Ohio, 1968.

I. Introduction

In this chapter we attempt to characterize the origin, nature, and magnitude of the forces which dictate the conformation of macromolecules and, for that matter, molecules in general. It should be clear that a complete *ab initio* quantum mechanical approach is out of the question for all but the simplest molecules such as HCl, H_2, etc. The major reason that this is the case follows from the fact that the size of the associated secular equation is so large that the time required to solve this equation, even with the fastest computer, is prohibitively long. As a result it is necessary to derive simplified energy expressions which adequately describe the interactions between nonbonded atoms in the molecules and which still can be evaluated rapidly. In attempting to perform conformational calculations one is constantly faced with the problem of maximizing the sophistication of the potential functions while minimizing the time required to evaluate these functions.

Over the last ten years many different levels of refinement have been used in performing conformational calculations. A general rule which seems to apply is that the level of sophistication used in the energy calculations is inversely proportional to the size of the molecule. For example, the conformational properties of pairs of peptide residues have been calculated via extended Hückel theory (1) a widely employed quantum mechanical approach while for long polypeptide chains similar to those found in proteins only crude "steric" potential functions have been employed (2). In many cases, especially for polymers with many degrees of freedom, the conformational calculation reduces to finding the minima of a function of many variables.

II. The Hard Sphere Model

The concept of spherical atoms of different size for different elements, packed together in ordered arrangement so that they are in mutual contact, is an old one (3). Once a sizable amount of data had been collected on the distance between atomic centers of atoms in various crystals, an attempt was made to deduce the corresponding size of the atomic spheres of different *bonded* atoms. Further, such work showed that these distances were relatively constant (and therefore additive) for the same pairs of bonded atoms regardless of the structural environment. Thus it was possible to define a set of atomic radii for certain standard types of bonding and environment which correctly predicted bond distances.

As data on the distance between nonbonded centers of pairs of atoms accumulated, it was observed that these distances for two atoms of an element or elements were also roughly the same in different structures (4a). The degree of constancy is not nearly as great as for bonded atoms and the interatomic distances are much larger. Nevertheless even roughly constant and additive "van der Waals radii" have proved useful in crystal structure analysis. Many sets of "van der Waals radii" have been proposed. Table 2-1 contains a

TABLE 2-1
Van der Waals Radii of Atoms and/or Groups of Atoms

Atom or group	van der Waals radius (Å)	Atom or group	van der Waals radius (Å)
H	1.05–1.25	O (sp)	1.30
C′ (sp^2)	1.50	O (sp^2)	1.35
C (sp^3)	1.65	OH	1.55
CH	1.85	F	1.40
CH^2	2.05	Cl	1.80
N (sp^2)	1.35	P	1.75
N (sp^3)	1.50	S	2.55
NH$_2$	1.75		

recently compiled listing of van der Waals radii. Within the last few years *contact distances* have come to be used in place of the van der Waals radii. A contact distance is the minimum distance between the centers of two atoms. Figure 2-1 schematically illustrates the van der Waals radii and the contact distance associated with a pair of atoms. In principle the contact distance should be equal to the sum of the van der Waals radii of the two atoms involved in the contact. In practice this does not turn out to be true. There is no single set of van der Waals radii which can be assigned to a set of atoms so that the correct contact distance is reproduced for every pairwise interaction. Thus one is forced to use contact distances rather than van der Waals radii in describing the closest packing of nonbonded atoms. Another way of saying this is that the van der Waals radii of nonbonded interactions are not strictly additive. Table 2-2 lists the most up-to-date set of contact distances. Note that

Table 2-2 lists two types of contact distances, "normally allowed" and "outer limit." Each outer limit value corresponds to shortest contact distance observed in the solid state. The "normally allowed" distance is the arithmetic average distance in which *all* recorded distances for each pair of atoms is included in the sum used to compute the average. These distances are most often determined from crystal structures. Molecular beam scattering may provide a new and elegant means to calculate contact distances.

If one defines an allowed conformation of macromolecule to be any configuration in which all variable nonbonded interaction distances are equal to or greater than the corresponding contact distances, then, in principle, all sterically allowed conformations of the macromolecule may be calculated. From the two types of contact distances defined in Table 2-2 it is possible to speak of "normally allowed conformations" and "outer limit conformations."

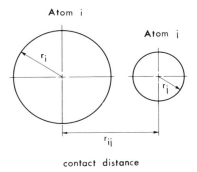

contact distance

Fig. 2-1. A two-dimensional demonstration of the concepts of van der Waals radii and contact distance for the pair of atoms *i* and *j*.

In normally allowed conformations all interaction distances are greater than or equal to the normally allowed contact distances, while in outer limit conformations at least one contact distance is less than the corresponding normal contact distance but greater than or equal to the outer limit contact distance.

Thus, if one had a molecule which possessed two degrees of freedom, i.e., two bond rotations are possible, he could construct a contact map which would show the regions of allowed conformations. An example of such a contact map for two planar peptide units of L-alanine is shown in Fig. 2-2.

$$\left[\begin{array}{c} \phi \overset{CH_3}{\underset{|}{\,}} \psi \\ -N \overset{\curvearrowright}{\,} C^\alpha \overset{\curvearrowright}{\,} C' - \\ | \quad\quad \| \\ H \quad\quad O \end{array} \right]_n$$

TABLE 2-2 Contact Distances of Many Pairwise Nonbonded Atomic Interactions to the Nearest Five-Hundredths of an Angstrom[a]

Atom	H		C'(sp²)		C(sp³)		N(sp²)		N(sp³)		O(sp)	
	Normal A	Outer A	Normal A	Outer A	Normal A	Outer A	Normal A	Outer A	Normal A	Outer A	Normal A	Outer A
H	2.15	1.90	2.75	2.70	2.85	2.80	2.60	2.50	2.65	—	2.55	2.35
C' (sp²)	2.75	2.70	2.95	2.90	3.10	2.95	2.85	2.70	3.00	—	2.75	2.70
C (sp³)	2.85	2.80	3.10	2.95	3.20	3.00	2.90	2.80	3.15	—	2.80	2.70
N (sp²)	2.60	2.50	2.85	2.70	2.90	2.80	2.70	2.60	2.95	3.00	2.65	2.60
N (sp³)	2.65	—	3.00	3.00	3.15	3.15	2.95	2.95	3.15	—	3.00	—
O (sp)	2.55	2.35	2.75	2.70	2.80	2.70	2.65	2.60	3.00	—	2.75	2.60
O (sp²)	2.65	2.55	2.85	2.75	2.80	2.70	2.70	2.60	3.10	2.95	2.75	2.65
F	2.50	2.40	2.95	2.80	3.10	2.95	2.75	2.65	2.90	—	2.70	2.55
Cl	2.90	2.75	3.25	3.15	3.45	3.30	3.20	3.05	3.35	—	3.20	3.05
P	2.90	2.70	3.25	3.10	3.45	3.35	3.15	3.05	3.30	—	3.10	3.05
S	3.60	3.45	4.00	3.95	4.20	3.95	4.00	3.90	4.10	—	3.90	3.75

Atom	O(sp²)		F		Cl		P		S	
	Normal A	Outer A	Normal A	Outer A	Normal A	Outer A	Normal A	Outer A	Normal A	Outer A
H	2.65	2.55	2.50	2.40	2.90	2.75	2.90	2.70	3.60	3.45
C' (sp²)	2.85	2.75	2.90	2.75	3.25	3.15	3.25	3.10	4.00	3.95
C (sp³)	2.80	2.70	2.90	2.70	3.45	3.30	3.45	3.35	4.20	3.95
N (sp²)	2.70	2.60	3.60	3.45	3.20	3.05	3.15	3.05	4.00	3.90
N (sp³)	3.10	2.95	2.90	—	3.35	—	3.30	—	4.10	—
O (sp)	2.75	2.65	2.70	2.55	3.20	3.05	3.10	3.05	3.90	3.75
O (sp²)	2.70	2.60	2.80	2.65	3.15	3.05	3.15	3.10	3.90	3.80
F	2.80	2.65	2.85	2.75	3.30	3.20	3.20	3.05	4.05	3.95
Cl	3.15	3.05	3.30	3.20	3.70	3.60	3.55	3.50	4.40	4.25
P	3.15	3.10	3.20	3.05	3.55	3.50	3.60	3.50	4.35	4.25
S	3.90	3.80	4.05	3.95	4.40	4.25	4.35	4.25	5.15	—

[a] Table positions which contain a dash indicate either insufficient data or excessive variability in the contact distance for that pair interaction.

Although such contact maps define regions of allowed conformations they cannot predict the most probable conformations. Further, it may well be that a total conformational energy lowering may result from more efficient packing if one, or perhaps even a few, contact distances are made slightly shorter than their normally allowed values.

Another difficulty which persists as long as the interactions are considered to be spherically symmetric, is the neglect of the true shape of the bonding

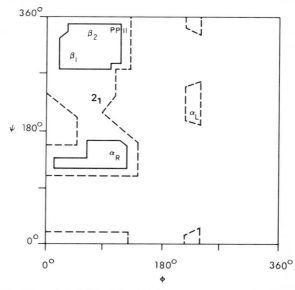

Fig. 2-2. Contact map of two planar peptide units of L-alanine (see text for details) using the contact distances given in Table 2-2. The dashed lines enclose allowed regions using the outer contact distances while the solid lines enclose allowed regions using the normal contact distances.

α_R	right-handed α-helix	β_2	parallel β-"helix"
2_1	twofold ribbon structure	PP II	left-handed 3_1-helix of poly-L-proline II
β_1	antiparallel β-"helix"	α_L	left-handed α-helix

See Bamford *et al.* (4b) for a detailed discussion of these polypeptide secondary structures.

orbitals. That is, the atoms cannot ultimately be treated as spheres in terms of their distribution of "bulk." Both the number of bonds in which the atom is involved as well as the bonding topology about the atom influence the distribution of "bulk" about the atom. Such a nonspherical distribution of "bulk" in atoms probably is a contributing factor to the variations found in the contact distances.

Inherently related to the shape of the atom is the electron density found on the atom. Since each atom has a unique characteristic electronegativity it follows that in a bonded system of atoms the distribution of charge is not uniform. Thus molecules are found to have dipole moments. Realistic conformational calculations should include the energy contributions due to the interaction of the charge distribution(s) on a given molecule or set of molecules.

Thus calculations involving the hard sphere model for atoms (or groups of atoms) can, at best, only differentiate between "allowed" and "forbidden" conformations. Even this distinction of states becomes uncertain for conformations in which only a few contact distances are slightly shorter than normally observed. However, when one is attempting to determine the conformational properties of a macromolecule with many free rotations, the hard sphere is very useful in reducing the size of the allowed conformational hyperspace.

III. Nonbonded Potential Functions

The total energy of any system of particles for which we neglect all possible vibrational complications can be given by

$$E_T = \sum_{i>j} V_{ij} + \sum_{i>j>k} V_{ijk} + \sum_{i>j>k>l} V_{ijkl} + \cdots \tag{2-1}$$

where the sum is over all particles in the system. If the particles make up a crystal, then the energy of the crystal is the value such that the above expression is minimized (this being rigorously true at absolute zero). There is still considerable debate as to whether or not this minimum is the global minimum or only a relative minimum which is not necessarily the global minimum. It can be seen that the energy is not simply the sum of the interaction of all possible pairs of interactions, V_{ij}, but also includes the effects which depend on the positions and properties of three (V_{ijk}), four (V_{ijkl}), and more particles. This result would not be expected if the charges on the particles were not polarizable. However, we do know that atoms, and groups of atoms, are polarizable and thus n-body ($n > 2$) interactions must be considered. To begin an analysis of the n-body interactions let us divide each of the types of interactions into two separate classes, namely, bonded interactions and nonbonded interactions. In other words we can write the terms in (2-1) as

$$V_{ij} = V_{ij}(\mathrm{B}) + v_{ij}(\mathrm{NB}) \tag{2-2}$$

$$V_{ijk} = V_{ijk}(\mathrm{B}) + v_{ijk}(\mathrm{NB}) \tag{2-3}$$

$$V_{ijkl} = V_{ijkl}(\mathrm{B}) + v_{ijkl}(\mathrm{NB}) \tag{2-4}$$

(B indicates bonded and NB nonbonded). In most conformational calculations the bonded interactions are normally assumed to be independent of conformation and thus make a constant contribution to the total energy. Hence they need not be considered when comparing two different conformations. For the time being let us adhere to the validity of this assumption and only investigate the nonbonded terms.

A. Analysis of v_{ij} (NB)

Generally only the pairwise additive terms are significant in conformational and configurational calculations although we will demonstrate, through the interaction equations, that higher order terms can be important when the particles have fixed interdependent positions. That is, potential energy curves relating the interaction energy of two nonbonded atoms with the distance between their centers involves the insufficiently tested assumption that the energy at a given distance is practically independent of (1) the orientation of the interatomic centerline relative to the orientations of the bonds by which these atoms are held to other atoms in the structure and (2) the nature and strengths of these bonds.

The general features of intermolecular potentials are very well known. At short distances the curve is repulsive and decreases roughly as an exponential in the separation. The repulsion arises from the coulombic nuclear–nuclear interaction and the electron–electron coulombic overlap interaction summing up to a greater repulsive value than the attractive interaction between the nuclear core and electrons on the two different atoms.

At distances near the minimum in energy the primary forces are the so-called London dispersion interactions which vary as R^{-6}, R^{-8}, and/or R^{-10} due to induced dipole, induced quadrupole, and induced octupole interactions. At very large distances (100–1000 Å) relativistic effects due to electron spin and the finite velocity of light change the form of this interaction greatly. In other words, we need a functional representation for the set of interactions described below:

$$E(R_{ij}) = \underbrace{\langle N(i)|N(j)\rangle}_{\text{nuclear–nuclear}} + \sum_k \underbrace{\langle N(i)|e_j(k)\rangle}_{\text{nuclear–electron}}$$

$$+ \sum_l \underbrace{\langle N(j)|e_i(l)\rangle}_{\text{nuclear–electron}} + \sum_k \sum_l \underbrace{\langle e_i(l)|e_j(k)\rangle}_{\text{electron–electron}} \qquad (2\text{-}5)$$

In practice the pairwise potential interactions are normally represented by one of three classical empirical potential functions.

1. *The Lennard-Jones 6-12 Potential Function*

$$P_{ij} = \epsilon_0[(R_0/R_{ij})^{12} - 2(R_0/R_{ij})^6] \qquad (2\text{-}6)$$

where ϵ_0 is the minimum energy at $R_{ij} = R_0$.

The attractive term, $-2(R_0/R_{ij})^6$, is rigorously derived for a pair of identical spherically symmetrical and chemically saturated molecules, while the repulsive term, $(R_0/R_{ij})^{12}$, is a "useful approximation" for the shape of repulsion which cannot be rigorously derived. The attractive term arises from an attraction between mutually induced fluctuating dipoles (distortion of electron orbitals) in the two freely rotating molecules. Notice that induced quadrupole and octupole moments are neglected. In principle, the parameters ϵ_0 and R_0 can be obtained for a given molecule from a quantum mechanical calculation, as suggested, but in practice they must be deduced from indirect experimental evidence. The second viral coefficient is the most useful source.

When the 6-12 potential function is used to describe the interaction between two nonbonded atoms which are part of a molecule and thus involved in a chemical bond the equation has the form

$$P_{ij} = \frac{B_{ij}}{R_{ij}^{12}} - \frac{A_{ij}}{R_{ij}^6} \qquad (2\text{-}7)$$

By applying Eq. (2-7) to the interaction between pairs of nonbonded atoms or groups of atoms we assume that the distortion of spherical symmetry and introduction of net residual charge on each of the pairs of atoms affect only the depth and position of the minimum in energy for the interaction. The reasonableness of this assumption is a subject of contention. The interactions due to the charge distribution on the molecule are assumed to be additively separable from the interaction described by the Lennard-Jones-type potential. These electrostatic interactions will be described in detail later.

The attractive term coefficients A_{ij} in Eq. (2-7) can be determined by applying the Slater–Kirkwood equation (5) in a slightly modified form as first suggested by Scott and Scheraga (6a);

$$A_{ij} = \frac{(\tfrac{3}{2}) e(\hbar/m^{1/2})\alpha_i \alpha_j}{C_{ij}} \qquad (2\text{-}8)$$

where

$$C_{ij} = (\alpha_i/N_{\text{eff}}^{(i)})^{1/2} + (\alpha_j/N_{\text{eff}}^{(j)})^{1/2} \qquad (2\text{-}9)$$

In these expressions m is the electron mass, e the electronic charge, α_i the atomic polarizability, and N_{eff} a correction factor which is supposed to take into account the nature, strengths, and orientations of the chemical bonds to which the pair of atoms belong. In essence, N_{eff} is the number of effective electrons surrounding the nucleus when an atom is chemically bound in some species.

Table 2-3 lists the values for α and N_{eff} for some atoms and groups of atoms.

TABLE 2-3

Atomic Polarizability (α) and Effective Number of Outer Shell Electrons (N_{eff}) for Various Atoms and Groups of Atoms[a]

Atom or group	$\alpha \times 10^{24}$ cm^3	N_{eff}
H	0.46	0.9
C' (sp^2)	1.20	4.8
C (sp^3)	1.30	5.0
CH$_2$	1.77	7.0
CH$_3$	2.05	7.8
N (sp^2)	1.15	6.0
O (sp^2)	0.84	7.0
F	1.93	8.5
Cl	24.5	14.0
P	23.0	13.0
S	23.9	13.5

[a] These parameters are used in the modified Slater–Kirkwood equation (Eq. 2-8). These values were taken from Pitzer (6b).

The repulsive term coefficients B_{ij} are calculated by insisting that the potential function P_{ij} have a minimum at the normal contact interaction distance R_0.

$$\frac{dP_{ij}}{dR} = \frac{6A_{ij}}{R_{ij}^7} - \frac{12B_{ij}}{R_{ij}^{13}} \tag{2-10}$$

$$\frac{6a_{ij}}{R_0^7} - \frac{12B_{ij}}{R_0^{13}} = 0 \tag{2-11}$$

$$\tfrac{1}{2}A_{ij}R_0^6 = B_{ij} \tag{2-12}$$

It should be noted that we can calculate an outer-limit contact distance R_0' for each potential function by determining that value of R_{ij} for which $P_{ij} = 0$. If the potential function is realistic the calculated R_0' should be similar to that found in crystals. We could have reversed this sequence of calculations and calculated B_{ij} from the R_0'. It is difficult to assert which of these two techniques is better. Tables 2-4A and B list a set of B_{ij} and A_{ij} for a large number of nonbonded atomic pair interactions.

TABLE 2-4A A_{IJ} and B_{IJ} for the Modified Lennard-Jones 6-12 Nonbonded Potential Function Given in Eq. (2-7)[a]

Atom	H		C'(sp²)		C(sp³)		N(sp²)		N(sp³)		O(sp)	
	A_{IJ}	B_{IJ}	A_{IJ}	B_{IJ}	A_{IJ}	B_{IJ}	A_{IJ}	B_{IJ}	A_{IJ}	B_{IJ}	A_{IJ}	B_{IJ}
H	see below[b]		165.8	1.924	127.4	3.743	124.9	2.710	128.3	1.995	128.5	1.391
C' (sp²)	165.8	1.924	600.2	19.62	465.3	19.70	571.4	14.50	453.7	12.16	452.5	9.753
C (sp³)	127.4	3.743	465.4	19.70	372.5	28.58	366.4	21.63	348.7	16.82	346.1	8.650
N (sp²)	124.9	2.710	571.4	14.50	366.4	21.63	362.9	16.09	346.1	10.58	344.3	6.085
N (sp³)	128.3	1.995	453.7	12.16	348.7	16.82	346.1	10.58	344.7	17.82	344.7	12.38
O (sp)	128.5	1.391	452.4	9.753	346.1	8.650	344.3	6.085	344.7	12.38	344.8	7.448
O (sp²)	123.9	2.503	461.9	11.48	367.2	20.52	365.0	15.28	349.5	15.75	348.7	7.538
F	79.80	0.9361	284.5	8.861	223.2	10.07	222.4	4.755	218.7	6.452	219.6	4.256
Cl	272.8	8.074	967.4	55.42	759.3	64.64	738.8	37.31	735.3	51.83	733.7	36.93
P	264.8	7.825	938.5	54.61	742.6	62.83	733.3	35.44	712.6	50.65	713.0	32.69
S	290.2	30.28	1076.0	268.8	845.9	225.1	1038.0	189.5	748.5	179.2	755.2	132.5

Atom	O(sp²)		F		Cl		P		S	
	A_{IJ}	B_{IJ}	A_{IJ}	B_{IJ}	A_{IJ}	B_{IJ}	A_{IJ}	B_{IJ}	A_{IJ}	B_{IJ}
H	123.9	2.503	79.80	0.9361	272.8	8.074	264.8	7.825	290.2	30.28
C' (sp²)	461.9	11.48	284.5	8.861	967.4	55.42	938.5	54.61	1076.0	268.8
C (sp³)	367.2	20.52	223.2	10.07	759.3	64.64	742.6	62.83	845.9	225.1
N (sp²)	365.0	15.28	222.4	4.755	738.8	37.31	733.3	35.44	1038.0	189.5
N (sp³)	349.5	15.75	218.7	6.452	735.3	51.83	712.6	50.65	748.5	179.2
O (sp)	348.7	7.538	219.6	4.256	733.7	36.93	713.0	32.69	755.2	132.5
O (sp²)	367.2	14.49	221.8	5.571	753.7	36.71	738.7	36.15	780.3	136.2
F	221.8	5.571	134.6	3.652	457.6	29.67	446.5	29.04	474.7	106.4
Cl	753.2	36.71	457.6	29.67	1562.0	200.5	1512.0	153.7	1608.0	585.1
P	738.7	36.15	446.5	29.04	1512.0	153.7	1473.0	170.7	1563.0	513.4
S	780.3	136.2	474.7	106.4	1608.0	585.1	1563.0	513.4	1685.0	1383.0

[a] The A_{IJ} are given in units of kcal $Å^6$/mole and the B_{IJ} 10^{-4} kcal $Å^{12}$/mole.

TABLE 2-4 B

Values of A_{ij} and B_{ij} for Various Distances and Interactions[a]

$R_0 A$	A_{ij} kcal Å⁶/mole	$B_{ij} \times 10^{-3}$ kcal Å¹²/mole	Type of H···H interaction
1.90	18.87	0.4436	Aromatic H···Aromatic H
2.00	25.58	0.8192	
2.10	34.17	1.469	
2.20	55.56	3.125	Aromatic H···Aliphatic H
2.30	64.77	5.352	
2.40	76.00	7.220	Aliphatic H···Aliphatic H

[a] The parameters for the H···H function are the least reliable of those reported in this table. The nature of the H···H interaction probably varies with the type of atom(s) to which each of the H's are bonded. We report a number of B_{ij} for various minimum energy distances and *roughly* designate which functions should be used for particular hydrogens.

2. The Exponential-6 Potential Function

$$P_{ij} = \frac{\epsilon_0}{1 - (6/\kappa)} \left[\frac{6}{\kappa} \exp\left(\kappa \left[1 - \frac{R_{ij}}{R_0} \right] \right) - \left(\frac{R_0}{R_{ij}} \right)^6 \right] \qquad (2\text{-}13)$$

where ϵ_0 and R_0 are the same as defined earlier and κ is an adjustable parameter (normally about 12.4) which introduces a shape dependence into the function.

When this function is used to describe the interaction between nonbonded interactions between atoms which are chemically bonded elements of a molecule it reduces to the familiar Buckingham potential function

$$P_{ij} = B_{ij} \exp(-C_{ij} r_{ij}) - (A_{ij}/r_{ij}) \qquad (2\text{-}14)$$

The Buckingham potential function differs from the Lennard-Jones function in the form of the repulsive term. A two parameter exponential function is used in place of a one parameter, inverse 12th power law. In principle this repulsive term should be more specific in describing the repulsive interaction because of the additional parameter. However, at this time no accurate values of C_{ij} are known for pair interactions. Best estimates are $4.35 \leqslant C_{ij} \leqslant 4.70$. In most circumstances $C_{ij} = 4.60$ for all types of pair interactions. Clearly the role of C_{ij} is to control the shape of the potential function. Once again the A_{ij} and B_{ij} are found in the same way as was described in the section on the Lennard-Jones 6-12 potential function.

3. The Kihara Potential Function

$$P_{ij} = 4\epsilon_0 \left[\left(\frac{1-\gamma}{(R_{ij}/\sigma) - \gamma} \right)^{12} - \left(\frac{1-\gamma}{(R_{ij}/\sigma) - \gamma} \right)^6 \right] \qquad (2\text{-}15)$$

where $P_{ij} = 0$ at σ. Through γ and σ the Kihara potential function introduces an effective hard core and a shape dependence.

We have already described some of the factors which have been neglected in computing these potential functions. However, we have completely over-looked any temperature dependence in the A_{ij}, B_{ij}, and C_{ij}. It seems reasonable that these coefficients should be temperature sensitive. After all, crystals melt, phase (and conformational) changes occur in solution and the solid state as a function of temperature. If one hopes to predict any of these thermodynamic properties of a system composed of macromolecules (or molecules for that matter) it is necessary to be able to describe A_{ij}, B_{ij}, and C_{ij} as a function of temperature. A consideration of temperature dependence in the potential functions is discussed in Chapter 3. Some possible means of determining temperature dependent parameters are discussed in detail.

IV. Electrostatic Energy

As stated earlier, atoms composing a molecule will have different electron densities owing to different electronegativities. Thus there must be some characteristic energy associated with the charge distribution within the molecule which is dependent upon the molecular conformation. The electrostatic energy is the major contribution to correct the nonbonded potential functions to take into account the fact that the nonbonded interactions occur between atoms or groups of atoms which are part of a molecule(s). Strictly speaking the nonbonded potential functions which take into account the steric forces are only valid for neutral "gas"-like atoms or molecules (i.e., He, CO_2, etc). Thus, when we include electrostatic energy contributions we account for the chemical bonding topology of the molecule.

A. The Dipole Model

One possible way of representing the nonuniform electron density within the molecule is by a set of bond and/or group moments. Table 2-5 lists some bond and/or group moments.

Given two point dipoles, μ_i and μ_j, separated by the vector $\mathbf{r}_{i,j}$, the mutual potential energy is

$$D_{i,j} = (\epsilon^{-1}[(\mu_i \cdot \mu_j)\, r_{i,j}^{-3} - 3(\mu_i \cdot \mathbf{r}_{i,j})(\mu_j \cdot \mathbf{r}_{i,j})\, r_{i,j}^{-5}] \tag{2-16}$$

where ϵ is the apparent bulk dielectric due to the molecular environment, to be discussed in greater length shortly.

The total electrostatic energy D_T by this technique is

$$D_T = \tfrac{1}{2} \sum_{i=1}^{n} \sum_{j \neq i}^{n} D_{i,j} \tag{2-17}$$

where n is the number of distinct bonds in the macromolecule.

TABLE 2-5

Bond and/or Group Moments Often Found in Macromolecules

Bond or group	Bond moment (debye units)	Bond or group	Bond moment (debye units)
O—H	1.53	C—O	0.86
N—H	1.31	C=O	2.48
C—H	0.35	C⋯N	0.20
C—F	1.50	C≡N	3.65
C—Cl	1.56	O—CH₃	0.65

B. Calculation of Residual Charges

The second means of computing the electrostatic energy depends upon being able to compute the actual set of residual charges residing on the atoms composing the molecule. These charges are then considered to be point charges located at the geometric centers of atoms in the molecule(s). For these charges to be realistic they must reproduce all associated dipole moments of bonds and groups of atoms.

There are many different means of calculating the residual charges using quantum mechanical schemes. In general each method yields different results than the others since each method defines electron density volume in a different way. For example, in some cases the electron density volume is defined in terms of cylinders along the bonds in which an atom participates while other techniques use volumes of spheres about atoms as effective electron density volumes. Since the total residual charge on an atom is directly proportional to the electron density volume these two approaches will yield different results.

There is one semiempirical technique of computing residual charges which has become popular. In this method one assumes that the saturated σ and π contributions to the residual charges are completely separable. The method has never been extended to d-orbital contributions. The saturated σ residual charges are calculated by applying the Del Re (7) technique. This is an application of the electronegativity concept in bonded systems.

From a mathematical point of view, the Del Re method may be considered as the counterpart, in saturated systems, to the simple Hückel method used for π electron calculations. In fact, it employs the same formalism as the simple Hückel procedure, but treats all the saturated bonds as strictly localized units and, instead of delocalized orbitals, introduces an inductive effect

TABLE 2-6

Set of Del Re Parameters for the Saturated Charge Distributions[a]

Bond	C—H	C—C	C—N	C—N[+]	C—O	C—O[1/2]
ϵ_{ab}	1.00	1.00	1.00	1.33	0.95	0.80
$\gamma_{a(b)}$	0.30	0.10	0.10	0.10	0.10	0.10
$\gamma_{b(a)}$	0.40	0.10	0.10	0.10	0.10	0.10
δ_a	0.07	0.07	0.07	0.07	0.07	0.07
$\delta_b{}^0$	0.00	0.07	0.24	0.31	0.40	0.33

Bond	N—H	N[+]—H	C—H	C—S	S—H	S—S
ϵ_{ab}	0.45	0.60	0.45	0.75	0.70	0.60
$\gamma_{a(b)}$	0.30	0.30	0.30	0.20	0.30	0.10
$\gamma_{b(a)}$	0.40	0.40	0.40	0.40	0.40	0.10
$\delta_a{}^0$	0.24	0.31	0.40	0.07	0.07	0.07
$\delta_b{}^0$	0.00	0.00	0.00	0.07	0.00	0.07

[a] Poland and Scheraga (8b) have suggested slightly different values for these parameters.

(similar to that originally proposed by Wheland and Pauling) (8a) as the main feature describing the influence of the neighboring atoms on each particular bond of a given molecule. The method, like the Hückel approach, is semiempirical and requires the choice of a suitable set of parameters. Those given in Table 2-6 have often been used with reasonable success.

The parameters used in the calculations are of three kinds. The bond parameters ϵ_{xy} correspond strictly to the exchange integrals of the usual Hückel method. The two other parameters, $\delta_x{}^0$ and $\gamma_{x(y)}$, are used to determine the coulomb integral δ_x of the various atoms according to the formula

$$\delta_x = \delta_x{}^0 + \sum_{y \text{ adj } x} \gamma_{x(y)} \delta_y \qquad (2\text{-}18)$$

If there are N atoms in the molecule, Eq. (2-18) leads to a set of N linear
equations in N unknowns which can be solved numerically. The inductive
parameter $\gamma_{x(y)}$ indicates the fraction of the coulomb integral of y that has to
be included in the coulomb integral x, in order to take into account the
inductive effect of y on x.

The residual charges can now be calculated as follows. For each bond x—y
one solves the secular equation:

$$\begin{pmatrix} \alpha_x - E & \beta_{xy} \\ \beta_{xy} & \alpha_y - E \end{pmatrix} = 0 \qquad (2\text{-}19)$$

where $\alpha_x = \alpha + \delta_x \beta$; $\beta_{xy} = \epsilon_{xy} \beta$; the lower root E_B, when inserted into the
linear system of equations

$$(\alpha_x - E) C_1 + \beta_{xy} C_2 = 0 \qquad (2\text{-}20)$$

$$\beta_{xy} C_1 + (\alpha_y - E) C_2 = 0 \qquad (2\text{-}21)$$

gives the coefficients C_1 and C_2 of the bonding orbital,

$$\psi = C_1 \chi_x + C_2 \chi_y \qquad (2\text{-}22)$$

The net charge appearing on x as a result of the formation of the x–y bond is
therefore

$$q_{x(y)} = 1 - 2C_1{}^2 \qquad (2\text{-}23)$$

Solving (2-20) and (2-21) and substituting into (2-23), one gets

$$q_{x(y)} = \frac{\delta_y - \delta_x}{2\epsilon_{xy}} \left[1 + \left(\frac{\delta_y - \delta_x}{2(\delta_x \delta_y)^{1/2}} \right)^2 \right]^{-1/2} \qquad (2\text{-}24)$$

Then the total residual charge of an atom is given by

$$Q_x = \sum_Y q_{x(y)} \qquad (2\text{-}25)$$

C. π Charge Calculation (Simple Hückel Method)

We assume that each molecular orbital, ϕ_i, can be written as a linear
combination of atomic orbitals χ_u.

$$\phi_i = \sum_{u=1}^{N} C_{iu} \chi_u \qquad (2\text{-}26)$$

where N is the number of $2P_z$ localized atomic orbitals χ_u. Since the ground
state electronic structure requires us to minimize the energy with respect to

C_{iu} we end up with the secular determinant

$$|H_{uv} - \epsilon_i S_{uv}| = 0 \tag{2-27}$$

where

$$
\begin{aligned}
H_{uv} &= \alpha & \text{if} \quad u = v \\
&= \beta & \text{if} \quad u \text{ and } v \text{ are bonded} \\
&= 0 & \text{if} \quad u \text{ and } v \text{ are not bonded} \\
S_{uv} &= 1 & \text{if} \quad u = v \\
&= 0 & \text{if} \quad u \neq v
\end{aligned}
$$

and ϵ_i is the energy of the ith molecular orbital and a solution of the secular equation.

We will discuss the meaning and methods of solving secular equations when we discuss quantum mechanical techniques later in this chapter. Once we have solved the secular determinant we can substitute back into the secular equation and find each set of $\{C_i\}$ associated with molecular orbital ϕ_i, which is associated with orbital energy ϵ_i.

$$(H_{11} - \epsilon_i S_{11}) C_{i1} + (H_{12} - \epsilon_i S_{12}) C_{i2} + \cdots = 0 \tag{2-28}$$
$$(H_{21} - \epsilon_i S_{21}) C_{i1} + (H_{22} - \epsilon_i S_2) C_{i2} + \cdots = 0 \tag{2-29}$$
$$\vdots$$

Then the π electron density on atom μ, q_μ, is defined as

$$q_\mu = \sum_i N_i C_{i\mu}^2 \tag{2-30}$$

where N_i is the number of electrons in the ith molecular orbital (N_i can have values 0, 1, 2) and $C_{i\mu}$ is the coefficient of atomic orbital χ_μ in the ith molecular orbital. For the ground state we, obviously, fill the molecular orbitals in the order of lowest to highest orbital energy. The net charge on atom μ, Q_μ, is defined as

$$Q_\mu = N_\mu - q_\mu \tag{2-31}$$

where N_μ is the effective nuclear π charge.

D. π Charge Calculation (Reproduction of Bond and/or Group Moments)

Although there is no absolute way to verify experimentally the accuracy of a calculated charge distribution it seems reasonable that one criterion required for a realistic charge distribution is that the charge distribution reproduce known experimental bond and/or group moments. Conversely if one knows the bond and/or group moments and a set of saturated charges he can solve for the set of π charges. There is no rigorous mathematical description of this technique since the distribution of groups having high group moments is a function of the bonding topology. In general the procedure taken in these calculations is as follows:

(1) Pick the M bonds or groups which have large moments and/or significant π charges.

(2) Consider each of the M bonds or groups as being isolated with respect to the π charges. That is, for each bond or group the sum of π charges is zero.

(3) Compute the set of π charges for each of the M groups.

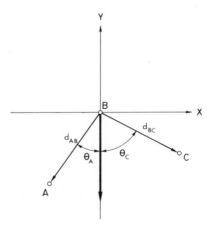

Fig. 2-3. An A, B, C set of atoms placed in an XY cartesian plane so that the bond moment, ρ_T, is in common with the negative y axis.

As an example let us derive expressions for the π charges on a group of atoms having the configuration

(examples of groups having this configuration are NH_2, COO^-, $COCH_3$, etc.). Let ρ_T be the group moment and $\{\sigma_A, \sigma_B, \sigma_C\}$ be the set of saturated charges on atoms (groups) $\{A, B, C\}$.

We begin by placing the configuration in the XY plane and arbitrarily assigning the origin of ρ_T at B. This is shown in Fig. 2-3.

Then the σ contributions ρ_{A_x}, ρ_{A_y}, ρ_{C_x}, and ρ_{C_y} to the total bond moment $\mathbf{\rho}_T$ are given by

$$\rho_{Ax} = -\sin \theta_A |\mathbf{d}_{AB}| \sigma_A k \tag{2-32}$$

$$\rho_{Ay} = -\cos \theta_A |\mathbf{d}_{AB}| \sigma_A k \tag{2-33}$$

$$\rho_{Bx} = \sin \theta_C |\mathbf{d}_{BC}| \sigma_C k \tag{2-34}$$

$$\rho_{By} = -\cos \theta_C |\mathbf{d}_{BC}| \sigma_C k \tag{2-35}$$

where k is a conversion constant. From the positioning of $\mathbf{\rho}_T$ it follows that

$$\rho_{Ax} + \rho_{Bx} + (-\sin \theta_A |\mathbf{d}_{AB}| \pi_A k) + (\sin \theta_C |\mathbf{d}_{BC}| \pi_C k) = 0 \tag{2-36}$$

and

$$\rho_{Ay} + \rho_{By} - \cos \theta_A |\mathbf{d}_{AB}| \pi_A k - \cos \theta_C |\mathbf{d}_{BC}| \pi_C k = -|\mathbf{\rho}_T| \tag{2-37}$$

From the isolation assumption,

$$\pi_A + \pi_B + \pi_C = 0 \tag{2-38}$$

Thus Eqs. (2-36), (2-37), (2-38) give us three equations in three unknowns π_A, π_B, π_C.

It should be obvious that if a particular group of atoms can take up different conformations, the above scheme is meaningless. This restriction can be circumvented by expressing (2-38) as

$$\pi_A + \pi_B + \pi_C = \delta_{BD} \tag{2-39}$$

where δ_{BD} is a parameter to account for electron exchange between atom B and an atom D, not belonging to the $\{A,B,C\}$ group. Each of the N groups have a characteristic $\delta_{\mu\nu}$ parameter for which

$$\sum_{\mu\nu} \delta_\mu = 0 \tag{2-40}$$

E. Calculation (Bonded) of the Monopole Electrostatic Energy

From elementary physics the electrostatic potential energy between atom i with residual charge Q_i and atom j with residual charge Q_j separated by a distance $r_{i,j}$ in which the medium has a bulk dielectric ϵ is given by

$$D_{i,j} = \frac{kQ_i Q_j}{\epsilon r_{i,j}} \tag{2-41}$$

where k is a conversion constant. The total electrostatic energy is given by

$$D_{\mathrm{T}} = \sum_{i=1}^{N-1} \sum_{j=i+1}^{N} D_{i,j} \qquad (2\text{-}42)$$

where $N \equiv$ number of atoms in the macromolecule, and additional restrictions on the sum can be imposed as desired (i.e., only nonbonded interactions might be considered).

V. Induced Dipole Interactions

The residual charge on an atom in a molecule arises because different atoms have different "electron pulling powers," e.g., different electronegativities. But it is also clear that each atom is, to some degree, polarizable. This is clear from the fact that spherical molecules which are chemically saturated, nevertheless, interact—the van der Waals attraction discussed earlier. Thus, the residual charge distribution on the molecule should produce an effective field in which each constituent polarizable atom (or group) finds itself. The net result is a set of interactions: induced dipole interactions, induced dipole–residual charge interactions, and induced dipole–induced dipole interactions. The dipole induction energy plus the induced dipole–induced dipole interaction energy should, in principle, be the precise stabilizing energy approximated by the attractive term in the nonbonded potential functions. This is strictly correct only for an isolated molecule *in vacuo* since the electric field is computed only from the charge distribution on the molecule.

We can derive a set of expressions to describe these interactions in the following way. Define $\mathbf{E}$ as the electric field, $\mathbf{m}$ the induced dipole, $\mathbf{r}$ the interaction vector, α the coefficient of polarizability, q the residual charge, and n the total number of atoms. We assume that the induced dipole of atom i is given by

$$m_x^{(i)} = \alpha_i E_x^{(i)} \qquad (2\text{-}43)$$
$$m_y^{(i)} = \alpha_i E_y^{(i)} \qquad (2\text{-}44)$$
$$m_z^{(i)} = \alpha_i E_z^{(i)} \qquad (2\text{-}45)$$

The field $\mathbf{E}^{(i)}$ at i due to all other polarizable atoms is

$$\mathbf{E}^{(i)} = \sum_{j \neq i=1}^{n} \left[-\nabla \frac{(\mathbf{m}_j \cdot \mathbf{r}_{ij})}{r_{ij}^3} + \frac{q_j \mathbf{r}_{ij}}{r_{ij}^3} \right] \qquad (2\text{-}46)$$

Let us define

$$X_{ij} \equiv x_i - x_j \tag{2-47}$$

$$Y_{ij} \equiv y_i - y_j \tag{2-48}$$

$$Z_{ij} \equiv z_i - z_j \tag{2-49}$$

where (x_k, y_k, z_k) are the coordinates of the kth atom. If we expand Eq. (2-46) in component form, we have

$$E_x^{(i)} = \sum_{j \neq i=1}^{n} \left[\frac{3X_{ij}^2 - r_{ij}^2}{r_{ij}^5} m_x^{(j)} + \frac{3Y_{ij}X_{ij}}{r_{ij}^5} m_y^{(j)} + \frac{3Z_{ij}X_{ij}}{r_{ij}^5} m_z^{(j)} + \frac{q_j X_{ij}}{r_{ij}^3} \right] \tag{2-50}$$

$$E_y^{(i)} = \sum_{j \neq i=1}^{n} \left[\frac{3X_{ij}Y_{ij}}{r_{ij}^5} m_x^{(j)} + \frac{3Y_{ij}^2 - r_{ij}^2}{r_{ij}^5} m_y^{(j)} + \frac{3Y_{ij}Z_{ij}}{r_{ij}^5} m_z^{(j)} + \frac{q_j Y_{ij}}{r_{ij}^3} \right] \tag{2-51}$$

$$E_z^{(i)} = \sum_{j \neq i=1}^{n} \left[\frac{3X_{ij}Z_{ij}}{r_{ij}^5} m_x^{(j)} + \frac{3Y_{ij}Z_{ij}}{r_{ij}^5} m_y^{(j)} + \frac{3Z_{ij}^2 - r_{ij}^2}{r_{ij}^5} m_z^{(j)} + \frac{q_j Z_{ij}}{r_{ij}^3} \right] \tag{2-52}$$

Substituting these expressions into Eqs. (2-43)–(2-45) gives

$$m_x^{(i)} = \alpha_i \sum_{j \neq i=1}^{n} \left[\frac{3X_{ij}^2 - r_{ij}^2}{r_{ij}^5} m_x^{(j)} + \frac{3Y_{ij}X_{ij}}{r_{ij}^5} m_y^{(j)} + \frac{3Z_{ij}X_{ij}}{r_{ij}^5} m_z^{(j)} + \frac{q_j X_{ij}}{r_{ij}^3} \right] \tag{2-53}$$

$$m_y^{(i)} = \alpha_i \sum_{j \neq i=1}^{n} \left[\frac{3X_{ij}Y_{ij}}{r_{ij}^5} m_x^{(j)} + \frac{3Y_{ij}^2 - r_{ij}^2}{r_{ij}^5} m_y^{(j)} + \frac{3Y_{ij}Z_{ij}}{r_{ij}^5} m_z^{(j)} + \frac{q_j Y_{ij}}{r_{ij}^3} \right] \tag{2-54}$$

$$m_z^{(i)} = \alpha_i \sum_{j \neq i=1}^{n} \left[\frac{3X_{ij}Z_{ij}}{r_{ij}^5} m_x^{(j)} + \frac{3Y_{ij}Z_{ij}}{r_{ij}^5} m_y^{(j)} + \frac{3Z_{ij}^2 - r_{ij}^2}{r_{ij}^5} m_z^{(j)} + \frac{q_j Z_{ij}}{r_{ij}^3} \right] \tag{2-55}$$

Since there is a set of equations like (2-53)–(2-55) for each atom i, we generate a system of $3n$ linear equations in $3n$ unknowns;

$$\mathbf{Ax = B} \tag{2-56}$$

where

$$\mathbf{x} = \begin{pmatrix} m_x^{(1)} \\ m_y^{(1)} \\ m_z^{(1)} \\ m_x^{(2)} \\ m_y^{(2)} \\ m_z^{(2)} \\ \vdots \\ m_x^{(n)} \\ m_y^{(n)} \\ m_z^{(n)} \end{pmatrix}, \quad \mathbf{b} = \begin{pmatrix} \alpha_1 \sum_{j=2}^{n} \dfrac{e_j X_{1j}}{r_{ij}^3} \\ \alpha_1 \sum_{j=2}^{n} \dfrac{e_j Y_{1j}}{r_{ij}^3} \\ \alpha_1 \sum_{j=2}^{n} \dfrac{e_j Z_{1j}}{r_{ij}^3} \\ \alpha_2 \sum_{j \ne (2)=1}^{n} \dfrac{e_j X_{2j}}{r_{2j}^3} \\ \alpha_2 \sum_{j \ne (2)=1}^{n} \dfrac{e_j Y_{2j}}{r_{2j}^3} \\ \alpha_2 \sum_{j \ne (2)=1}^{n} \dfrac{e_j Z_{2j}}{r_{2j}^3} \\ \vdots \\ \alpha_n \sum_{j=1}^{n-1} \dfrac{e_j X_{nj}}{r_{nj}^3} \\ \alpha_n \sum_{j=1}^{n-1} \dfrac{e_j Y_{nj}}{r_{nj}^3} \\ \alpha_n \sum_{j=1}^{n-1} \dfrac{e_j Z_{nj}}{r_{nj}^3} \end{pmatrix}, \quad \text{and}$$

$$\mathbf{A} = \begin{pmatrix} \tilde{A}_{11} & \tilde{A}_{12} & \tilde{A}_{13} & \cdots & \tilde{A}_{1n} \\ \tilde{A}_{21} & \tilde{A}_{22} & \tilde{A}_{23} & \cdots & \tilde{A}_{2n} \\ \vdots & \vdots & \vdots & \vdots & \vdots \\ \tilde{A}_{n1} & \tilde{A}_{n2} & \tilde{A}_{n3} & \cdots & \tilde{A}_{nn} \end{pmatrix}$$

where

$$\tilde{A}_{ii} = \begin{pmatrix} 1 & 0 & 0 \\ 0 & 1 & 0 \\ 0 & 0 & 1 \end{pmatrix}, \quad \tilde{A}_{ij} = -\alpha_i \begin{pmatrix} \dfrac{3X_{ij}^2 - r_{ij}^2}{r_{ij}^5} & \dfrac{3Y_{ij}X_{ij}}{r_{ij}^5} & \dfrac{3Z_{ij}X_{ij}}{r_{ij}^5} \\ \dfrac{3X_{ij}Y_{ij}}{r_{ij}^5} & \dfrac{3Y_{ij}^2 - r_{ij}^2}{r_{ij}^5} & \dfrac{3Y_{ij}Z_{ij}}{r_{ij}^5} \\ \dfrac{3X_{ij}Z_{ij}}{r_{ij}^5} & \dfrac{3Y_{ij}Z_{ij}}{r_{ij}^5} & \dfrac{3Z_{ij}^2 - r_{ij}^2}{r_{ij}^5} \end{pmatrix}$$

We can solve these simultaneous equations by any number of methods, some of which will be discussed later, for the set of triplets $\{m_x^{(i)}, m_y^{(i)}, m_z^{(i)}\}$. Then the various interaction energies can be calculated as follows.

Induced dipole energy

$$E_{\text{IND}} = \frac{1}{\epsilon} \sum_{i=1}^{n} \frac{\mathbf{m}_i \cdot \mathbf{m}_i}{2\alpha_i} \tag{2-57}$$

Induced dipole–residual charge interaction energy

$$E_{\text{ID-RC}} = \frac{1}{\epsilon} \sum_{i=1}^{n} \sum_{j \neq i=1}^{n} \frac{e_i(\mathbf{m}_j \cdot \mathbf{r}_{ij})}{r_{ij}^3} \tag{2-58}$$

Induced dipole–induced dipole interaction energy

$$E_{\text{ID-ID}} = \frac{1}{\epsilon} \sum_{i=1}^{n} \sum_{j \neq i=1}^{n} \left[\frac{\mathbf{m}_i \cdot \mathbf{m}_j}{r_{ij}^3} - \frac{3(\mathbf{m}_i \cdot \mathbf{r}_{ij})(\mathbf{m}_j \cdot \mathbf{r}_{ij})}{r_{ij}^5} \right] \tag{2-59}$$

VI. Discussion of the Concept of a Dielectric on the Molecular Level

Virtually all the effort has been applied to the accurate calculation of bond moments and partial charges, and little or no consideration has been given to the accuracy of the dielectric factor in pairwise electrostatic calculations. However, an error in the dielectric is as critical as an error in the bond moments and partial charges. Moreover, an accurate calculation of the bond moments and partial charges is meaningless in the electrostatic energy calculation if there is a large error in the dielectric term.

The major reason for the lack of information on the dielectric constant's role in macromolecular conformation is that the problem is not well understood. The very idea of a constant bulk valued dielectric constant ϵ which is used in all of these calculations, is, at best, a crude representation of reality. The concept of a bulk dielectric constant arises from a model where the two interacting charged species are immersed in a homogeneous medium which has an associated constant dielectric effect. For macromolecules in any solvent, the associated medium is heterogeneous with respect to both the direction of the interaction vector of the two atoms, as well as the magnitude of this interaction vector. Qualitatively, the dielectric effect arises because of shielding and coupling effects of the solvent molecules as well as the atoms of the solute molecules composing the system. In this respect the dielectric constant reflects the magnitude of 3-, 4-, ..., N-body interactions which we have neglected. At interaction distances near the contact distance, the dielectric effect of the pair interaction should be near unity regardless of the direction of the interaction vector. This should be the case because the probability of shielding or coupling from other constituents would be low due to the short interaction distance. At moderate interaction distances the dielectric term would be sensitive to both the magnitude and direction of the interaction vector owing to the competing dielectric effects of the solvent and solute molecules. At large interaction distances the value of the dielectric term should assume the value of the macroscopic bulk dielectric constant associated with the solution.

Since we are able to qualitatively describe the behavior of the dielectric constant with distance we can assume a model which is consistent with the qualitative scheme. We know at the contact distance r_{con}, $\epsilon(r_{ij}) = 1$ and for some particular interaction distance r_{eff}, the dielectric constant $\epsilon(r_{ij}) = \epsilon_0$, the bulk dielectric constant of the system. However, we do not know the functional behavior of the dielectric between r_{con} and r_{eff}. Let us assume a linear dependence of dielectric on distance in this interval and compare the equation which results to the classic coulombic potential function in which we use a constant dielectric constant, ϵ_0. In our model we will assume $\epsilon(r_{ij}) = \epsilon_0$ for $r_{ij} > r_{eff}$, that is, we assume the coulomb potential is valid in this region. Hence the linear dielectric model we propose here will differ from the coulomb potential only in the region $r_{con} \leqslant r_{ij} < r_{eff}$. Figure 2-4 shows our linear dielectric model. For $r_{con} \leqslant r_{ij} < r_{eff}$ the electrostatic potential is

$$\phi_{ij} = -k \int_{r_{eff}}^{r_{ij}} \frac{Q_i Q_j}{(at+b)\,t^2}\,dt - k \int_{\infty}^{r_{eff}} \frac{Q_i Q_j}{\epsilon_0 t^2}\,dt \tag{2-60}$$

$$\phi_{ij} = -k \left[\frac{-1}{bt} + \frac{a}{b^2}\ln\frac{(at+b)}{t} \right]\Bigg|_{r_{eff}}^{r_{ij}} Q_i Q_j + \frac{KQ_i Q_j}{\epsilon_0 r_{eff}} \tag{2-61}$$

$$\phi_{ij} = -kQ_i Q_j \left[\frac{-1}{br_{ij}} + \frac{1}{br_{eff}} + \frac{a}{b^2}\ln\frac{(ar_{ij}+b)}{r_{ij}} - \frac{a}{b^2}\ln\frac{(ar_{eff}+b)}{r_{eff}} \right] + \frac{kQ_i Q_j}{\epsilon_0 r_{eff}} \tag{2-62}$$

$$\phi_{ij} = \frac{kQ_i Q_j}{b}\left\{ \frac{1}{r_{eff}} - \frac{1}{r_{ij}} \right\} + kQ_i Q_j \frac{a}{b^2}\left\{ \ln\frac{(ar_{eff}+b)}{r_{eff}} - \ln\frac{(ar_{ij}+b)}{r_{ij}} \right\} + \frac{kQ_i Q_j}{\epsilon_0 r_{eff}} \tag{2-63}$$

Calculation of a and b for $y = ax + b$ in the interval (r_{con}, r_{eff}) is achieved by the following boundary equation:

$$y - \epsilon_c = \frac{\epsilon_0 - \epsilon_c}{r_{eff} - r_{con}} (x - r_{con}) \tag{2-64}$$

$$y = \left(\frac{\epsilon_0 - \epsilon_c}{r_{eff} - r_{con}} \right) x + \epsilon_c - \left[\frac{\epsilon_0 - \epsilon_c}{r_{eff} - r_{con}} \right] r_{con} \tag{2-65}$$

$$a = \frac{\epsilon_0 - \epsilon_c}{r_{eff} - r_{con}} \tag{2-66}$$

$$b = \epsilon_c - \left[\frac{\epsilon_0 - \epsilon_c}{r_{eff} - r_{con}} \right] r_{con} \tag{2-67}$$

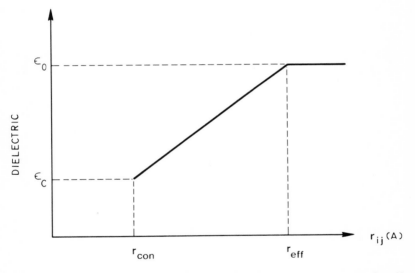

Fig. 2-4. The dependence of the dielectric $\epsilon(r_{ij})$ on the interaction distance, r_{ij}, for a linear dielectric model. The ϵ_c is the value of the dielectric at the contact distance, r_{con}, and ϵ_0 is the effective bulk value of the dielectric at some effective interaction distance, r_{eff}.

Let us compare (2-63) to the coulomb potential for some realistic values of r_{eff}, r_{con}, ϵ_0, ϵ_c, for polypeptides in solution for $r_{con} \leqslant r_{ij} < r_{eff}$.

$$r_{eff} = 7.0 \text{ Å}, \qquad r_{con} = 3.0 \text{ Å}, \qquad \epsilon_c = 1.0, \qquad \epsilon_0 = 4.0$$

Plots of $\phi_{ij}(\text{constant})/kQ_iQ_j$ and $\phi_{ij}(\text{linear})/kQ_iQ_j$ versus r_{ij} are shown in Fig. 2-5.

Thus we see that the electrostatic potential between two charged species is very sensitive to the spatial dielectric behavior of the medium which separates them. If we were to go one step further in the refinement of the theory to describe the electrostatic potential between two charged species, it would be necessary to take into account the possible heterogeneous nature of the solution or crystal. Let us concentrate the discussion on the dielectric properties of macromolecular solutions. Similar statements can be made about

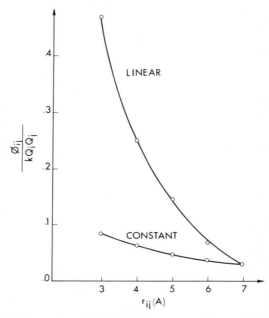

Fig. 2-5. A plot of $\phi_{ij}(\text{constant})/kQ_iQ_j$ and $\phi_{ij}(\text{linear})/kQ_iQ_j$ versus r_{ij}, the interaction distance between species i and j, which demonstrates the critical role of dielectric behavior in estimating the electrostatic pair potential energy.

the nature of the dielectric medium in crystals as those given here for solutions. The interaction vector between two nonbonded atoms in a macromolecule which is in solution might traverse a medium composed only of macromolecular atoms or, for the other extreme, intersect a medium composed only of solvent molecules. Obviously, most interactions occur across mediums composed of varying amounts of macromolecule and solvent. Unless the dielectric behavior of the solvent molecules is very similar to that of the macromolecule, sizable errors could result if the precise nature of the atomic heterogeneity is not taken into account in the theory to describe the dielectric medium between each pair of nonbonded interactions occurring in the macromolecule. However, from the computational point of view, the "bookkeeping" required in order to estimate the dielectric nature of medium between two nonbonded pairs of atoms in the macromolecule is far too time consuming to be practical. Hence, the dielectric medium has been considered homogeneous with respect to polymer and solvent. It has been only recently, in fact, that distance-dependent dielectric functions (9) have been introduced into the derivation of the electrostatic potential. Normally a dielectric constant (usually that of the bulk material) was inserted into the classical coulombic expression for pair interactions.

From what has been said, one would expect that using a constant dielectric to describe the medium between all pairs of nonbonded interactions would lead to gross errors in the total pairwise electrostatic potential energy. However, this does not turn out to be the case for macromolecules possessing a net zero charge. The results of such calculations indicate, in the majority of cases, that the constant dielectric model is nearly identical to the distance-dependent dielectric models with respect to the total pairwise electrostatic potential energy. Why is this so? Well, the total electrostatic energy, as we have said, is simply equal to the sum of the individual pairwise interactions of the constituent pairs of atoms in the macromolecule. For a neutral macromolecule there is a balanced number of attractive and repulsive interactions. In this case the total electrostatic potential energy is a measure of the nonuniformity of the charge distribution over the molecule as well as the set of associated interaction distances between all nonbonded pairs of atoms in the macromolecule. For each pairwise attractive electrostatic interaction there is a corresponding repulsive electrostatic interaction whose energy is approximately the same but opposite in sign. Thus very different individual electrostatic pair-potential functions will yield rather similar total electrostatic energies for neutrally charged macromolecules. However, when the total charge on the macromolecule is no longer equal to zero (the macromolecule possesses ionized groups) the total electrostatic potential energy becomes very sensitive to the form of the electrostatic pair-potential functions. In this case there is no longer a balance in the number of attractive and repulsive energy interactions. Past calculations were performed on neutral macromolecules so that the electrostatic potential energy calculations did not seriously affect the results. Workers are only now beginning conformational potential energy calculations on charged macromolecules (10). Hopefully, these studies will help to identify what direction will need to be taken in order to refine the electrostatic pair-potential functions.

VII. Polymer–Solvent Interactions

A. Introduction

The interaction between polymer and solvent is not well understood on the molecular level. This is mainly the result of the complex nature of this type of interaction. The extent to which solvent species "bind" to the polymer and the degree to which such "binding" can modify polymer conformation has eluded quantitative investigation. However, as we continue to probe deeper into investigating the forces that dictate polymer conformation in solution it is becoming ever more evident that polymer–solvent forces play a major role in conformational control. This fact is perhaps most evident in protein chemistry where the conformation of a protein is altered by minor changes in solvent composition.

There have been some ambitious efforts to study the ordering of solvent molecules about a polymer. Most notably are the pioneering efforts of Berendson (11) who used NMR spectroscopy to attempt to determine the structuring of water molecules about collagen and the infrared studies of Swenson and Formanek (12) who showed that water molecules are bonded to the carbonyl oxygens of poly-L-proline II in aqueous solution. Indeed the crystal structure studies of many proteins provides information concerning the arrangement of solvent molecules about the protein molecule (13). Unfortunately these structural investigations provide only a minimum amount of information concerning the molecular thermodynamics of the polymer–solvent interaction. What would be desirable is a relationship between polymer conformation and polymer–solvent free energy. Such a relationship would make it possible to determine the distribution of conformational states for a given polymer interacting with a particular solvent.

B. Models to Describe the Molecular Thermodynamics of Polymer–Solvent Interactions

One of the early attempts to develop a model to describe polymer–solvent interactions was carried out by Nemethy and Scheraga (14) as part of their comprehensive treatment of the structure of water. These workers carried out an extensive investigation of the hydrophobic bond–water interacting with hydrocarbons. Using available data they were able to estimate the experimental values of enthalpy $\Delta H°$, entropy $\Delta S°$, and free energy $\Delta F°$, for aqueous solutions containing hydrocarbons. These quantities are listed in Table 2-7. Also listed in Table 2-7 is a set of computed values of $\Delta H°$, $\Delta S°$, and $\Delta F°$ made by Nemethy and Scheraga using a solvent-ordering model in which ordered clustering of water molecules was assumed to occur about hydrocarbon molecules. The change in $\Delta F°$ in going from methane to ethane should correspond roughly to the $\Delta F°$ of a methyl group while the change in $\Delta F°$ in going from ethane to propane, or propane to butane, should correspond to the $\Delta F°$ of a methylene group. This, of course, presupposes conformational variations in the hydrocarbon molecules make constant contributions to $\Delta H°$ and $\Delta S°$.

Later Gibson and Scheraga (15) extended this approach to include polar solute species in aqueous solution and estimated the $\Delta F°$ of several atoms and groups. These quantities were then used in a polymer–solvent model very similar to the hydration shell model which is discussed in detail in the next section.

TABLE 2-7

Observed and Calculated Thermodynamic Parameters for the Transfer of Normal Aliphatic Hydrocarbons from the Liquid (or Nonpolar Solution) to Aqueous Solution at 25°C

	Observed			Calculated		
Substance	$\Delta F°$ (kcal/mole)	$\Delta H°$ (kcal/mole)	$\Delta S°$ (e.u.)	$\Delta F°$ (kcal/mole)	$\Delta H°$ (kcal/mole)	$\Delta S°$ (e.u.)
Methane	+2.51 to +3.15	−2.86 to −2.25	−18.4 to −16.8	+2.60	−1.97	−15.3
Ethane	+3.32 to +3.86	−2.37 to −1.27	−19.5 to −16.8	+3.77	−1.85	−18.8
Propane	+4.90 to +4.91	−2.09 to −1.45	−23.5 to −21.3	+4.96	−1.41	−21.2
Butane	+5.82 to +6.00	−0.96 to −0.72	−22.7 to −21.9	+5.84	−1.18	−23.0

Glasel (16) carried out a very interesting series of NMR experiments for a variety of polymers, with various types of side chains (see Table 2-8), in D_2O. By plotting the relaxation rate $(1/T_1)$ as a function of any of a number of solution variables, i.e., concentration, pH, and viscosity, he was able to make some deductions about the characteristic polymer–solvent interactions.

Poly(methacrylic acid), poly(L-lysine), and poly(L-glutamic acid) each undergo a conformational transition as a function of pH. This is reflected in the $(1/T_1)$ versus pH curves presented in Figs. 2-6a to 2-6c. There is, for all three polymers, a definite change in the relaxation rate over a relatively small range of pH's. (More will be said about these transitions later.) Nevertheless, Glasel, using data obtained from plots of the type given in

TABLE 2-8

Polymers Used in the NMR Polymer–Solvent Investigation

(I) Poly(vinyloxazolidinone methyl)
(PVO)

(II) Poly(vinylpyrrolidone)
(PVP)

(III) Poly(methacrylic acid)

(IV) Poly(L-glutamic acid)

(V) Poly(L-lysine)

(VI) Poly(adenylic acid)

(VII) Poly(uridylic acid)

Fig. 2-6, has postulated the following phenomenological rules concerning D_2O–polymer interactions:

(1) The following groups, where M^+ and X^- are counterions,

$$>C{=}O, \quad >NH, \quad -\overset{\overset{\displaystyle O}{|}}{C}-O^- \, M^+, \quad C-NH_3^+ \, X^-$$

do not form strong interactions with water.

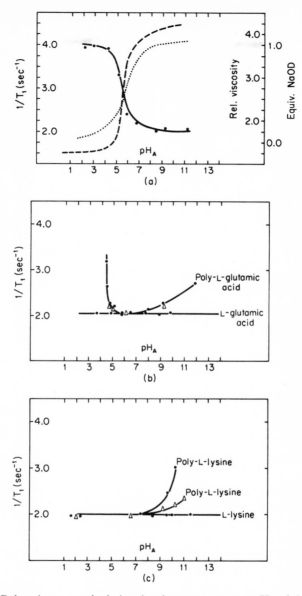

Fig. 2-6. (a) Relaxation rate and relative viscosity versus apparent pH and titration curve of a 3% solution of poly(methacrylic acid) (mol wt = 300,000 in D_2O): solid line, relaxation rate; dashed line, viscosity; dotted line, titration curve. (b) Relaxation rate versus apparent pH for 3% wt/v solutions of poly(L-glutamic acid) and glutamic acid, 0.1 N NaCl in D_2O: ●, mol wt = 100,000; △, mol wt = 3000. (c) Relaxation rate versus apparent pH for 3% wt/v solutions of poly(L-lysine) and L-lysine, 1 M KBr in D_2O: ●, mol wt = 75,000; △, mol wt = 5000. Reprinted from Glasel, *J. Amer. Chem. Soc.*, **92**, 375 (1970). Copyright 1970 by the American Chemical Society. Reprinted by permission of the copyright owner.

(2) $\overset{\text{O}}{\overset{\|}{—C}}\!—\text{OH},\ \text{CN}\!\!<^{\text{H}}_{\text{H}}$ *do* form strong interactions with water.

(3) Polymers which have side chains having formal charges will strongly interact with solvent only if the charges are intra- or intermolecularly wholly or partly neutralized; that is, counterion effects are eliminated.

(4) When the conformational fluctuations are large, and have characteristic times of the order of 10^{-3} sec, the D_2O–polymer interactions are destroyed.

The most ambitious theoretical approach to studying the effect of water molecule–polymer interactions has been launched by Krimm and Venkatachalam (17). These workers have brought water molecules into the vicinity of the carbonyl groups in poly-L-proline using the geometry shown in Fig. 2-7. By varying the parameters ω, ψ, α, and θ they were able to compute

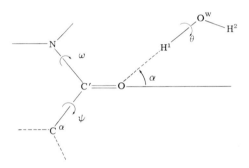

Fig. 2-7. Geometry of the imide group with a hydrogen-bonded water molecule. From Krimm and Venkatachalam (17).

the conformational free energies for poly-L-proline. This biopolymer can exist in two forms; *trans* and *cis* which are defined in detail in Chapter 4. The free energy maps for these two forms of poly-L-proline with and without water–poly-L-proline interactions are shown in Figs. 2-8a to 2-8d. These results indicate that the cis form is nearly independent of interaction with water while the trans form of the biopolymer is moderately sensitive to interaction with water. This is in agreement with experiment (see Chapter 4).

The disadvantages of this approach to investigating polymer–solvent interactions is (a) the excessive time required in the computations to compute free energies due to the large number of degrees of freedom, (b) the neglect of interactions between all groups and/or atoms in the polymer with water molecules.

A general method which circumvents both of these problems is the use of a hydration shell model. We will describe this model in considerable detail in

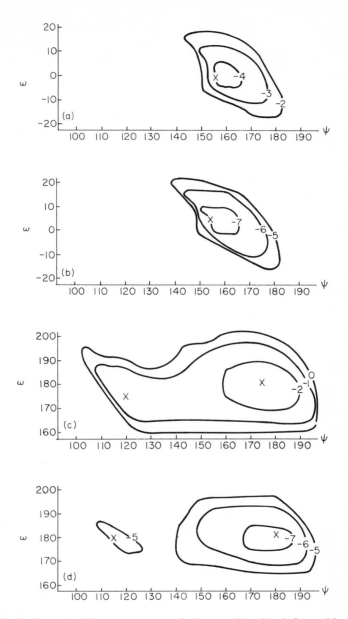

Fig. 2-8. Conformational free energy maps of poly-L-proline: (a) cis form without water; (b) cis form with water; (c) trans form without water; (d) trans form with water. Energy contours are in kcal/mole of residues and the × denote minima. From Krimm and Venkatachalam (17).

the next section. We state in advance that the hydration shell model has the disadvantage of (a) not accounting for possible long range ordering of solvent molecules about a polymer, (b) introduces several debatable assumptions concerning the molecular energetics of polymer–solvent interactions. Nevertheless, the hydration shell model represents the most effective means of dealing with the molecular aspects of polymer–solvent interactions at present.

C. The Hydration Shell Model

1. Description of the Model

The concept of a hydration shell to describe the behavior of solvent molecules near a solute species has been used for many years (18). As mentioned earlier, Gibson and Scheraga (15) modified existing hydration shell models so as to be applicable to the atoms of a solute macromolecule. The model described here is a further modification of the basic hydration shell concept. It differs from the Gibson–Scheraga model in the size and properties of the hydration shells as well as the criteria for calculating excluded volumes in the hydration shells.

In adopting a hydration shell model we assume that a characteristic sphere can be centered about each atom of the macromolecule. The size of the sphere, which defines the hydration shell, is dependent upon the solvent molecule and solute atom of the macromolecule. A particular change in free energy is associated with the removal of a solvent molecule from the hydration shell. The size of the hydration shell and the shape of the solvent molecule dictates how many solvent molecules can occupy the hydration shell. The sum of the intersections of the van der Waals volumes of the atoms of the macromolecule with the hydration shell results in an excluded volume which determines how many solvent molecules are removed from the hydration shell when the macromolecule is in a particular conformation. Thus the hydration shell is sensitive to conformation via excluded hydration shell volumes.

The hydration shell model is a four-parameter system: n, the maximum number of solvent molecules which can occupy the hydration shell; Δf, the change in free energy associated with the removal of one solvent molecule from the hydration shell; R_v, the effective radius of the hydration shell; and V_f, the free volume of packing associated with one solvent molecule in the hydration shell.

Table 2-9 contains the values of the parameters n, Δf, R_v, and V_f for several different atoms solvated in aqueous solution. Also given in Table 2-9 are the values for each V_i', which is defined as the effective volume of the hydration shell of atom i. This quantity must be known in order to compute V_f and, in itself, is a useful parameter for measuring the "amount of room" which solvent molecules can occupy about atom i of the macromolecule. Perhaps the best way to explain how the values of the hydration shell parameters are calculated is to describe the computational procedure for a representative case.

For example, the values of n, R_v, V_f, and V' were calculated for $N(sp^2)$ as follows: First, a bonding geometry was assigned to the $N(sp^2)$ atom by choosing a sphere having the van der Waals radius of nitrogen (1.35 Å) to

TABLE 2-9

Macromolecule–Solvent Interaction Parameters R_v, V_f, n, Δf for Some Atoms and Groups of Atoms in Aqueous Solution (Also Given is V', the Effective Hydration Shell Volume)[a]

Atom or group	n	Δf (kcal/mole)	R_v (Å)	V_f (Å³)	V' (Å³)
N (sp²)	2	0.63	4.33	35.8	114.0
C (sp²)	2	0.63	3.90	14.3	71.0
O (carbonyl)	2	1.88	3.94	67.6	177.6
H (amide)	2	0.31	3.54	31.3	105.0
CH₃ (aliphatic)	8	−0.13	5.50	41.8	498.2
CH₂ (aliphatic)	4	−0.10	5.50	60.8	328.0
CH (aliphatic)	2	−0.13	5.50	104.8	252.0
CH (aromatic)	3	0.11	3.90	3.3	76.6
O (hydroxyl)	2	1.58	3.94	55.2	152.8
H (hydroxyl)	2	0.31	3.54	54.7	151.8
O⁻ (carboxyl)	4	4.20	4.10	42.5	255.0
O (carboxyl)	2	4.20	4.10	64.1	170.6
H (carboxyl)	2	0.31	3.54	54.7	151.8

[a] The volume of the H_2O molecule, $V_s = 21.2$ Å³.

represent the bulk of the atom. Then, to account for the trigonal covalent bonds in which the $N(sp^2)$ atom participates, three cylinders of indefinite length and having a radius equal to three-quarters of the radius of the sphere were positioned around the van der Waals sphere in threefold symmetry. The radius of the cylinders were determined from studying CPK molecular models of $N(sp^2)$. The sphere and cylinders represented a steric potential barrier to the solvent molecules. The geometry is shown in Fig. 2-9. Solvent molecules were assigned initial positions about this model for the $N(sp^2)$ atom and the total interaction energy of the system, including interactions between the $N(sp^2)$ atom and solvent molecules as well as solvent molecule–solvent molecule interactions, was minimized using the Davidson (19) technique. The conformational potential functions and parameters described in other sections of this chapter were employed to calculate the total interaction energy for each iteration in the minimization. The values of the residual charges on the atoms or groups vary from macromolecule to macromolecule. However, this variation is rather small (<0.09 a.u.) in all groups reported here. Thus, adopting the residual charges found in polypeptide chains as universally representative of all situations is not too bad an assumption. Whenever possible, use was made of the symmetry of the model in order to reduce the number of variables to be minimized. Since the $N(sp^2)$ atomic geometry

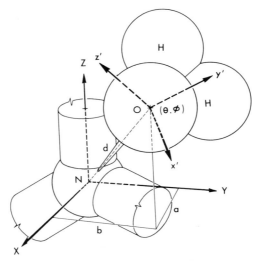

Fig. 2-9. Geometry of an $N(sp^2)$ atom in a molecule used to determine the characteristic polymer–solvent interaction parameters using the hydration shell model. d is the distance of the water molecule from the N, and a and b define the position of the water molecule in the XY plane of the N cartesian frame. θ_s and ϕ_s define the relative orientation of the water molecule in the cartesian frame associated with the water molecule.

possessed a twofold plane of symmetry (the plane being defined by the axes of the cylinders), solvent molecules were added to the system in pairs of two—one on each side of the plane, and having mirror positions and orientations relative to one another. In order to restrict the calculations to a single hydration shell, the constraint was imposed that only those minima in potential energy were considered which corresponded to positions and orientations of the solvent molecules within a distance equal to the sum of the van der Waals radius for the $N(sp^2)$ atom and the maximum steric diameter of a water molecule (3.7 Å from CPK models). This distance represents an initial upper-limit approximation for the radius of the hydration shell of the $N(sp^2)$ atom.

The volume V_s of the solvent molecule (in this case water) was calculated by assigning van der Waals spheres to represent the oxygen and two hydrogens and determining how much volume these three spheres manifested when they were allowed to overlap in a manner consistent with the bond lengths and bond angle of the H_2O molecule. Equation (2-72) was used to make this volume calculation and V_s for H_2O is given in Table 2-9.

Values of n equal to 2, 4, 6, 8 were used in the energy minimization. The results indicated that $n = 2$ yielded the deepest energy minimum and the two water molecules positioned and orientated themselves about the $N(sp^2)$ atom in such a manner that the effective hydration shell radius R_v is 4.33 Å. Next the value of V' was computed by subtracting the volumes of the van der Waals sphere and the three cylinders of the $N(sp^2)$ model from the hydration shell volume of the $N(sp^2)$ atom using R_v as the hydration shell radius. Then V_f could be computed from

$$V_f = (V' - nV_s)/n \tag{2-68}$$

The Δf's listed in Table 2-9 are taken from Gibson and Scheraga (15) rather than using the free energies computed from the energy minimization data necessary in the determination of the values of n, R_v, and V_f. The Gibson and Scheraga parameters were found experimentally and, hence, are probably more accurate than the theoretical predictions. It is interesting to note, however, that no theoretical interaction free energy between polymer atom and solvent molecule differs from the corresponding free energy interaction as given by Gibson and Scheraga by more than 12%. Using analogous procedures to those presented for the $N(sp^2)$ atom it was possible to evaluate the necessary polymer–solvent interaction parameters for the atomic species listed in Table 2-9.

To complete the polymer–solvent interaction model the following properties are assigned to the hydration shells:

(1) Each solvent molecule occupies an identical volume, $(V_s + V_f)$, in a hydration shell which is independent of how many other solvent molecules are present at any instant.

(2) The dynamic characteristics of the polymer–solvent interaction is included by *assuming* a *linear* relationship between the total change in free energy and the amount of excluded volume in the hydration shell.

(3) There is no additional contribution to the polymer–solvent free energy from a hydration shell once n solvent molecules have been removed from the hydration shell.

(4) All energy parameters are temperature independent and valid only near room temperature.

There are three unique types of intersections between the hydration shell of atom i and the van der Waals sphere of atom j having volume V_j and radius r_j:

(1) $V_i' \cap v_j = 0$ when $R_{v_i} + r_j \leqslant r_{ij}$, where r_{ij} is the distance between the centers of atom i and atom j.

(2) $V_i' \cap v_j = \frac{4}{3}\pi r_j^3$ when $r_j + r_{ij} \leqslant R_{v_i}$.

(3) The third case is $R_{v_i} + r_j > r_{ij}$. With the aid of Fig. 2-10 we can derive an expression for $V_i' \cap v_j$. First we compute the volume associated with $V_i' \cup v_j$ by the solid of revolution technique which is described in any elementary calculus text.

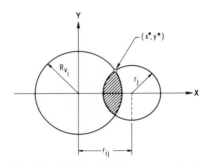

Fig. 2-10. A two-dimensional projection of the geometry of the intersection of the solvation shell of atom i and the van der Waals sphere of atom j.

From simple analytic geometry it is found that the x coordinate, x^*, of the intersection of the solvation shell and the van der Waals sphere is given by

$$x^* = (R_{v_i}^2 - r_j^2 + r_{ij}^2)/2r_{ij} \tag{2-69}$$

Then for $r_j + R_{v_i} > r_{ij}$

$$V_i' \cup v_j = \pi \left[\int_{-R_{v_i}}^{x^*} (R_{v_i}^2 - x^2)\, dx + \int_{x^*}^{r_j + r_{ij}} (r_j^2 - [x - r_{ij}]^2)\, dx \right] \tag{2-70}$$

and

$$V_i' \cap v_j = \frac{4}{3}\pi (R_{v_i}^3 + r_j^3) - V_i' \cup v_j \tag{2-71}$$

or

$$V_i' \cap v_j = \frac{1}{3}\pi [2R_{v_i}^3 + 2r_j^3 + r_{ij}^3] - \pi [R_{v_i}^2 x^* + (r_{ij} - x^*)(r_j^2 + r_{ij} x^*)] \tag{2-72}$$

Then for some particular conformation K of the macromolecule having a total of N atoms the total excluded volume of the ith hydration shell due to all other nonbonded atoms is

$$V_i^0 = \sum_{i \neq j}^{N} (V_i' \cap v_j) \tag{2-73}$$

and i and j not bonded. Since the solvation free energy is, in terms of this model, the change in free energy in going from a completely solvated state to some partially solvated state dictated by conformation K, the total solvation free energy $F(K)_i$ associated with atom i in the Kth conformation is

(a) $\quad F(K)_i = \Delta f \left[n - \left(\dfrac{V_i^0}{V_s + V_f} \right) \right] \qquad$ when $\quad V_i^0 < n(V_s + V_f) \quad$ (2-74)

(b) $\quad F(K)_i = 0 \qquad\qquad\qquad\qquad$ when $\quad V_i^0 \geqslant n(V_s + V_f) \quad$ (2-75)

where n, Δf, and V_f are chosen according to which type of atom is indicated by the index i.

From the way $F(K)_i$ is defined in Eqs. (2-74) and (2-75), it is clear that $F(K)_i$ is discontinuous for $V_i^0 = n(V_s + V_f)$. However, since we are considering the intersection of spheres with spheres V_i^0 will never equal $n(V_s + V_f)$ unless there is a steric overlap of the van der Waals spheres. Under such conditions the conformational energy will be exceedingly high and the entire structure will be rejected. Hence, this polymer–solvent model has continuous energy functions for all "sterically" allowed polymer conformations and can be used without fear of "blowups" in any minimization routine.

The total solvation free energy $F_T(K)$ for the macromolecule in the Kth conformation is simply

$$F_T(K) = \sum_{i=1}^{N} F(K)_i \tag{2-76}$$

2. Polymer–Solvent Studies Using the Hydration Shell Model

In this section we report the results of two types of polymer–solvent studies using the hydration shell model. In the first investigation aqueous solution–polypeptide interactions were considered as part of a conformational study of a variety of polypeptides having chemically and structurally different sidechains. In the second study the hydration shell parameters were *theoretically* computed for some solvents in addition to water. A series of conformational energy investigations were carried out for oliogomeric and polymeric forms of poly-L-alanine. The available polymer–solvent interactions were included in the energy calculations.

(a) *In Table 2-10* we report the values of the free energies $\langle A \rangle$, internal

TABLE 2-10

Free Energies $\langle A \rangle$, Average Internal Energies $\bar{E}$, and Entropies $\langle S \rangle$ Associated with Different Conformations of Several Homopolypeptides[a]

	$\langle A \rangle$, kcal/mol of residue			E, kcal/mol of residue			$\langle S \rangle$, cal/(mol of residue deg)		
	Aqueous solution	Vacuum	ΔA (PS)	Aqueous solution	Vacuum	$\Delta\bar{E}$ (PS)	Aqueous solution	Vacuum	ΔS (PS)
			Poly-L-alanine						
α_R[b]	−5.5	−5.4	−0.1	−4.3	−4.4	+0.1	3.56	3.40	0.16
$\beta_{\#}$[c]	−3.4	−3.1	−0.3	−1.6	−1.2	−0.4	5.94	6.57	−0.63
PP II[d]	−4.2	−3.6	−0.6	−2.5	−1.6	−0.9	5.68	6.31	−0.63
			Poly-L-proline						
PP II	−1.7	−1.1	−0.6	0.0	0.5	−0.5	5.74	5.29	−0.45
			Poly-L-valine						
α_R	12.3	11.7	+0.6	12.3	11.7	+0.6	0.00	0.00	0.00
$\beta_{\#}$	−2.0	−2.3	+0.3	−0.9	−1.3	+0.4	3.75	3.78	−0.03
PP II	0.9	0.0	+0.9	1.9	0.8	+1.1	3.03	2.50	+0.53
ECF[e]	−2.7	−2.9	+0.2	−1.6	−1.8	+0.2	3.74	3.73	+0.01
			Poly-L-serine						
α_R	−0.3	+2.5	−2.8	−0.0	2.8	−2.8	0.90	1.03	−0.13
$\beta_{\#}$	−3.9	−0.3	−3.6	−3.2	0.3	−4.1	2.29	2.05	+0.24
PP II	−2.7	+0.8	−3.5	−2.1	1.3	−3.3	1.85	1.47	+0.38
			Poly(L-aspartic acid) neutral form						
α_R	−19.5	−5.1	−14.4	−19.3	−4.8	−14.5	0.67	1.03	−0.36
$\beta_{\#}$	−18.8	+0.5	−19.3	−18.4	+1.3	−19.7	1.57	2.86	−1.29
PP II	−19.3	−1.6	−17.7	−18.9	−1.1	−17.8	1.39	1.88	−0.49
ECF	−14.2	+1.2	−15.4	−13.9	+1.4	−15.3	0.94	0.73	−0.12
			Poly(L-aspartic acid) charged form						
α_R	−7.8	+19.8	−27.6	−7.5	+20.2	−27.7	1.05	1.25	−0.20
$\beta_{\#}$	−1.4	+22.3	−23.7	−0.7	+22.9	−23.6	2.09	2.19	−0.10
PP II	+4.7	+26.7	−22.0	−5.3	+27.6	−22.3	1.95	3.08	−1.13
ECF	+3.6	+24.5	−20.9	+4.3	+25.1	−20.8	2.44	1.95	+0.49
			Poly(L-glutamic acid) neutral form						
α_R	−0.3	+12.1	−12.4	+0.2	12.7	−12.9	1.46	2.45	−0.99
$\beta_{\#}$	1.7	+17.1	−18.9	+0.6	20.1	−20.7	3.55	4.03	−0.48
PP II	1.6	−17.1	−18.7	+0.4	20.2	−20.6	3.61	5.27	−1.66
ECF	1.4	+16.8	−18.2	+0.2	20.0	−20.2	3.77	5.95	−2.18
			Poly(L-glutamic acid) charged form						
α_R	−4.1	+20.0	−24.1	−3.6	+21.0	−25.6	1.50	3.40	−1.90
$\beta_{\#}$	−12.4	+11.8	−24.2	−11.5	+13.1	−24.6	3.22	4.15	−0.93
PP II	−15.7	+10.6	−26.3	−14.6	+12.2	−26.8	3.50	5.40	−1.90
ECF	−20.9	+9.0	−20.9	−19.8	+10.7	−30.5	3.51	5.60	−2.09
			Poly-L-histidine neutral form						
α_R	−8.8	−7.6	−1.2	−8.3	−7.1	−1.2	1.41	1.41	0.00
$\beta_{\#}$	−6.8	−4.8	−2.0	−6.2	−4.2	−2.0	2.00	2.11	−0.11
PP II	−6.5	−4.5	−2.0	−6.0	−4.1	−1.9	1.74	1.65	+0.09

TABLE 2-10 (continued)

	⟨A⟩, kcal/mol of residue			E, kcal/mol of residue			⟨S⟩, cal/(mol of residue deg)		
	Aqueous solution	Vacuum	ΔA (PS)	Aqueous solution	Vacuum	$\Delta\bar{E}$ (PS)	Aqueous solution	Vacuum	ΔS (PS)
				Poly-L-histidine charged form					
α_R	+2.3	+7.4	−5.1	2.9	8.0	−5.1	1.95	1.90	0.05
$\alpha_{\#}$	−2.8	+4.2	−7.0	−2.2	4.9	−7.1	2.10	2.50	−0.40
PP II	−3.5	−3.1	−6.6	−2.9	3.8	−6.7	1.98	2.25	−0.27
ECF	−1.6	+5.3	−6.9	−0.9	6.0	−6.9	2.40	2.43	−0.03
				Poly-L-lysine neutral form					
α_R	−5.8	−5.2	−0.6	−4.5	−4.0	−0.5	4.15	3.95	−0.20
$\beta_{\#}$	−0.9	−0.2	−0.7	+0.7	1.5	−0.8	5.41	5.68	−0.27
PP II	+0.3	+1.0	−1.0	+2.0	3.0	−1.0	4.85	6.40	−0.55
ECF	−0.4	+0.7	−1.1	+1.3	2.4	+1.1	5.76	6.13	−0.37
				Poly-L-lysine charged form					
α_R	−1.1	+6.5	−7.6	0.0	7.6	−7.6	3.65	3.50	+0.15
$\beta_{\#}$	−2.3	+5.8	−8.1	−0.8	7.5	−8.3	5.13	5.71	−0.58
PP II	−3.7	+4.3	−8.0	−2.1	6.0	−8.1	5.41	5.83	−0.42
ECF	−1.9	+6.0	−7.9	−0.5	7.5	−8.0	4.70	5.02	−0.32

[a] ⟨A⟩, $\bar{E}$, and ⟨S⟩ are given for both aqueous solution and vacuum. The changes in free energies, ΔA (PS), average internal energies, $\Delta\bar{E}$ (PS), and entropies, ΔS (PS), in going from the vaccum to aqueous solution are also listed. The temperature was fixed at 298°K. Reprinted from Hopfinger, *Macromolecules* **4**, 731 (1971). Copyright 1971 by the American Chemical Society. Reprinted by permission of the copyright owner.

[b] Right-handed α helix. [c] Antiparallel β sheet. [d] Poly-L-proline II helix.

[e] Extended charged-form helix.

energies $\bar{E}$, and entropies ⟨S⟩ for a number of polypeptides in and out of aqueous solution (20). Some conclusions which can be considered valid only for polar solvents have been postulated from these calculations. Further, since the size and shape of the solvent molecule appears to be as important as its polarity (see next section) in defining the characteristic interaction with solute molecules, care must be taken in adopting the findings reported here to *all* types of polar solvents. The postulated conclusions are:

(*i*) *For polar solvents*, such as water, the favorable interactions of polar groups with solvent are, in magnitude, larger than the unfavorable interactions of nonpolar groups with solvent. Poly-L-valine has a destabilization free energy in aqueous solution (as compared to a vacuum) of about 0.5 kcal/mole/residue while poly-L-aspartic acid in the neutral form has a stabilization free energy in aqueous solution (as compared to a vacuum) of around 17 kcal/mole/residue.

(*ii*) *The change in polymer–solvent free energy* in going from a polar solvent to a vacuum (neutral solvent) is roughly proportional to the solubility of the polymer. Those homopolypeptides which have charged sidechains are most soluble. The higher the hydrophobic character of the homopolypeptide the lower the solubility. Poly-L-proline in form II probably should have a higher $\Delta A(\text{PS})$ in view of its known high solubility in aqueous solution (21). This suggests that the Δf's used to describe the carbonyl carbon and oxygen polymer–solvent interactions should be different for imino acids from the Δf's used for carbonyl and oxygen polymer–solvent interactions used for amino acids.

(*iii*) *The polymer–solvent interaction free energy* is dependent upon the conformation of the polymer's backbone and the size and chemical nature of the polymer side chain. Poly-L-alanine would be more soluble (in aqueous solution) in a PP II conformation than a right-handed α-helix. However, from a comparison of the conformational free energies it can be seen that a right-handed α-helix is statistically more probable. Thus, maximum water solubility of poly-L-alanine is diminished by conformational restrictions.

(*iv*) *The polymer–solvent interactions* (in this case strictly for aqueous solution and homopolypeptides) do not change the ranking of the conformations of nonionizable homopolypeptides with respect to stability relative to the vacuum calculations. However, the statistical weights associated with observing the various conformations is sensitive to polymer–solvent interactions. In general, the right-handed α-helix becomes less probable while the other conformations become more probable when aqueous solution–homopolypeptide interactions are included in the calculations. Figures 2-11d and 2-14d in the next section are conformational (ϕ, ψ) maps for poly-L-alanine based on vacuum calculations and aqueous solution–homopolypeptide calculations respectively. The region in the upper left hand corner of the map increases in stability for aqueous solution–poly-L-alanine interactions at the expense of the right-handed α-helical region, $\phi = 130°$, $\psi = 120°$. Thus, if one attempts to use conformational calculations to predict Boltzmann average properties of a macromolecule (i.e., spectra, transition temperatures) in aqueous solution, it is important to be sure that the polymer–solvent interactions are included in the computations. Hence, Aebersold and Pysh (22) were not able to accurately predict the statistical average CD spectra of various homopolypeptides perhaps because they neglected polymer–solvent interactions, and not because of major errors in the conformational potential functions. Lastly, polymer–solvent interactions may cause changes in the conformations of heteropolypeptide chains which contain a mixture of polar and hydrophobic groups (i.e., proteins) and are in aqueous solution. The effects of polymer–solvent interactions are maximized under these conditions.

(*v*) *Aqueous solution–homopolypeptide interactions* for homopolypeptides having ionizable sidechains are extremely stabilizing (10–30 kcal/mole/residue).

Hence we should expect that these interactions are extremely important in dictating the conformational properties of such macromolecules.

(*b*) *The values of the hydration shell parameters* n, R_v, V_f, *and* V' were computed for formic acid, acetic acid, methanol, and ethanol. This was accomplished using the same procedures employed in calculating the hydration shell parameters for aqueous solution. For these solvents the Δf computed from the energy minimization data required to compute n, R_v, V_f, and V' were adopted in the conformational energy calculations since no experimental values of Δf were available as in the case of aqueous solvent.

For formic acid and acetic acid it is conceivable that in some cases the neutral form of these solvent molecules interact with some groups in the solute macromolecule, and the negatively charged form interacts with other macromolecular groups. In the calculation of the hydration shell parameters it has been assumed that for both solvent species the neutral form interacts with the nonpolar and negatively charged polar atoms or groups of macromolecules, and the negatively charged form interacts with the positively charged polar solute atoms or groups.

For ethanol and acetic acid, where internal bond rotations are possible, the following conformations were adopted for the solvent molecules and held fixed during the calculations:

(1) in ethanol the CH_3 group was held fixed in a *gauche* position to the CH_2OH group;
(2) in acetic acid the chosen geometry, looking down the C—C bond from the methyl group, was

The favorable O^----H interaction is responsible for the stability of this conformation.

The values of the hydration shell parameters are given in Table 2-11(a–d).

Conformational energy calculations were carried out for two, three, and four planar L-alanine peptide units joined at the common C^α atoms and poly-L-alanine (e.g., up to 20th residue neighbor interactions were considered in simulating the polymer) in ethanol, methanol, formic acid, acetic acid, water, and vacuum. In every case the equivalence condition was invoked. Calculations upon di-, tri-, and tetra-L-alanine planar peptide unit configurations were carried out to determine the effect of nearest neighbor, first and second neighbor, and first, second, and third neighbor planar peptide

TABLE 2-11

Polymer–Solvent Parameters for the Hydration Shell Model for a Variety of Solvents

Atom or group	n	Δf (kcal/mole)	R_v (Å)	V_f (Å^3)	V' (Å^3)	Molecular species
		(a) Formic acid[a]				
N (sp^2)	2	1.90	5.80	24.40	116.40	b
C (sp^2)	2	2.85	5.25	16.50	100.60	a
O (carbonyl)	2	3.40	5.40	43.80	145.20	b
H (amide)	1	5.95	4.30	22.70	56.50	a
CH$_3$ (aliphatic)	6	−0.09	5.95	30.75	387.30	a
CH$_2$ (aliphatic)	4	−0.08	5.95	28.60	249.60	a
CH (aliphatic)	1	0.06	5.75	15.75	49.55	a
CH (aromatic)	2	0.08	4.45	19.00	105.60	a
O (hydroxyl)	1	2.55	4.85	40.60	74.40	b
H (hydroxyl)	1	4.75	4.30	43.95	77.75	a
O$^-$ (carboxyl)	3	5.05	5.85	29.75	190.65	b
O (carboxyl)	1	3.55	5.20	18.50	52.30	b
H (carboxyl)	1	3.55	5.20	18.50	52.30	b
		(b) Acetic acid[b]				
N (sp^2)	2	1.95	6.60	48.75	200.70	b
C (sp^2)	2	2.85	5.60	39.80	182.80	a
O (carbonyl)	2	3.35	5.80	96.50	286.2	b
H (amide)	1	5.85	4.90	47.35	197.90	a
CH$_3$ (aliphatic)	6	−0.08	6.20	31.60	499.20	a
CH$_2$ (aliphatic)	4	−0.08	6.20	35.83	349.72	a
CH (aliphatic)	1	0.06	5.95	18.98	70.58	a
CH (aromatic)	2	0.08	4.65	22.07	147.34	a
O (hydroxyl)	1	2.55	5.80	48.60	100.20	b
H (hydroxyl)	1	4.55	4.90	67.40	119.00	a
O$^-$ (carboxyl)	3	4.96	6.60	49.50	303.00	b
O (carboxyl)	1	3.45	6.10	33.70	85.3	b
H (carboxyl)	1	3.45	6.10	33.70	85.3	b

[a] a, HCOO$^-$; b, HCOOH; $V_s = 33.8$ Å^3.
[b] a, H$_3$CCOO$^-$; b, H$_3$CCOOH; $V_s = 51.6$ Å^3.

unit interactions, respectively, on the conformation of this polymer. In other words we can assess the importance of long range interactions upon the conformational properties of poly-L-alanine. Flory (23) has postulated that two planar peptide units joined at the common C$^\alpha$ atom is a good model for the secondary structural properties of a homopolypeptide when in the random state. In so far as this is correct, one can consider the results presented here as demonstrating the conformational properties of poly-L-alanine when it

TABLE 2-11 (continued)

Atom or group	n	Δf (kcal/mole)	R_v (Å)	V_f (Å³)	V' (Å³)
		(c) Methanol[c]			
N (sp²)	2	0.23	4.70	41.6	115.8
C (sp²)	2	0.18	5.20	53.6	185.5
O (carbonyl)	2	1.45	4.80	43.9	216.8
H (amide)	1	0.30	4.10	30.5	98.5
CH₃ (aliphatic)	4	0.38	5.90	19.5	258.5
CH₂ (aliphatic)	3	0.32	5.70	21.6	195.0
CH (aliphatic)	2	0.32	5.70	21.6	195.0
CH (aromatic)	2	0.46	4.70	28.5	145.5
O (hydroxyl)	1	0.85	3.65	38.8	107.3
H (hydroxyl)	1	0.85	3.65	38.8	107.3
O⁻ (carboxyl)	2	2.80	4.80	48.5	223.1
O (carboxyl)	1	1.30	4.35	28.6	95.5
H (carboxyl)	1	1.15	4.35	29.3	106.6
		(d) Ethanol[d]			
N (sp²)	2	0.18	6.20	49.8	218.3
C (sp²)	2	0.15	5.80	57.5	173.5
O (carbonyl)	2	1.18	6.20	49.1	246.3
H (amide)	1	0.28	5.50	45.6	189.5
CH₃ (aliphatic)	4	0.41	7.10	31.3	533.1
CH₂ (aliphatic)	3	0.39	7.00	31.8	383.5
CH (aliphatic)	2	0.39	6.90	33.0	247.5
CH (aromatic)	2	0.40	6.35	42.9	165.0
O (hydroxyl)	1	0.57	4.15	51.6	118.7
H (hydroxyl)	1	0.57	4.20	51.6	124.4
O⁻ (carboxyl)	2	2.45	6.35	60.7	290.0
O (carboxyl)	1	1.10	5.90	43.5	173.5
H (carboxyl)	1	0.88	5.90	43.5	173.5

[c] CH_3OH; $V_s = 42.8$ Å³.
[d] CH_3CH_2OH; $V_s = 63.7$ Å³.

possesses virtually no locally ordered secondary structure, the random coil, up through a nearly completely ordered secondary structure—the simulated polymer calculation invoking the equivalence condition. Thus, in the most general sense, we are able to describe the secondary structural properties of poly-L-alanine as a function of solvent and extent of local order. We now proceed to discuss the results of these calculations and to make some correlations with experimental findings.

(*i*) *Conformational properties of poly-L-alanine in various solvents*   Poly-L-alanine (PLA) indicates a preference to adopt the right-handed α-helical conformation in nonpolar solvents, as well as precipitates and films made from nonpolar solutions (24). In polar solvents, and precipitates and films made from polar solutions, PLA is often found to exist in the β structure (25). The ordered polytripeptide, (Gly-Ala-Ala)$_n$, has been found to be in the left-handed 3_1-helical conformation in aqueous solution (26). This suggests that alanine residues can adopt the 3_1-helical conformation as a stable state.

The conformational energy maps of L-alanine chains have been computed in a number of solvents. Note that these calculations are for an isolated chain in a solvent near room temperature. Figures 2-11 to 2-16 show the digital (ϕ, ψ) conformational energy maps of (a) two, (b) three, (c) four, and (d) twenty planar peptide units of L-alanine (the model for PLA) in vacuum (2-11), methanol (2-12), ethanol (2-13), water (2-14), formic acid (2-15), and acetic acid (2-16) mediums. The equivalence condition†was employed in these calculations and rotation of the methyl side chain group about the C^{α}–C^{β} bond was chosen to minimize the total conformational energy at each scan point. The bond rotation $\phi(N–C^{\alpha})$ is plotted along the horizontal axis and the rotation $\psi(C^{\alpha}–C')$ is plotted on the vertical axis. The figures appear at the end of this section.

The energy maps of PLA in methanol and ethanol are nearly identical. The right-handed α-helix, located near $\phi = 120°$, $\psi = 140°$, is the preferred conformation for both solvents. Relative minima at $\phi = 20°$, $\psi = 280°$ and $\phi = 100°$, $\psi = 320°$, on both maps, are indicative of the possible existence of isolated stable β and left-handed 3_1-helical conformations, respectively. For ethanol, in fact, the right-handed α-helix and the left-handed 3_1-helix are approximately equally stable.

The conformational map of PLA *in vacuo* indicates that the biopolymer is more restricted in conformational freedom than in methanol and ethanol. This map indicates less flexibility of the biopolymer than computed by other workers (27, 28). The reason for this is due to an increase in the stabilization energy given to the hydrogen bond. In these calculations the ΔH of breaking the hydrogen bonds of the α-helix in water was chosen to be 1.5 kcal/mole (29, 30). This required an intrachain maximum hydrogen bonding energy of -4.6 kcal/mole as compared to the -3.5 kcal/mole normally used. These calculations indicate that a completely ordered chain of PLA *in vacuo* would overwhelmingly prefer the right-handed α-helix.

PLA in aqueous solution also is most stable in the right-handed α-helix. However, the conformational flexibility of the polymer is much enhanced over what it was *in vacuo*. There are relative minima located at the points on the map

† See Semen (34) for a discussion of the equivalence condition.

corresponding to the β and left-handed 3_1-helical structures just as for methanol and ethanol.

The conformational map of PLA in acetic acid indicates a radical departure in the conformational properties of the biopolymer. Only conformations in the upper left-hand corner of the map, β and left-handed 3_1-helical-type structures are permitted. The global minimum at $\phi = 80°$, $\psi = 300°$ corresponds to a structure very similar to the left-handed 3_1-helix as well as the "extended helix" proposed for charged polypeptides such as poly(-L-glutamic acid) (31, 32). A relative minimum is noted at the β position on the map. This radical change in the conformational properties is to be expected since one expects the polar groups of the solvent molecules to strongly interact with PLA. Interactions involving the carbonyl oxygen and amide hydrogen with solvent would be especially stabilizing (33). The net effect of such interactions would be to "pull" the α-helix apart in order to expose the carbonyl oxygens and amide hydrogens to solvent. Thus extended conformations should be preferred.

However, what is most surprising is the conformational map of PLA in formic acid. The first thought one might have is to expect PLA in formic acid to behave nearly the same as in acetic acid. An inspection of the map of PLA in formic acid indicates that the biopolymer has conformational properties very similar to PLA in aqueous solution. From an inspection of the polymer–solvent parameters presented in Tables 2-9 and 2-11 it is seen that the ΔF of formic and acetic acid are nearly identical while the R_v of formic acid are about midway between the R_v of water and acetic acid. Further, the V_f of water and formic acid are quite similar and smaller than the V_f of acetic acid.

We conclude from these observations that in this case the size and shape of the solvent molecule overrides the free energy interactions between solvent and solute species in dictating conformation. The additional bulk of a methyl group in acetic acid has the capacity to pull the α-helix apart whereas formic acid maintains the α-helical structure. Remember these findings are true only for *isolated* PLA chains which are *perfectly ordered*. Aggregation of PLA chains in the β form or chain-folding of a single chain in the β form could very likely be more stable than the isolated and perfectly ordered structures. Also, the absence of perfect ordering, as in a random coil, would reduce the stability of α-helical segments and favor the extended structures. We will discuss this in more detail in the next section.

(*ii*) *Ordering and conformational stability* An interesting question which can be dealt with using conformational energy calculations is the sensitivity of conformation to segmental ordering of a macromolecule in various solvents. Specifically, we are able to compare the stable conformations of two, three, four, and twenty (model for the polymer) planar peptide units of alanine, each obeying the equivalence condition, in a variety of solvents. This analysis makes it possible to study the local chain conformations of PLA as a function of increasing segmental ordering beginning with the random coil (two planar peptide alanine units) (28). (Use Figs. 2-11 to 2-16 as references in the discussion to follow.)

In methanol a β structure near $\phi = 20°$, $\psi = 320°$ is the most stable structure for the dimer, trimer, and tetramer of alanine. The right-handed α-helix is the most stable structure for the polymer. An approximate left-handed 3_1-helix is preferred for the dimer, trimer, and tetramer of alanine in ethanol. This global minimum is located at $\phi = 100°$, $\psi = 300°$. Again the right-handed α-helix is the most stable structure for the polymer. The 2_1-helix located at $\phi = 80°$, $\psi = 260°$ is the preferred conformation for two planar peptide units of alanine *in vacuo*. A structure near the right-handed α-helix, $\phi = 120°$, $\psi = 160°$, namely the 3_{10}, threefold helix (35) is the most stable structure for the trimer. The right-handed α-helix is the preferred conformation for the tetramer and polymer in vacuo. The β structure at $\phi = 20°$, $\psi = 320°$, is the most stable conformation of the dimer, trimer, and tetramer of alanine in aqueous solution. The right-handed α-helix is the most likely structure of the polymer in aqueous solution. The dimer, trimer, and tetramer of alanine in acetic acid indicates the near 3_1-helix located at $\phi = 100°$, $\psi = 300°$ is the most stable conformation. The polymer also prefers the near 3_1 left-handed helical conformation in acetic acid. Apparently this structure allows the maximum exposure of the polymer to the solvent. However, for PLA, the global minimum shifts slightly to $\phi = 80°$, $\psi = 300°$. The dimer of alanine in formic acid is about equally stable in the β- or 3_1-helical conformations. The trimer indicates a definite preference for a β structure located at $\phi = 20°$, $\psi = 280°$ as does the tetramer. The polymer is most stable in the right-handed α-helix.

(*iii*) *Conclusions*   We can make the following statements about polymer–solvent interactions from the calculations reported here.

a. The conformation of an ordered segmental unit is very sensitive to solvent. Methanol and ethanol, which differ only by a CH_2 group, promote different extended structures for short-ordered alanine chain segments. The same is true for acetic and formic acid. Apparently the additional bulk of acetic acid and ethanol compared to formic acid and methanol, respectively, results in a preference of the left-handed 3_1-helix over the β structures in these two solvents.

b. The added stability of the hydrogen bond occurring in the formation of α-helix has the cumulative effect of making the right-handed α-helix the most stable conformation in polar solvents where the opposite might be expected. The long range ordering, e.g., long chain segments of uniform conformation, is the major factor in deciding if the α-helix or some extended conformation is the most stable structure. There appears to be some critical length of the ordered chain segments for which a transition from an extended structure to a right-handed α-helix might be expected to take place.

c. The size and shape of the solvent molecule can be more important in dictating polymer conformation than the characteristic free energy interactions between solvent molecules and atoms in the solute macromolecule.

d. The solvent-dependent conformational energies are a *relative* measure of the solubility of a polymer. PLA would, according to the calculations, be most soluble in acetic acid, then formic acid, then water, then methanol and lastly ethanol. This is in approximate agreement with experiment. Formic and acetic acid should be interchanged in the above sequence. The values of the solvent dependent conformational energies are not realistic *absolute* measures of solubility. These calculations suggest that PLA is more soluble *in vacuo* than in formic acid or water!

	0	20	40	60	80	100	120	140	160	180	200	220	240	260	280	300	320	340	360
360	0.0	-1.0	-1.0	-0.5	-0.8	-0.9	0.4					0.9	4.7					4.9	0.0
340	-0.3	-1.2	-1.2	-0.8	-1.1	-1.4	-1.1	6.3				5.1	5.0					4.6	-0.3
320	-0.6	-1.5	-1.3	-0.9	-1.3	-1.7	-1.7	-0.2					5.6					4.6	-0.6
300	-0.4	-1.5	-1.3	-1.0	-1.5	-2.0	-2.0	-1.2	5.9				6.0					4.8	-0.4
280	-0.2	-1.3	-1.2	-0.9	-1.7	-2.3	-1.8	0.9	5.1			4.6	6.0					5.1	-0.2
260	0.1	-1.0	-0.9	-0.8	-2.7	-1.6				6.7	3.4	1.5	6.0					5.3	0.1
240	0.2	-0.8	-0.7	-0.6	-1.7	1.0				5.3	0.1	0.6	5.8					5.5	0.2
220	0.5	-0.6	-0.5	-0.4	-1.4	0.3					-0.4	0.0	5.5					5.8	0.5
200	2.8	0.3	0.0	0.2	-0.6	-0.9					6.8	-0.3	5.4						2.8
180		4.3	1.1	0.8	0.1	-0.7	1.8					2.4	5.0						
160	2.4	2.6	1.2	0.5	-0.1	-0.8	-1.2	6.6					4.5					5.9	2.4
140	-0.2	-0.9	-0.5	-0.2	-0.6	-1.0	-1.4	-1.0					5.3					4.6	-0.2
120	-0.6	-1.2	-0.8	-0.3	-0.5	-0.8	-0.9	-0.5	4.6				5.8					4.3	-0.6
100	-0.2	-0.9	-0.4	0.2	0.0	-0.1	0.7	3.4	6.6				3.8					4.7	-0.2
80	0.8	0.2	0.7	1.3	1.2	1.3	5.1			6.0	1.6	-0.3	4.3					5.7	0.8
60	1.6	0.9	1.4	2.1	1.9	2.2					0.5	0.4	5.7					6.4	1.6
40	0.8	0.1	0.6	1.2	1.0	1.2					0.9	0.2	5.4					5.8	0.8
20	0.1	-0.8	-0.5	0.1	-0.2	-0.2	5.1					6.5	-0.2	4.8				5.1	0.1
0	0.0	-1.0	-1.0	-0.5	-0.8	-0.9	0.4					0.9	4.7					4.9	0.0
	0	20	40	60	80	100	120	140	160	180	200	220	240	260	280	300	320	340	360

Fig. 2-11a. The lowest energy is −2.7 kcal/mole/dipeptide, located at (80, 260), for the dialanine chain *in vacuo*.

	0	20	40	60	80	100	120	140	160	180	200	220	240	260	280	300	320	340	360
360	0.9	-0.8	-1.8	-1.2	-2.0	-2.4	-3.2					0.8	6.0					8.0	0.9
340	0.1	-2.2	-2.2	-1.7	-2.6	-3.2	-3.1					9.2	6.6					7.4	0.1
320	-1.4	-2.8	-2.6	-2.0	-3.0	-3.8	-4.2	-1.4					7.6					6.3	-1.4
300	-1.3	-2.8	-2.7	-2.3	-3.2	-4.4	-4.9	-3.4					8.5					6.7	-1.3
280	-0.6	-2.4	-2.4	-2.2	-3.6	-5.0	-4.2	0.8	8.8			8.7	8.9					7.4	-0.6
260	-0.1	-1.9	-1.9	-1.9	-3.7	-3.3					5.5	2.3	8.9					8.0	-0.1
240	0.4	-1.5	-1.5	-1.6	-3.7	1.6					-1.6	-0.3	8.5					8.3	0.4
220	0.6	-1.2	-1.3	-1.4	-3.4	0.0						-2.8	7.2					9.0	0.6
200	5.3	0.6	-0.3	-0.5	-2.4	-3.2						-4.0	5.8						5.3
180		8.6	2.0	0.7	-1.2	-3.7	1.0					1.8	4.8						
160	4.6	5.4	2.4	0.5	-1.2	-3.6	-5.6						4.6					8.7	4.6
140	-0.8	-1.5	-1.0	-0.6	-1.5	-2.9	-5.2						7.0					6.2	-0.8
120	-1.4	-2.4	-1.6	-0.9	-1.2	-1.9	-2.9	-3.0					8.4					5.7	-1.4
100	-0.8	-1.5	-0.9	0.2	-0.1	-0.3	0.8	5.2					4.6					6.5	-0.8
80	1.4	0.6	1.4	2.1	1.9	2.3					2.4	-1.5	5.5					8.4	1.4
60	2.9	1.9	2.8	3.8	3.2	4.0					0.1	-0.2	8.1						2.9
40	1.4	0.2	1.0	2.1	1.4	1.8					0.8	-0.7	7.4					8.8	1.4
20	1.1	-0.6	-1.1	-0.2	-0.7	-1.0	9.3						-1.4	6.2				7.6	1.1
0	0.9	-0.8	-1.8	-1.2	-2.0	-2.4	-3.2					0.8	6.0					8.0	0.9
	0	20	40	60	80	100	120	140	160	180	200	220	240	260	280	300	320	340	360

Fig. 2-11b. The lowest energy is −5.6 kcal/mole/tripeptide, located at (120, 160), for the trialanine chain *in vacuo*.

	0	20	40	60	80	100	120	140	160	180	200	220	240	260	280	300	320	340	360
360	1.7	-0.7	-2.8	-2.0	-3.3	-2.6	0.8					1.5							1.7
340	0.5	-3.2	-3.4	-2.6	-4.8	-5.1	-4.0						8.0						0.5
320	-2.3	-4.2	-3.9	-3.0	-5.4	-7.0	-6.5	-2.7										7.8	-2.3
300	-2.2	-4.2	-4.2	-3.7	-5.8	-6.8	-6.6	-5.8											-2.2
280	-0.3	-4.4	-4.4	-4.2	-5.6	-6.6	-5.6	1.2											-0.3
260	-0.7	-2.0	-2.1	-4.1	-5.5	-4.3						1.5							-0.7
240	5.5	-2.1	-1.5	-0.2	-3.3	4.6					-5.7								5.5
220	1.0	-1.7	-2.2	-2.8	-4.8	0.2						-8.0							1.0
200	7.7	0.8	-0.7	-1.3	-4.3	-5.6						-8.4	4.8						7.7
180			2.4	0.1	-2.9	-7.0	0.0					0.9	4.3						6.6
160	6.6	8.1	3.3	-0.3	-3.9	-7.7	-10.6						4.5						6.6
140	-1.4	-2.2	-1.5	-1.2	-3.6	0.4	-11.5											7.7	-1.4
120	-2.3	-3.3	-2.3	-1.2	-2.0	-4.6		-7.7										7.2	-2.3
100	-1.9	-2.7	-2.0	0.2	-0.4	-0.7	-0.7						6.1						-1.9
80	2.7	0.9	0.2	4.0	2.0	3.0					3.8	-1.8	7.7						2.7
60	4.1	3.4	4.1	3.4	6.1	6.4					-0.6	0.2							4.1
40	1.8	1.8	2.0	2.8	-0.5	2.2					0.6	-1.5	8.4						1.8
20	1.9	-0.4	-1.9	-0.6	-1.4	-1.9						-1.7	7.4						1.9
0	1.7	-0.7	-2.8	-2.0	-3.3	-2.6	0.6					1.5							1.7
	0	20	40	60	80	100	120	140	160	180	200	220	240	260	280	300	320	340	360

Fig. 2-11c. The lowest energy is −11.5 kcal/mole tetrapeptide, located at (120, 140), for the tetraalanine chain *in vacuo*.

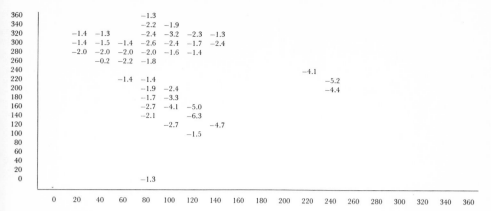

Fig. 2-11d. The lowest energy is −6.3 kcal/mole/residue, located at (120, 140), for the infinite alanine chain *in vacuo*.

	0	20	40	60	80	100	120	140	160	180	200	220	240	260	280	300	320	340	360
360	−7.5	−9.0	−8.7	−8.0	−7.9	−7.9	−6.6						−5.6	−1.8				−2.7	−7.5
340	−8.0	−9.4	−8.8	−8.3	−8.3	−8.3	−8.0	−0.5					−1.5	−1.8				−3.1	−8.0
320	−8.4	−9.8	−9.1	−8.4	−8.4	−8.4	−8.3	−6.9						−1.3				−3.2	−8.4
300	−8.0	−9.5	−8.7	−8.1	−8.3	−8.5	−8.4	−7.5	−0.3					−0.7				−2.7	−8.0
280	−7.6	−9.1	−8.4	−7.9	−8.4	−8.9	−8.0	−5.5	−1.2				−1.7	−0.6				−2.3	−7.6
260	−7.3	−8.8	−8.1	−7.7	−8.5	−7.9						−3.1	−4.9	−0.5				−2.0	−7.3
240	−7.1	−8.5	−7.9	−7.6	−8.5	−5.3					−1.1	−6.6	−5.6	−0.6				−1.9	−7.1
220	−6.9	−8.4	−7.7	−7.4	−8.1	−6.0						−7.2	−6.2	−0.9				−1.5	−6.9
200	−4.5	−7.4	−7.2	−6.8	−7.3	−7.2							−6.5	−1.0					−4.5
180		−3.4	−6.1	−6.2	−6.6	−7.1	−4.3						−3.8	−1.4					
160	−5.0	−5.2	−6.2	−6.6	−7.0	−7.3	−7.6							−2.0				−1.6	−5.0
140	−7.8	−8.9	−8.0	−7.4	−7.6	−7.7	−7.8	−7.4						−1.3				−2.9	−7.8
120	−8.0	−9.1	−8.1	−7.3	−7.4	−7.4	−7.3	−6.8	−1.7					−0.6				−3.1	−8.0
100	−7.5	−8.5	−7.5	−6.7	−6.7	−6.7	−5.8	−2.7						−2.4				−2.6	−7.5
80	−6.5	−7.5	−6.5	−5.6	−5.6	−5.4	−1.4					−4.9	−6.1	−1.9				−1.6	−6.5
60	−5.9	−6.9	−5.8	−4.9	−5.0	−4.6						−5.9	−5.4	−0.4				−1.0	−5.9
40	−6.8	−7.9	−6.8	−5.9	−6.0	−5.7						−5.6	−5.7	−0.8				−1.8	−6.8
20	−7.3	−8.7	−8.1	−7.2	−7.2	−7.1	−1.7					0.0	−6.2	−1.5				−2.2	−7.3
0	−7.5	−9.0	−8.7	−8.0	−7.9	−7.9	−6.6						−5.6	−1.8				−2.7	−7.5

Fig. 2-12a. The lowest energy is −9.8 kcal/mole/dipeptide, located at (20, 320), for the dialanine chain in methanol.

	0	20	40	60	80	100	120	140	160	180	200	220	240	260	280	300	320	340	360
360	−9.7	−12.0	−12.7	−11.3	−11.4	−11.2	−8.4						−6.7	−2.1				−2.6	−9.7
340	−10.9	−13.7	−12.9	−12.1	−12.2	−12.1	−11.4							−1.9				−3.6	−10.9
320	−12.8	−14.8	−13.6	−12.2	−12.3	−12.3	−12.1	−8.9						−0.8				−4.9	−12.8
300	−12.1	−14.3	−13.2	−12.0	−12.1	−12.4	−12.3	−10.3										−4.0	−12.1
280	−11.2	−13.7	−12.6	−11.6	−12.2	−13.0	−11.4	−6.0										−3.1	−11.2
260	−10.6	−13.0	−12.0	−11.3	−12.4	−10.9						−0.3	−3.9					−2.4	−10.6
240	−10.1	−12.5	−11.6	−11.1	−12.4	−5.7						−7.6	−5.8					−2.0	−10.1
220	−9.7	−12.2	−11.3	−10.6	−11.7	−7.1							−8.5					−1.2	−9.7
200	−4.9	−10.3	−10.2	−9.4	−10.1	−9.7							−9.6	−0.7					−4.9
180		−2.1	−7.9	−8.3	−8.9	−10.3	−4.6						−3.8	−2.0					
160	−5.9	−5.7	−7.9	−9.0	−9.4	−10.3	−11.4							−2.3				−1.7	−5.9
140	−11.5	−12.9	−11.4	−10.5	−10.4	−10.0	−11.1							−0.5				−4.4	−11.5
120	−11.8	−13.4	−11.8	−10.3	−10.1	−9.4	−8.6	−8.3										−4.5	−11.8
100	−10.8	−12.2	−10.7	−9.1	−8.8	−8.1	−5.6							−2.5				−3.4	−10.8
80	−8.7	−10.0	−8.0	−7.1	−6.7	−5.8						−4.6	−8.0	−1.9				−1.6	−8.7
60	−7.4	−8.9	−7.0	−5.4	−5.5	−4.3						−6.7	−6.9					−0.2	−7.4
40	−9.2	−10.9	−9.0	−7.3	−7.5	−6.6						−6.0	−7.5	0.1				−1.7	−9.2
20	−9.2	−11.4	−11.4	−9.8	−9.7	−9.5							−8.4	−1.4				−2.6	−9.2
0	−9.7	−12.0	−12.7	−11.3	−11.4	−11.2	−8.4						−6.7	−2.1				−2.6	−9.7

Fig. 2-12b. The lowest energy is −14.8 kcal/mole/tripeptide, located at (20, 320), for the trialanine chain in methanol.

Fig. 2-12c

	0	20	40	60	80	100	120	140	160	180	200	220	240	260	280	300	320	340	360
360	−10.7	−13.9	−15.5	−13.5	−13.7	−12.2	−8.3						−6.6	−0.1				−1.3	−10.7
340	−12.5	−16.9	−15.8	−14.7	−15.5	−15.0	−13.1							−1.3				−2.9	−12.5
320	−16.0	−18.6	−16.9	−14.8	−15.8	−16.3	−15.2	−10.5						0.5				−5.5	−16.0
300	−15.1	−17.9	−16.5	−14.7	−15.4	−15.3	−14.5	−12.9										−3.6	−15.1
280	−12.9	−17.8	−16.3	−15.0	−15.0	−13.1	−5.7											−2.8	−12.9
260	−13.1	−15.2	−13.9	−14.8	−15.0	−12.0							−2.4					−2.0	−13.1
240	−11.7	−15.2	−13.3	−10.9	−12.8	−2.8						−10.6						−0.7	−11.7
220	−11.1	−14.6	−13.7	−13.0	−13.4	−6.8							−12.3					0.5	−11.1
200	−3.7	−11.7	−11.7	−10.7	−11.9	−11.8							−13.1	−0.0					−3.7
180	0.9	−7.8	−8.3	−9.4	−12.6	−5.0							−4.0	−1.6					
160	−5.3	−4.5	−7.7	−9.1	−10.1	−12.9	−15.6							−2.1				−0.4	−5.3
140	−13.9	−15.5	−13.3	−11.6	−10.8	−4.6	−16.1											−4.5	−13.9
120	−14.4	−16.1	−13.9	−11.8	−10.8	−9.4			−11.7									−4.7	−14.4
100	−13.5	−15.1	−11.3	−10.0	−9.5	−7.2	−4.4							−1.0				−2.3	−13.5
80	−8.9	−11.4	−10.8	−6.1	−7.1	−4.6						−3.4	−8.4	0.2				−1.1	−8.9
60	−7.8	−9.1	−7.0	−6.7	−3.1	−1.8						−7.7	−6.8						−7.8
40	−10.5	−12.7	−9.4	−7.6	−10.0	−6.4						−6.5	−8.7	0.6				−0.5	−10.5
20	−9.9	−12.9	−13.7	−11.4	−11.2	−10.9							−9.2	−0.7				−1.7	−9.9
0	−10.7	−13.9	−15.5	−13.5	−13.7	−12.2	−8.3						−6.6	−0.1				−1.3	−10.7

Fig. 2-12c. The lowest energy is −18.6 kcal/mole/tetrapeptide, located at (20, 320), for the tetraalanine chain in methanol.

Fig. 2-12d

	0	20	40	60	80	100	120	140	160	180	200	220	240	260	280	300	320	340	360
360	6.6	5.8	4.8	5.4	5.3	6.6													6.6
340	5.9	4.4	4.7	5.1	4.3	4.7	5.9												5.9
320	4.4	3.8	4.3	5.0	4.2	3.6	4.5	6.0										7.0	4.4
300	4.6	4.0	4.3	4.9	4.3	4.7	5.4	4.9											4.6
280	6.0	3.5	3.9	4.3	4.8	5.6	5.9												6.0
260	5.1	5.4	5.8	4.1	5.0	6.5													5.1
240	6.0	5.0	6.0	7.8	7.2							4.6							6.0
220	6.2	5.1	5.2	5.2	5.9								3.8						6.2
200			6.1	6.3	5.9	5.5							4.1						
180					7.1	5.2	7.1						7.4						
160					6.9	5.0	3.4												
140	5.2	5.0	5.7				2.6											7.5	5.2
120	5.1	4.9	5.4	6.1				4.2										7.5	5.1
100	5.0	4.7																	5.0
80													7.2						
60												6.7	7.7						
40													6.4						
20		6.1	5.4	6.1	6.1	6.2							6.9						
0	6.6	5.8	4.8	5.4	5.3	6.6													6.6

Fig. 2-12d. The lowest energy is 2.6 kcal/mole/residue, located at (120, 140), for the infinite alanine chain in methanol.

Fig. 2-13a

	0	20	40	60	80	100	120	140	160	180	200	220	240	260	280	300	320	340	360
360	−7.3	−8.7	−8.4	−8.2	−9.0	−9.8	−7.4						−6.4	−3.5				−2.4	−7.3
340	−7.7	−8.9	−8.6	−8.4	−9.2	−9.8	−9.1	−1.7					−2.1	−2.5				−2.8	−7.7
320	−7.8	−9.0	−8.5	−8.4	−9.2	−9.8	−10.5	−8.0						−1.8				−2.7	−7.8
300	−7.9	−9.0	−8.7	−8.7	−9.7	−11.0	−10.3	−9.1	−2.3					−1.6				−2.7	−7.9
280	−7.9	−9.3	−8.9	−9.0	−10.2	−10.7	−10.1	−7.8	−3.3				−2.7	−1.9				−3.1	−7.9
260	−8.3	−9.8	−9.4	−9.6	−10.6	−10.2					−1.8	−5.2	−6.4	−2.2				−3.0	−8.3
240	−7.8	−9.4	−9.0	−9.3	−10.2	−7.4					−2.6	−8.3	−7.3	−2.7				−2.5	−7.8
220	−7.5	−9.0	−8.6	−9.0	−9.9	−7.9						−8.6	−8.7	−2.6				−2.1	−7.5
200	−5.1	−8.1	−8.1	−8.2	−9.0	−9.1						−1.7	−8.0	−2.9					−5.1
.180		−3.9	−6.8	−7.3	−8.4	−9.1	−6.3						−5.2	−2.9					
160	−5.1	−5.3	−6.4	−7.3	−8.3	−9.2	−9.1	−1.8						−3.0				−1.7	−5.1
140	−8.0	−9.0	−8.4	−8.2	−8.9	−9.5	−10.3	−9.1						−2.3				−3.2	−8.0
120	−8.4	−9.5	−8.8	−8.4	−9.0	−9.7	−9.0	−8.9	−3.7					−1.9				−3.6	−8.4
100	−8.3	−9.4	−8.7	−8.3	−9.0	−8.7	−5.2	−5.2	−2.2					−4.1				−3.5	−8.3
80	−6.8	−8.3	−7.5	−7.2	−7.8	−7.2	−2.8				−1.8	−6.9	−8.2	−3.5				−1.9	−6.8
60	−5.9	−7.1	−6.4	−6.4	−7.1	−6.1						−7.6	−7.7	−2.9				−1.0	−5.9
40	−6.7	−7.8	−7.2	−6.9	−8.1	−7.1						−7.2	−8.6	−2.6				−1.7	−6.7
20	−7.1	−8.5	−8.0	−7.8	−8.5	−8.3	−2.6					−1.6	−7.6	−3.1				−2.1	−7.1
0	−7.3	−8.7	−8.4	−8.2	−9.0	−9.8	−7.4						−6.4	−3.5				−2.4	−7.3

Fig. 2-13a. The lowest energy is −11.0 kcal/mole/dipeptide, located at (100, 300), for the dialanine chain in ethanol.

Fig. 2-13b.

φ\ψ	0	20	40	60	80	100	120	140	160	180	200	220	240	260	280	300	320	340	360
360	−5.7	−7.9	−8.5	−8.2	9.7	−11.3	−6.4						−4.4	−1.6				1.4	−5.7
340	−6.6	−9.3	−8.8	−8.5	−10.0	−11.3	−9.8							0.3				0.5	−6.6
320	−7.9	−9.6	−8.7	−8.3	−9.8	−11.2	−12.5	−7.1										−0.4	−7.9
300	−8.3	−10.1	−9.0	−9.1	−10.6	−13.2	−11.9	−9.2										−0.5	−8.3
280	−8.0	−10.2	−9.5	−9.5	−11.5	−12.4	−11.2	−6.5										−0.9	−8.0
260	−8.7	−11.2	−10.5	−10.6	−12.1	−11.2						−2.2	−3.2					−0.5	−8.7
240	−7.6	−10.0	−9.3	−9.7	−11.2	−5.5						−8.1	−5.9					0.6	−7.6
220	−7.1	−9.5	−8.8	−9.1	−10.7	−6.5							−9.9	0.8				1.4	−7.1
200	−2.5	−7.7	−7.8	−7.9	−9.4	−9.8							−9.8	−1.2					−2.5
180		0.6	−5.3	−6.3	−8.7	−10.7	−5.2						−3.6	−1.3					
160	−2.5	−2.2	−4.5	−6.1	−7.7	−10.6	−11.2							−0.8					−2.5
140	−8.2	−9.5	−8.3	−7.7	−8.4	−9.8	−12.9							1.5				−1.1	−8.2
120	−8.8	−10.3	−9.0	−8.0	−8.5	−9.0	−8.9	−9.3										−1.6	−8.8
100	−8.7	−9.9	−8.8	−7.9	−8.6	−7.4	−5.1	−2.4					−2.0					−1.6	−8.7
80	−5.5	−7.8	−6.5	−6.1	−6.8	−5.1						−4.4	−7.8	−1.1				1.6	−5.5
60	−4.0	−5.6	−4.4	−4.6	−5.9	−3.3						−5.7	−6.9	−0.1					−4.0
40	−5.5	−7.3	−6.2	−5.6	−8.0	−5.7						−5.0	−8.9	0.3					−5.5
20	−5.5	−7.7	−7.9	−7.4	−8.8	−8.4							−7.0	−0.7				1.1	−5.5
0	−5.7	−7.9	−8.5	−8.2	−9.7	−11.3	−6.4						−4.4	−1.6				1.4	−5.7

Fig. 2-13b. The lowest energy is −13.2 kcal/mole/tripeptide, located at (100, 300), for the trialanine chain in ethanol.

Fig. 2-13c.

φ\ψ	0	20	40	60	80	100	120	140	160	180	200	220	240	260	280	300	320	340	360
360	−5.5	−8.5	−10.0	−9.5	−11.8	−12.5	5.4						−2.8	0.5					−5.5
340	−7.0	−11.0	−10.4	−9.9	−12.9	−14.0	−10.4							1.9				2.4	−7.0
320	−9.4	−11.6	−10.3	−9.5	−12.6	−14.8	−15.4	−7.2										0.6	−9.4
300	−10.1	−12.4	−11.3	−10.8	−13.7	−16.6	−13.5	−11.0										1.0	−10.1
280	−8.8	−13.3	−12.2	−12.1	−14.3	−14.4	−12.5	−5.7										−0.4	−8.8
260	−11.0	−13.1	−12.1	−14.1	−14.7	−12.6							−0.7					0.1	−11.0
240	−8.6	−11.9	−10.2	−9.1	11.1	−2.5						−9.0						2.6	−8.6
220	−7.5	−10.9	−10.1	−10.6	−11.8	−5.7							−12.5						−7.5
200	−0.2	−7.7	−7.8	−7.8	−10.2	−11.1							−12.1	0.2					−0.2
180		−3.5	−4.9	−8.5	−11.9	−4.5							−2.4	0.5					
160	−0.2	0.7	−2.4	−4.7	−7.9	12.2	−13.5							1.1					−0.2
140	−9.2	−10.5	−8.5	6.7	−8.2	−3.5	−16.6											0.1	−9.2
120	−10.3	−11.9	−9.8	−8.0	7.1	−8.9		−10.9										−0.8	−10.3
100	−10.8	−12.2	−8.8	−8.1	−8.4	−4.7	−3.1						−0.3					−0.0	−10.8
80	−4.8	−8.7	−8.6	−4.8	−7.0	−2.9						−2.1	−7.6	1.1				3.0	−4.8
60	−3.4	−4.8	−0.7	−5.8	−3.8	−0.4						−5.0	−6.2	1.3					−3.4
40	−5.8	−8.2	−6.0	−5.7	−11.1	−5.1						−3.5	−10.3	0.9					−5.8
20	−5.2	−8.2	−9.2	−8.5	−10.4	−9.4							−6.6	0.2				3.0	−5.2
0	−5.5	−8.5	−10.0	−9.5	−11.8	−12.5	−5.4						−2.8	0.5					−5.5

Fig. 2-13c. The lowest energy is −16.6 kcal/mole/tetrapeptide, located at (100, 300), for the tetraalanine chain in ethanol.

Fig. 2-13d.

φ\ψ	0	20	40	60	80	100	120	140	160	180	200	220	240	260	280	300	320	340	360
360	7.1	6.3	5.4	5.6	4.8	5.7													7.1
340	6.6	5.2	5.3	5.5	4.0	4.2	6.3												6.6
320	5.4	5.0	5.4	5.7	4.2	3.3	4.1	6.9										7.9	5.4
300	5.2	4.6	5.0	5.2	3.9	3.5	5.3	5.1											5.2
280	6.2	3.9	4.2	4.3	4.2	5.0	5.7	7.8										7.5	6.2
260	4.7	5.0	5.3	3.5	4.3	5.5												7.6	4.7
240	6.0	5.0	6.1	7.6	7.1							6.1							6.0
220	6.6	5.5	5.7	5.4	5.8	7.8								4.5					6.6
200			7.0	7.0	7.0	6.2	5.7							4.6					
180					8.3	7.2	5.7	7.6						8.2					
160						6.8	5.4	4.6											
140	6.0	5.9	6.7					3.3										8.2	6.0
120	5.5	5.4	6.1	7.0					5.3									7.8	5.5
100	4.9	4.7																	4.9
80													7.1						
60												7.7	7.6						
40													5.6	7.5					
20	7.2	6.4	5.6	5.9	5.3	6.0							7.4	7.9					7.2
0	7.1	6.3	5.4	5.6	4.8	5.7													7.1

Fig. 2-13d. The lowest energy is 3.3 kcal/mole/residue, located at (100, 320) and (120, 140), for the infinite alanine chain in ethanol.

Fig. 2-14a

	0	20	40	60	80	100	120	140	160	180	200	220	240	260	280	300	320	340	360
360	-6.5	-7.8	-7.8	-7.2	-7.2	-7.0	-5.2						-3.5	-0.2				-1.6	-6.5
340	-6.9	-8.0	-7.5	-7.6	-7.4	-6.6	1.1						0.5	0.0				-1.8	-6.9
320	-7.1	-8.2	-8.1	-7.6	-7.8	-7.6	-7.1	-5.2						0.5				-1.8	-7.1
300	-6.9	-8.1	-8.0	-7.5	-7.8	-7.7	-7.2	-5.9	1.4					0.9				-1.5	-6.9
280	-6.6	-7.8	-7.7	-7.3	-7.8	-7.9	-6.8	-3.7	0.8				0.0	1.0				-1.2	-6.6
260	-6.3	-7.5	-7.4	-7.1	-7.8	-7.1						-0.8	-3.0	0.9				-1.0	-6.3
240	-6.1	-7.2	-7.2	-6.9	-7.6	-4.4					1.3	-4.0	-3.8	0.8				-0.8	-6.1
220	-5.9	-7.1	-7.0	-6.6	-7.4	-5.2						-4.4	-4.3	0.6				-0.6	-5.9
200	-3.6	-6.3	-6.5	-6.1	-6.6	-6.4							-4.5	0.6					-3.6
180		-2.3	-5.5	-5.6	-6.1	-6.3	-3.1						-1.8	0.3					
160	-4.0	-4.0	-5.5	-6.0	-6.4	-6.5	-6.2							-0.2				-0.5	-4.0
140	-6.7	-7.6	-7.3	-6.8	-6.9	-6.8	-6.5	-5.4						0.6				-1.7	-6.7
120	-7.0	-7.9	-7.5	-6.8	-6.7	-6.7	-6.2	-5.1	0.6					1.3				-1.9	-7.0
100	-6.6	-7.5	-7.1	-6.3	-6.4	-6.1	-4.8	-1.6						-0.8				-1.6	-6.6
80	-5.7	-6.5	-6.0	-5.3	-5.3	-4.8	-0.6						-2.1	-0.3				-0.7	-5.7
60	-5.0	-5.8	-5.3	-4.5	-4.5	-3.9							-3.2	0.9				0.0	-5.0
40	-5.8	-6.7	-6.2	-5.4	-5.5	-4.9						-3.0	-4.0	0.5				-0.7	-5.8
20	-6.4	-7.6	-7.3	-6.6	-6.6	-6.3	-0.7						-4.5	-0.2				-1.4	-6.4
0	-6.5	-7.8	-7.8	-7.2	-7.2	-7.0	-5.2						-3.5	-0.2				-1.6	-6.5

Fig. 2-14a. The lowest energy is −8.2 kcal/mole/dipeptide, located at (20, 320), for the dialanine chain in water.

Fig. 2-14b

	0	20	40	60	80	100	120	140	160	180	200	220	240	260	280	300	320	340	360
360	-7.4	-9.4	-10.3	-9.2	-9.4	-9.1	-5.6						-3.4	0.7				-0.1	-7.4
340	-8.3	10.8	-10.8	-9.9	-10.2	-9.9	-8.7							1.2				-0.8	-8.3
320	-9.7	-11.4	-11.2	-10.2	-10.6	-10.4	-9.7	-5.8						2.3				-1.8	-9.7
300	-9.5	-11.3	-11.2	-10.4	-10.7	-10.8	-10.2	-7.6						3.3				-1.3	-9.5
280	-8.6	-10.7	-10.6	-10.0	-10.7	-11.1	-9.3	-3.2						3.6				-0.5	-8.6
260	-8.1	-10.1	-9.9	-9.5	-10.6	-9.4						3.3	-1.3	3.6				0.1	-8.1
240	-7.6	-9.6	-9.4	-9.0	-10.5	-4.1						-3.3	-3.3	3.4				0.5	-7.6
220	-7.2	-9.2	-9.1	-8.8	-10.1	-5.8							-5.3	3.0				1.2	-7.2
200	-2.4	-7.4	-8.1	-7.6	-8.7	-8.2							-6.4	2.1					-2.4
180		0.7	-5.9	-6.5	-7.3	-8.3	-2.5						-1.0	1.1					
160	-3.2	-2.6	-5.6	-7.0	-7.7	-8.2	-8.8							0.5				1.1	-3.2
140	-8.8	-9.8	-9.2	-8.3	-8.6	-8.2	-8.5							2.3				-1.5	-8.8
120	-9.5	-10.7	-9.9	-8.6	-8.5	-8.1	-6.8	-5.0										-2.0	-9.5
100	-9.0	-10.0	-9.3	-7.9	-7.6	-6.9	-4.1	2.4						-0.4				-1.4	-9.0
80	-6.9	-7.9	-7.0	-6.0	-5.6	-4.4						-0.4	-5.3	0.5				0.3	-6.9
60	-5.5	-6.6	-5.6	-4.2	-4.3	-2.7						-2.8	-4.0	3.0				1.8	-5.5
40	-7.0	-8.4	-7.5	-5.9	6.1	-4.8						-2.3	-4.7	2.1				0.6	-7.0
20	-7.3	-9.1	-9.6	-8.2	-8.1	-7.6							-5.6	0.8				-0.6	-7.3
0	-7.4	-9.4	-10.3	-9.2	-9.4	-9.1	-5.6						-3.4	0.7				-0.1	-7.4

Fig. 2-14b. The lowest energy is −11.4 kcal/mole/tripeptide, located at (20, 320), for the trialanine chain in water.

Fig. 2-14c

	0	20	40	60	80	100	120	140	160	180	200	220	240	260	280	300	320	340	360
360	-8.5	-11.1	-13.0	-11.4	-11.9	-10.0	-5.2						-2.6	2.9				1.2	-8.5
340	-9.8	-13.8	-13.8	-12.4	-13.7	-12.6	-10.0							2.2				0.1	-9.8
320	-12.5	-14.8	-14.4	-12.9	-14.3	-14.5	-12.4	-6.8						3.9				-2.0	-12.5
300	-12.2	-14.6	-14.5	-13.3	-14.5	-14.0	-12.3	-9.8										-0.6	-12.2
280	-10.0	-14.4	-14.3	-13.4	-13.8	-13.5	-11.0	-2.6										0.0	-10.0
260	-10.4	-11.9	-11.7	-13.0	-13.6	-11.2							0.2					0.6	-10.4
240	-9.0	-11.8	-10.9	-8.9	-11.3	-1.9						-6.0						1.9	-9.0
220	-8.4	-11.3	-11.5	-11.4	-12.6	-6.4							-9.3					3.1	-8.4
200	-1.3	-8.6	-9.9	-9.7	-11.6	-11.2							-10.2	1.3					-1.3
180		3.8	-6.5	-8.0	-9.9	-11.9	-3.4						-1.6	0.2					
160	-2.6	-1.1	-5.5	-7.8	-10.2	-12.2	-13.1							0.0				2.4	-2.6
140	-10.9	-12.1	-11.0	-9.8	-10.2	-3.6	-13.5							3.5				-1.4	-10.9
120	-11.9	-13.3	-12.2	-10.4	-10.2	-9.6		-8.6										-2.0	-11.9
100	-11.8	-13.0	-10.3	-9.3	-9.2	-7.7	-3.4							0.3				-0.4	-11.8
80	-7.4	-9.6	-10.0	-5.5	-6.7	-4.3						1.4	-5.7	2.0				0.6	-7.4
60	-6.1	-7.0	-6.1	-6.0	-2.5	-1.0						-2.9	-3.6	4.8				3.6	-6.1
40	-8.4	-10.2	-8.2	-6.6	-9.1	-5.1						-1.9	-5.5	2.5				1.8	-8.4
20	-8.3	-10.8	-12.0	-10.0	-10.0	-9.2							-5.8	1.5				0.0	-8.3
0	-8.5	-11.1	-13.0	-11.4	-11.9	-10.0	-5.2						-2.6	2.9				1.2	-8.5

Fig. 2-14c. The lowest energy is −14.8 kcal/mole/tetrapeptide, located at (20, 320), for the tetraalanine chain in water.

90

Fig. 2-14d

φ\ψ	0	20	40	60	80	100	120	140	160	180	200	220	240	260	280	300	320	340	360
360	-1.1	-1.7	-2.7	-2.2	-2.5	-0.9													-1.1
340	-1.5	-3.0	-3.0	-2.5	-3.5	-2.7	-1.3												-1.5
320	-2.8	-3.4	-3.2	-2.7	-3.7	-4.1	-2.7	-1.0										-0.2	-2.8
300	-2.7	-3.3	-3.3	-2.9	-3.8	-3.2	-2.1	-2.2											-2.7
280	-1.4	-3.7	-3.7	-3.4	-3.1	-2.4	-1.7												-1.4
260	-2.3	-1.8	-1.8	-3.5	-3.0	-1.8													-2.3
240	-1.4	-2.2	-1.5									-2.7							-1.4
220	-1.2	-2.1	-2.4	-2.6	-2.5	-0.6							-4.0						-1.2
200		-1.2	-1.8	-2.1	-2.9	-3.0							-3.8	-0.8					
180				-1.5	-2.6	-3.6	-0.9						-0.6	-0.9					
160					-2.5	-4.0	-4.3						-0.5						
140	-2.1	-2.3	-1.8				-5.0												-2.1
120	-2.4	-2.6	-2.3	-1.8	-1.7				-3.6									0.0	-2.4
100	-2.8	-3.0	-1.0	-1.4	-1.6														-2.8
80													-0.4						
60												-0.1							
40													-0.8						
20		-1.7	-2.4	-1.8	-1.9	-1.6													
0	-1.1	-1.7	-2.7	-2.2	-2.5	-0.9													-1.1

Fig. 2-14d. The lowest energy is −5.0 kcal/mole/residue, located at (120, 140), for the infinite alanine chain in water.

Fig. 2-15a

φ\ψ	0	20	40	60	80	100	120	140	160	180	200	220	240	260	280	300	320	340	360
360	-12.8	-13.0	-13.6	-12.4	-13.2	-13.7	-12.3						-10.6	-6.6				-7.4	-12.8
340	-12.9	-12.9	-13.3	-12.1	-12.8	-13.0	-14.5						-5.8	-7.0				-7.5	-12.9
320	-13.8	-13.7	-13.9	-12.6	-13.2	-13.3	-14.6	-12.7						-7.8				-8.2	-13.8
300	-14.8	-13.6	-13.8	-13.4	-13.1	-13.7	-15.1	-13.9	-5.3					-7.5				-8.9	-14.8
280	-15.0	-15.0	-15.2	-14.0	-14.3	-14.6	-13.7	-10.8	-7.9				-8.9	-7.8				-9.2	-15.0
260	-14.6	-14.6	-14.8	-13.9	-14.0	-13.8						-8.5	-11.6	-7.2				-9.0	-14.6
240	-12.9	-13.0	-14.4	-13.7	-14.1	-10.6					-5.7	-13.6	-13.0	-6.4				-7.2	-12.9
220	-11.2	-11.4	-12.0	-12.4	-11.8	-9.3						-11.9	-8.4	-5.4				-5.4	-11.2
200	-9.3	-10.9	-11.8	-11.2	-12.1	-11.6								-6.7					-9.3
180			-6.8	-10.6	-10.3	-11.9	-12.3	-7.8						-7.0					
160	-9.6	-8.5	-10.5	-10.4	-10.5	-12.2	-13.1	-9.4						-6.6				-5.7	-9.6
140	-12.5	-12.2	-12.4	-11.2	-11.0	-12.0	-13.3	-11.9										-7.1	-12.5
120	-14.6	-14.4	-14.5	-13.2	-13.8	-13.7	-14.3	-14.1	-6.0				-6.4					-9.1	-14.6
100	-15.4	-15.3	-15.4	-14.0	-13.6	-13.8	-12.3	-8.4					-8.7					-9.7	-15.4
80	-14.0	-13.9	-14.0	-12.6	-12.5	-13.0	-8.3				-6.1	-8.8	-11.7	-7.4				-8.5	-14.0
60	-12.9	-12.6	-12.6	-11.3	-11.3	-12.0						-10.7	-12.0	-5.3				-6.2	-12.9
40	-12.5	-12.6	-13.8	-12.3	-12.3	-13.2						-9.6	-11.6					-6.8	-12.5
20	-12.4	-12.6	-13.0	-12.8	-12.6	-13.4	-6.8						-12.1	-5.4				-6.7	-12.4
0	-12.8	-13.0	-13.6	-12.4	-13.2	-13.7	-12.3						-10.6	-6.6				-7.4	-12.8

Fig. 2-15a. The lowest energy is −15.2 kcal/mole/dipeptide, located at (40, 280), for the dialanine chain in formic acid.

Fig. 2-15b

φ\ψ	0	20	40	60	80	100	120	140	160	180	200	220	240	260	280	300	320	340	360
360	-8.5	-8.7	-9.7	-7.6	-8.1	-7.3	-3.2						-0.8					-0.5	-8.5
340	-9.3	-10.0	-9.9	-7.9	-8.3	-7.3	-6.9											-1.2	-9.3
320	-12.4	-12.2	-11.7	-9.0	-9.7	-8.4	-8.0	-1.7										-3.7	-12.4
300	-14.4	-12.4	-11.9	-11.2	-9.6	-9.3	-9.6	-6.1										-4.9	-14.4
280	-14.4	-14.9	-14.4	-12.0	-11.8	-11.3	-7.8	-1.6										-5.0	-14.4
260	-13.2	-13.6	-13.2	-11.3	-10.8	-8.5												-3.9	-13.2
240	-8.9	-9.7	-11.2	-9.2	-9.2	-2.1					-3.1	-1.2							-8.9
220	-6.4	-6.9	6.8	-5.0	-4.7	-0.6						-2.6							-6.4
200	-2.0	-5.3	-6.0	-4.0	-4.1	-4.0						-5.1							-2.0
180			-3.4	-2.4	-2.6	-4.7													
160	-2.1	0.0	-2.7	-2.8	-2.2	-4.0	-5.7												-2.1
140	-7.6	-7.0	-6.2	-4.1	-3.2	-2.8	-5.5												-7.6
120	-11.2	-10.8	-9.5	-6.4	-6.5	-3.9	-4.2	-5.2										-3.1	-11.2
100	-13.4	-12.8	-11.8	-8.4	-6.1	-4.3	-0.3								0.0			-4.9	-13.4
80	-11.0	-10.2	-9.0	-6.3	-4.3	-3.0							-4.2					-2.8	-11.0
60	-8.8	-8.0	-6.6	-3.6	-2.5	-1.4					-0.8	-4.0							-8.8
40	-8.6	-8.0	-9.3	-6.1	-5.3	-4.7						-4.0						-0.1	-8.6
20	-7.7	-7.9	-8.4	-7.7	-6.4	-6.2						-4.1						-0.2	-7.7
0	-8.5	-8.7	-9.7	-7.6	-8.1	-7.3	-3.2						-0.8					-0.5	-8.5

Fig. 2-15b. The lowest energy is −14.9 kcal/mole/tripeptide, located at (20, 280), for the trialanine chain in formic acid.

91

	0	20	40	60	80	100	120	140	160	180	200	220	240	260	280	300	320	340	360
360	−7.5	−8.3	−10.4	−8.2	−9.2	−7.0	−2.1						0.2					2.8	−7.5
340	−8.6	−10.6	−10.7	−8.6	−10.2	−9.0	−6.7											1.8	−8.6
320	−13.4	−13.2	−12.7	−9.8	−11.7	−11.3	−9.4	−2.2										−1.9	−13.4
300	−16.5	−13.6	−13.2	−12.3	−11.8	−11.4	−10.1	−7.4										−2.9	−16.5
280	−14.6	17.0	−16.2	−13.1	−12.6	−11.9	−8.2	−0.1										−2.5	−14.6
260	−13.9	−13.4	−12.6	−12.3	−11.6	−8.3							2.7					−1.2	−13.9
240	−8.5	−10.0	−9.8	−6.6	−7.5	2.1						−4.4							−8.5
220	−6.1	−7.2	−7.1	−6.2	−5.8	−0.2							−7.0						−6.1
200	0.6	−4.8	−5.3	−3.8	−5.6	−6.4							−8.2						0.6
180			−0.3	−0.9	−3.7	−7.4	−0.2						0.9						
160	0.3	0.2	−0.8	−4.0	−7.5	−10.3													0.3
140	−8.3	−7.3	−5.3	−3.2	−3.0	1.5	−10.5											2.3	−8.3
120	−11.2	−10.3	−8.0	−0.9	−2.3	−3.5		−6.8										−0.5	−11.2
100	−14.1	−13.4	−10.1	−6.0	−2.4	0.2	0.3							2.6				−1.5	−14.1
80	−9.2	−9.1	−9.2	−2.7	−2.1	1.5							−3.8					−0.2	−9.2
60	−8.0	−6.1	−4.9	−3.4	1.2	2.8						−0.7	−3.2						−8.0
40	−7.9	−8.0	−8.1	−5.1	−6.5	−2.9						1.6	−4.4					3.0	−7.9
20	−6.7	−7.4	−8.9	−7.9	−6.8	−6.5							−4.0					2.4	−6.7
0	−7.5	−8.3	−10.4	−8.2	−9.2	−2.1	−2.1						0.2					2.8	−7.5

Fig. 2-15c. The lowest energy is −17.0 kcal/mole/tetrapeptide, located at (20, 280), for the tetraalanine chain in formic acid.

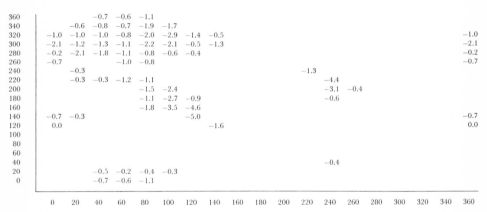

	0	20	40	60	80	100	120	140	160	180	200	220	240	260	280	300	320	340	360
360			−0.7	−0.6	−1.1														
340		−0.6	−0.8	−0.7	−1.9	−1.7													
320	−1.0	−1.0	−1.0	−0.8	−2.0	−2.9	−1.4	−0.5											−1.0
300	−2.1	−1.2	−1.3	−1.1	−2.2	−2.1	−0.5	−1.3											−2.1
280	−0.2	−2.1	−1.8	−1.1	−0.8	−0.6	−0.4												−0.2
260	−0.7			−1.0	−0.8														−0.7
240		−0.3										−1.3							
220		−0.3	−0.3	−1.2	−1.1								−4.4						
200					−1.5	−2.4							−3.1	−0.4					
180					−1.1	−2.7	−0.9						−0.6						
160					−1.8	−3.5	−4.6												
140	−0.7	−0.3				−5.0													−0.7
120	0.0						−1.6												0.0
100																			
80																			
60																			
40													−0.4						
20			−0.5	−0.2	−0.4	−0.3													
0			−0.7	−0.6	−1.1														

Fig. 2-15d. The lowest energy is −5.0 kcal/mole/residue, located at (120, 140), for the infinite alanine chain in formic acid.

	0	20	40	60	80	100	120	140	160	180	200	220	240	260	280	300	320	340	360
360	−22.9	−24.2	−23.9	−23.8	−25.1	−26.9	−23.7						−22.3	−19.7					−22.9
340	−23.0	−24.2	−23.7	−23.7	−24.9	−26.6	−25.8						−18.2	−19.6					−23.0
320	−24.4	−24.8	−24.8	−24.1	−25.4	−27.2	−27.9	−23.8						−18.9				−18.5	−24.4
300	−24.3	−25.1	−24.4	−24.0	−26.4	−28.0	−27.8	−26.7	−18.3					−19.1				−18.7	−24.3
280	−23.9	−24.6	−23.6	−23.4	−25.8	−26.5	−24.6	−21.8					−19.6						−23.9
260	−21.6	−22.4	−22.6	−22.4	−24.8	−23.7						−19.0	−21.7						−21.6
240	−22.5	−23.2	−22.4	−23.6	−24.7	−20.9						−23.1	−25.0	−18.0					−22.5
220	−21.6	−22.7	−23.0	−23.0	−23.9	−21.9						−24.1	−25.2	−19.1					−21.6
200	−18.7	−21.3	−21.2	−21.4	−23.5	−22.6							−23.6	−19.1					−18.7
180		−20.4	−20.8	−23.0	−24.0	−20.1							−20.0	−18.4					
160	−19.4	−19.2	−20.2	−20.8	−22.5	−25.4	−24.1												−19.4
140	−22.0	−22.6	−21.8	−21.2	−22.4	−25.2	−25.4	−23.2											−22.0
120	−23.6	−23.2	−22.2	−21.5	−24.2	−24.5	−25.5	−22.5											−23.6
100	−21.8	−23.0	−22.3	−21.6	−22.7	−24.3	−22.1	−18.1											−21.8
80	−22.6	−22.6	−21.8	−22.1	−23.3	−23.0	−18.8					−20.8	−22.4	−18.3					−22.6
60	−22.0	−23.0	−22.8	−22.1	−23.5	−23.2						−24.0	−24.0	−18.5					−22.0
40	−22.8	−23.8	−23.0	−23.0	−24.3	−23.9						−21.5	−24.5	−18.0					−22.8
20	−23.0	−24.4	−24.3	−23.6	−25.6	−26.5	−19.2						−24.1	−19.1					−23.0
0	−22.9	−24.2	−23.9	−23.8	25.1	−26.9	−23.7						−22.3	−19.7					−22.9

Fig. 2-16a. The lowest energy is −28.0 kcal/mole/dipeptide, located at (100, 300), for the dialanine chain in acetic acid.

Fig. 2-16b

	0	20	40	60	80	100	120	140	160	180	200	220	240	260	280	300	320	340	360
360	-25.8	-27.6	-27.7	-26.8	-28.7	-31.0	-22.9						-20.5	-20.3					-25.8
340	-26.5	-28.8	-27.8	-26.9	-28.7	-31.0	-27.8							-19.6					-26.5
320	-30.0	-30.7	-29.8	-28.2	-30.1	-32.6	-32.5	-22.2										-21.3	-30.0
300	-30.2	-31.2	-29.8	-28.8	-32.5	-34.4	-32.5	-28.4										-21.5	-30.2
280	-28.8	-30.2	-28.8	-27.9	-31.4	-30.6	-26.3												-28.8
260	-23.3	-25.2	-26.2	-25.7	-28.6	-25.2													-23.3
240	-23.4	-25.2	-24.6	-26.8	-26.8							-20.7	-23.4						-23.4
220	-22.2	-24.1	-23.2	-23.3	-23.4	-19.6							-25.5						-22.2
200		-22.5	-22.0	-21.5	-22.7	-21.7							-22.5						
180			-20.9	-20.8	-24.7	-24.1													
160	-19.6		-20.6	20.9	-23.1	-26.0	-23.2												-19.6
140	-23.3	-24.2	-22.5	-20.7	-21.3	-22.7	-25.9												-23.3
120	-24.5	-25.4	-23.5	-21.4	-21.8	-21.2	-24.3	-19.7											-24.5
100	-23.1	-23.7	-22.1	-20.0	-20.6	-21.8													-23.1
80	-25.5	-24.2	-22.4	-22.9	-23.8	-20.8							-22.4						-25.5
60	-24.2	-25.4	-24.8	23.0	-24.7	-22.0						-22.7	-25.6						-24.2
40	-25.9	-27.1	-25.1	-24.6	-26.6	23.8							-26.5						-25.9
20	-25.8	-27.6	-27.7	-25.9	-28.9	-29.3							-24.7	-19.7					-25.8
0	-25.8	-27.6	-27.7	-26.8	-28.7	-31.0	-22.9						-20.5	-20.3					-25.8

Fig. 2-16b. The lowest energy is −34.4 kcal/mole/tripeptide, located at (100, 300), for the trialanine chain in acetic acid.

Fig. 2-16c

	0	20	40	60	80	100	120	140	160	180	200	220	240	260	280	300	320	340	360
360	-28.8	-31.2	-31.8	-30.0	-32.8	-34.1													-28.8
340	-30.2	-33.6	-32.1	-30.4	-33.7	-35.8	-28.7												-30.2
320	-35.8	-36.8	-35.1	-32.5	-36.0	-39.4	-37.1											-24.2	-35.8
300	-36.4	-37.8	-35.8	-33.8	-39.6	-41.1	-36.2	-30.4										-24.0	-36.4
280	-33.3	-37.0	-34.9	33.3	-37.2	-34.2	-26.8												-33.3
260	-26.2	-27.6	-29.3	-30.2	-32.6	-26.4													-26.2
240	-25.1	-27.9	-26.2	-27.7	-27.3														-25.1
220	-22.2	-25.4	-24.3	-24.4	-23.0								-21.7						-22.2
200				-21.3															
180						-21.6													
160						-22.7													
140	-23.7	24.0						-22.2											-23.7
120	-26.5	-27.1	-23.4																-26.5
100	-26.1	-26.6	-22.1																-26.1
80	-28.2	-26.6	-25.4	-22.6	-23.8								-21.3						-28.2
60	-27.0	-27.8	-27.4	-26.3	-24.1							-22.2	-26.2						-27.0
40	-29.2	-30.7	-27.2	-27.1	-31.4	-23.6							-28.5						-29.2
20	-28.7	-31.1	-31.5	-28.9	-33.1	32.5							-24.6						-28.7
0	-28.8	-31.2	-31.8	-30.0	-32.8	-34.1													-28.8

Fig. 2-16c. The lowest energy is −41.1 kcal/mole/tetrapeptide, located at (100, 300), for the tetraalanine chain in acetic acid.

Fig. 2-16d

	0	20	40	60	80	100	120	140	160	180	200	220	240	260	280	300	320	340	360
360	-3.0	-3.6	-4.1	-3.2	-4.1	-3.1													-3.0
340	-3.7	-4.8	-4.3	-3.5	-5.0	-4.8													-3.7
320	-5.8	-6.1	-5.3	-4.3	-5.9	-6.8	-4.6											-2.9	-5.8
300	-6.2	-6.6	-6.0	-5.0	-7.1	-6.7	-3.7											-2.5	-6.2
280	-4.5	-6.8	-6.1	-5.4	-5.8	-3.6													-4.5
260	-2.9	-2.4	-3.1	-4.5	-4.0														-2.9
240		-2.7																	
220																			
200																			
180																			
160																			
140																			
120																			
100	-3.0	-2.9																	-3.0
80		-2.4	-3.0																
60																			
40																			
20	-2.9	-3.5	-3.8	-3.0	-4.2														-2.9
0	-3.0	-3.6	-4.1	-3.2	-4.1	-3.1													-3.0

Fig. 2-16d. The lowest energy is −7.1 kcal/mole/residue, located at (80, 300), for the infinite alanine chain in acetic acid.

93

VIII. Semiempirical Quantum Mechanical Calculations

A. One-Electron Models

For any predescribed conformation of a molecule one is faced with the formidable task of solving the N-electron Schrödinger equation

$$H(N)\psi(N) = E(N)\psi(N) \qquad (N > 1) \tag{2-77}$$

where $H(N)$ is the N-electron Hamiltonian which contains nonseparable terms $1/r_{ij}$ which represent the relative interaction positions of all electrons i and j. $\psi(N)$ is a nonspecified N-electron wave function which characterizes the properties of the electrons in the molecule. $E(N)$ is the total energy of the molecule. Because $H(N)$ is nonseparable and $\psi(N)$ is nonspecific, Eq. (2-77) cannot be solved exactly. However, let us introduce the concept of molecular orbitals, MO's, which assume that each of the N electrons in the molecule moves throughout the molecule in a path dictated by the potential generated by all other electrons and nuclei composing the molecule. Then $H(N)$ can be expressed as

$$H(N) = \sum_{i=1}^{N} \mathscr{H}_{eff}(i) \tag{2-78}$$

where $\mathscr{H}_{eff}(i)$ is the effective Hamiltonian associated with the ith molecular orbital. That is, we can rewrite the N-electron Hamiltonian as a sum of N-effective one-electron Hamiltonians. If ϕ_i is the wave function associated with the ith molecular orbital, then Eq. (2-77) reduces to N one-electron equations of the form

$$\mathscr{H}_{eff}(i)\,\phi_i = \epsilon_i\,\phi_i \tag{2-79}$$

where

$$E(N) = \sum_{i=1}^{N} \epsilon_i \tag{2-80}$$

and

$$\psi(N) = \prod_{i=1}^{N} \phi_i \tag{2-81}$$

in its simplest form.

Now if we make the additional assumption that each MO, ϕ_i, can be written as a linear combination of atomic orbitals, LCAO, then ϕ_i takes the form

$$\phi_i = \sum_{j=1}^{N} C_{ij} \chi_j \tag{2-82}$$

The LCAO approximation states that the behavior of an electron in a MO near a particular atom is approximately the same as it would be in the isolated atom.

In Eq. (2-82) χ_j is the jth atomic orbital and C_{ij} is the weighting coefficient. Since the sum in Eq. (2-82) extends over all N electrons, the equation

$$\mathscr{H}_{\text{eff}}(i) \sum_{j=1}^{N} C_{ij} \chi_j = \epsilon_i \sum_{j=1}^{N} C_{ij} \chi_j \tag{2-83}$$

will yield all N possible MO's. Let us now proceed to solve Eq. (2-83) subject to the normalization constraint

$$\int_\tau \phi_i^* \, \phi_i \, d\tau = 1 \tag{2-84}$$

First we premultiply each side of Eq. (2-83) by the appropriate expression for ϕ_i

$$\int_\tau \left[\sum_{k=1}^{N} C_{ik} \chi_k \mathscr{H}_{\text{eff}} (i) \sum_{j=1}^{N} C_{ij} \chi_j \right] d\tau = \epsilon_i \int_\tau \left[\sum_{k=1}^{N} C_{ik} \chi_k \sum_{j=1}^{N} C_{ij} \chi_j \right] d\tau \tag{2-85}$$

after some algebra

$$\sum_{k=1}^{N} \sum_{j=1}^{N} C_{ik} C_{ij} \int_\tau \chi_k \mathscr{H}_{\text{eff}} (i) \chi_j \, d\tau = \epsilon_i \sum_{k=1}^{N} \sum_{j=1}^{N} C_{ik} C_{ij} \int_\tau \chi_k \chi_j \, d\tau \tag{2-86}$$

We define

$$H_{kj}^{(i)} = \int_\tau \chi_k \mathscr{H}_{\text{eff}} (i) \chi_j \, d\tau \tag{2-87}$$

$$S_{kj}^{(i)} = \int_\tau \chi_k \chi_j \, d\tau \tag{2-88}$$

Since our LCAO sum extends over all N electrons, we can be sure that we will get all N molecular orbitals from (2-86) and, therefore, we will drop the i subscript. This leads to a simple expression for the orbital energy

$$\frac{\sum_{k=1}^{N} \sum_{j=1}^{N} C_k C_j H_{kj}}{\sum_{k=1}^{N} \sum_{j=1}^{N} C_k C_j S_{kj}} = \epsilon \tag{2-89}$$

We now minimize ϵ with respect to the C_u. This is known as imposing the variation principle which states:

Given any approximate wave function satisfying the boundary conditions of the problem, the expectation value of the energy calculated from this function will always be higher than the true energy of the ground state.

We must now find the set of

$$\left(\frac{\partial \epsilon}{\partial C_u}\right)_{C_k} = 0 \qquad (k = 1, 2, \ldots, N = u)$$

After some algebraic manipulation we arrive at

$$\sum_{i=1}^{N} (H_{ij} - S_{ij}\,\epsilon)\, C_i = 0 \qquad (j = 1, 2, \ldots, N) \qquad (2\text{-}90)$$

For Eq. (2-90) to lead to a nontrivial solution it follows that

$$|H_{ij} - S_{ij}\,\epsilon| = 0 \qquad (j = 1, 2, \ldots, N) \qquad (2\text{-}91)$$

Equation (2-90) is known as the secular equation of the molecule and Eq. (2-91) is known as the secular determinant.

Now how do we calculate the H_{ij} and the S_{ij} so that we can calculate the ϵ and thus arrive at the C_i? The methods used to calculate the H_{ij} and the S_{ij} distinguish the various one-electron schemes from one another.

B. The Extended Hückel (EH) Method

In this method each H_{ii} is taken as the ionization potential for the appropriate electronic state in the appropriate isolated atom. The S_{ij} are computed according to Eq. (2-88) in which the χ_u are chosen to be Slater wave functions (36). The Slater wave functions are semiempirical approximations for more accurate wave functions calculated from refined two-electron schemes. It should be noted that most two-electron schemes are too complex to apply to large molecules (fifty orbitals or more). The forms of the Slater wave functions for several different electronic states are given in Table 2-12. The M represents the appropriate normalization constant which makes the probability χ^2 over all space unity. The constant, c, is the effective nuclear charge which is computed by modifying the atomic number, Z, with a shielding constant, S, $c = Z - S$. The shielding constants are derived from contributions from the environment of the electron according to the following rules:

(1) $S = 0$ for all electrons outside the principal quantum number being considered.

(2) $S = 0.35$ for each electron with the same principal quantum number. A value of $S = 0.30$ is used for consideration of the 1s orbital.

(3) If the orbital being considered is an s or a p, then $S = 0.85$ from each electron in the next inner shell and $S = 1.00$ for each electron further in.

Since we can determine the χ_u, the S_{ij} may be calculated, as stated earlier, from

$$S_{ij} = \int_\tau \chi_i \chi_j \, d\tau \tag{2-92}$$

TABLE 2-12

Slater Wave Functions[a]

n	l	State	χ
1	0	1s	$M \exp(-cr)$
2	0	2s	$Mr \exp(-cr/2)$
2	1	$2p_x$	$Mx \exp(-cr/2)$
3	0	3s	$Mr^2 \exp(-cr/3)$
3	1	$3p_x$	$Mxr \exp(-cr/3)$

[a] M is the appropriate normalization constant, c the shielding factor, x a variable distance, and r the internuclear distance.

The off-diagonal elements, H_{ij} $(i \neq j)$, normally called the exchange or resonance integrals, are approximated by the expression:

$$H_{ij} = \tfrac{1}{2}[k_1(H_{ii} + H_{jj}) S_{ij}] \tag{2-93}$$

or

$$H_{ij} = -k_2 (H_{ii} H_{jj})^{1/2} S_{ij} \tag{2-94}$$

where the k_1 and k_2 are calculated constants which reproduce some experimental value (e.g., the rotational barrier height in ethane). Commonly used values for k_1 are near 1.75. It should be noted that in some cases the H_{ii} are corrected ionization potentials which take into account electron–electron interactions. Such corrections are, at the very best, crude.

One must solve Eq. (2-79) for each trial conformation of the molecule tested. Stable conformations of the molecule are those for which Eq. (2-80) yields a relative minimum in total energy. It should be clear that this technique minimizes the bonded and nonbonded energy simultaneously. That is, up to now we have implicitly assumed that the bonded energy remains constant (e.g., there are no bond angle or bond length distortions) as different molecular conformations are realized. While this is a rather good approximation for the majority of cases, one can expect some small perturbations in bond lengths and bond angles of the molecule as conformational changes take place.

C. Applications of Semiempirical Quantum Mechanics

Depending upon the quantum mechanical technique employed, such calculations are more or less useful in identifying the functional forms of several types of energy interactions. Among the interactions which are

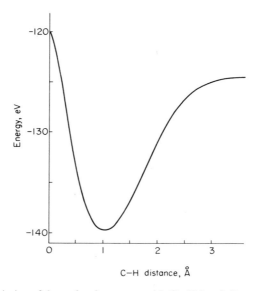

Fig. 2-17. Variation of the molecular energy with C—H bond distance in methane.

difficult to *theoretically* characterize by any means other than those involving quantum mechanics are:

(1) *Calculation of bond lengths and angles.* By minimizing the total molecular energy as a function of the constituent bond lengths and angles it is possible to determine the molecular structure of simple molecules. An example of this type of calculation is shown in Fig. 2-17 for the C—H bond lengths in methane using the extended Hückel approach.

(2) *Calculation of force constants.* From Fig. 2-17 it is seen that the shape of the potential of bond energy versus bond distance is part of the information obtained in determining equilibrium bond lengths in simple molecules. It is possible to fit a function, f, to the potential curve. The force constant, K, can

then easily be computed,

$$K = \frac{\partial^2 f}{\partial r^2}\bigg|_{r_{eq}} \tag{2-95}$$

where r_{eq} is the equilibrium bond distance. Once K is known, the bond length (the same holds bond angles) distortion potential energy, E_{dis}, may be determined using a Hookean type function,

$$E_{dis} = (K/2)\,(r - r_{eq})^2 \tag{2-96}$$

(3) *Calculation of hydrogen-bonding functions.* Perhaps the most important application of quantum mechanical schemes is in the elucidation of the nature of hydrogen bonding. A great number of investigations involving hydrogen bonds have been carried out and in the next section we summarize the results of two of these investigations.

(4) *Characterization of torsional potential functions.* We have discussed a number of energy contributions which contribute to hindering and restricting rotation about bonds in this chapter. In addition to these contributions are intrinsic torsional potential functions which arise because of the orbital configuration of the electrons about the atoms involved in the bonds. Since this type of energy contribution results from orbital configurations it is completely quantum mechanical in nature. Later in this chapter we will discuss this type of energy contribution in detail paying particular attention to experimental methods of characterizing this phenomenon.

IX. Hydrogen Bond Potential Functions

The one definite fact about hydrogen bonds is that there does not appear to be any definite rules which govern their geometry. This is demonstrated in Fig. 2-18 where we plot the NH$\cdots$O (Å) distance versus the NH$\cdots$O=C angle for observed NH$\cdots$O=C hydrogen bonds found in a variety of crystals. The question as to whether or not the variation in hydrogen bond geometry is completely due to variations in the strength of the donor–acceptor interaction cannot be resolved because of the restrictions and conditions imposed by the bonding geometries of the groups near the donor and the acceptor. The existence of a hydrogen bond can be inferred whenever the distance between the donor and the acceptor is less than the associated outer contact distance for the pair interaction. The violation of the outer contact distance criteria is due to the fact that there is a strong attractive electrostatic interaction between donor and acceptor—the acceptor has a large negative residual charge and the donor has a large positive residual charge. Thus the hydrogen bond is largely, but not exclusively, electrostatic in nature. Table 2-13 contains a list of the geometric parameters for a number of different hydrogen bonds.

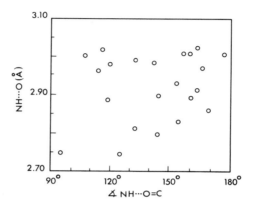

Fig. 2-18. A plot of the NH$\cdots$O distance versus the NH$\cdots$O=C angle in various crystals.

A large number of different models have been proposed to describe the interaction potential energy of a hydrogen bond as a function of the geometry of the bond. In Fig. 2-19 is shown the general geometry of the hydrogen bonding interaction. The following equations, based upon the geometry shown in Fig. 2-19, have been relatively successful in describing certain types of hydrogen bonds.

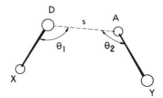

Fig. 2-19. The geometry of the hydrogen bond. D denotes the donor and the A denotes the acceptor atom. The symbol s indicates the distance between the donor and the acceptor.

A. The N — H $\cdots$ O = C(sp^2) Hydrogen Bond Function

This is a hybrid expression for the hydrogen-bond potential energy as a function of a specific set of geometric parameters. The basic form of the function was worked up by Lippincott and Schroeder (37) and later modified by Schroeder, Lippincott, Moulton, and Kromhout (38). The final equation,

TABLE 2-13

Geometry of a Selected Number of Hydrogen Bonds[a]

Compound	O⋯O	O—H	H⋯O	∠H—O⋯O	Donor	Acceptor
			O—H⋯O hydrogen bonds			
Boric acid	2.69	0.95	1.75	0°	acid OH	acid OH
Acetylglycine	2.56	0.94	1.63	2°	carboxyl OH	carbonyl O
Iodic acid	2.69	0.99	1.70	5°	acid OH	acid O
Xanthazole monohydrate	2.83	1.02	2.02	18°	H_2O	carbonyl O

Compound	N⋯O	N—H	H⋯O	∠H—N⋯O	Donor	Acceptor
			N—H⋯O hydrogen bonds			
L-alanine	2.81	0.90	1.91	5°	—NH_3^+	carboxyl O
Cytosine	2.98	0.86	2.14	9°	—NH_2	carbonyl O
Glycine	2.77	0.92	1.87	8°	—NH_3^+	carboxyl O
	2.85	0.85	2.03	12°	—NH_3^+	carboxyl O

Compounds	N⋯N	N—H	H⋯N	H—N⋯N	Donor	Acceptor
			N—H⋯N hydrogen bonds			
Cytosine	2.84	0.88	2.02	17°	ring NH	ring N
Purine	2.85	0.86	2.00	8°	ring NH	ring N

Compound	F—H⋯F	F⋯H	F—H	H-bond enthalpy	F—H⋯F
			Fluorine hydrogen bonds		
$(HF)_6$	2.55	1.55	1.00	−6.7 kcal/mole-bond	—
$(HF)_n$	2.49	—	—	—	120.1°

[a] Distances are in angstroms.

given here, also contains the modifications made by Scheraga and co-workers (39).

$$E(s, \theta_1, \theta_2) = -D^* \cos^2(\theta_1) \exp\left[\frac{-n^*(R - s - r_0^*)^2}{2(R - s)} \right]$$

$$-D^* \cos^2(\theta_2) \exp\left[\frac{-n^*(R - s - r_0^*)^2}{2(R - s)} \right] + A \exp(-bR)$$

$$-\tfrac{1}{2}A \left(\frac{R_0}{R} \right)^m \exp(-bR_0) \qquad (2-97)$$

where D^* is the strength of the O⋯H interaction, m an adjustable exponent,

n^* the variable parameter proportional to the ionization potential of the hydrogen, R the N to O distance, r_0^* the O—H bond length in the equilibrium configuration, A the measure of the strength of the N$\cdots$O potential, b the measure of the "hardness" of the N$\cdots$O potential (normally is assigned a value of 4.6), and R_0 the N to O distance of the equilibrium configuration. Both θ_1 and θ_2 must be greater than or equal to 90° for this function to be validly employed. When this condition is not satisfied the classical potential energy interactions given earlier in this chapter are normally used to describe the interaction.

B. The Corrected Sum Hydrogen Bond Function

This is a completely empirical potential function in which it is assumed that the interaction between the donor and the acceptor is given by the sum of the London–van der Waals and electrostatic potential functions (just like all pairwise classical potential interactions) plus some angle-dependent correction term (40). That is, it is assumed that any hydrogen bond interaction energy can be expressed by an equation of the general form

$$E(s, \theta_1, \theta_2) = \frac{-a}{s^6} + \frac{b}{s^{12}} + k \frac{Q_d \, Q_a}{\epsilon s} - \frac{G}{s^n} f(\theta_1, \theta_2) \qquad (2\text{-}98)$$

For the N—H$\cdots$O=C(sp²) hydrogen bond, the angular dependence is shown in Figs. 2-20a and 2-20b. These are empirical functions which have been chosen because they yield results (from a statistical basis) which are most consistent with the data found in hydrogen bond plots of the type shown in Fig. 2-18. In general, the angular dependent function $f(\theta_1, \theta_2)$ takes on the form

$$\begin{aligned} f(\theta_1, \theta_2) &= 0 \qquad \text{if} \quad \theta_1 < \theta_1^* \quad \text{and/or} \quad \theta_2 < \theta_2^* \\ &= \cos^u k\theta_1 \cos^v j\theta_2 \qquad \text{if} \quad \theta_1 \geqslant \theta_1^* \quad \text{and} \quad \theta_2 \geqslant \theta_2^* \qquad (2\text{-}99) \end{aligned}$$

The parameters θ_1^* and θ_2^* are known as the critical angle parameters. The value of θ_1^* is usually near 135° and θ_2^* is in the range around 90°. The value of G is computed by insisting that $E(s, \theta_1, \theta_2)$ is minimized at $s = R_0$ (the optimum hydrogen bond distance). That is, G is computed from the constraint that

$$\left(\frac{\partial E(s, \theta_1, \theta_2)}{\partial s} \right) \bigg|_{s = R_0} = 0$$

with $f(\theta_1, \theta_2) = 1$, the maximum value of $f(\theta_1, \theta_2)$.

The parameter n dictates the strength of the hydrogen bond interaction for a particular choice of R_0. If $n = 12$, then G takes the form

$$G = b + k \frac{Q_d Q_a}{\epsilon} \left(\frac{s^{11}}{12}\right) - a \left(\frac{s^6}{2}\right) \tag{2-100}$$

Then at $s = R_0$ and $\theta_1 = 180°$ and $\theta_2 = 180°$ we have

$$E(R_0, 180°, 180°) = -\left(\frac{1}{2}\right)\frac{a}{R_0^6} + \left(\frac{11}{12}\right)k\frac{Q_d Q_a}{\epsilon R_0} \tag{2-101}$$

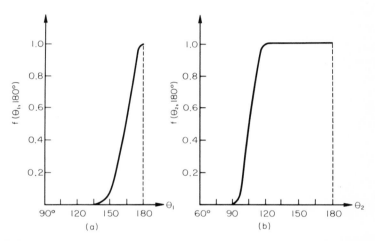

Fig. 2-20. (a) A plot of $f(\theta_1, 180°)$ versus θ_1. This is the θ_1-dependent plot of the angular dependence in a N—H$\cdots$O=C(sp²) hydrogen bond. The specific functional dependence is
$$f(\theta_1, 180°) = 0 \qquad \text{for} \quad \theta_1 \leqslant 135°$$
$$= \cos^4 2\theta_1 \qquad \text{for} \quad 135° < \theta_1 \leqslant 180°$$
(b) A plot of $f(180°, \theta_2)$ versus θ_2. This is the θ_2 angular-dependent plot of the angular dependence in a N—H$\cdots$O=C(sp²) hydrogen bond. The specific functional dependence is
$$f(180°, \theta_2) = 0 \qquad \text{for} \quad \theta \leqslant 90°$$
$$= \cos^2 3\theta_2 \qquad \text{for} \quad 90° < \theta_2 \leqslant 120°$$
$$= 1 \qquad \text{for} \quad 120° < \theta_2 \leqslant 180°$$

We can get some idea of the hydrogen-bonding energies predicted by Eq. (2-101) by calculating $E(R_0, 180°, 180°)$ versus R_0 for an O$\cdots$H interaction for which the residual charges are $Q_a = -0.450$ e and $Q_d = 0.210$ e, and $\epsilon = 3.5$. This particular set of parameters would describe the hydrogen bond which occurs between backbone carbonyl oxygens and amide hydrogens in polypeptides. Table 2-14 contains the values of $E(R_0, 180°, 180°)$ as a function of R_0 for this particular hydrogen bond. For a general measure of the reasonableness of the hydrogen bond energies given in Table 2-14 we note that experimental values for hydrogen bond energies range between 2 and 10 kcal/mole.

A slightly different form for Eq. (2-98) which has the capacity to be extremely flexible and adjustable in terms of both energy and equilibrium distance has the form for $f(\theta_1, \theta_2) = 1$ given by

$$E(s, 180°, 180°) = \frac{-(a - a_0)}{s^6} + \frac{(b - b_0)}{s^{12}} + \frac{kQ_d Q_a}{\epsilon s} \qquad (2\text{-}102)$$

That is, we assume the perfect linear hydrogen bond is identical in mathematical form to the usual nonbonded interaction in which the coefficients of

Table 2-14
H-Bond Energy Using Eq. (2-101)[a]

R_0 (Å)	$E(R_0, 180°, 180°)$ (kcal/mole)	R_0 (Å)	$E(R_0, 180°, 180°)$ (kcal/mole)
1.60	−8.2	1.90	−6.2
1.65		1.95	
1.70	−7.9	2.00	−5.4
1.75		2.05	
1.80	−7.1	2.10	−4.7
1.85			

[a] Parameters used are those of backbone polypeptide hydrogen bonds.

the Lennard-Jones 6-12 potential function have been modified. If E_0 is the desired minimum interaction energy, then

$$E_0 = E(R_0, 180°, 180°), \qquad 0 = \frac{\partial}{\partial s} E(R_0, 180°, 180°) \qquad (2\text{-}103)$$

yield two linear equations in two unknowns, a_0 and b_0. This set of equations is easily solved. Once a_0 and b_0 have been determined the general form of the potential function becomes

$$E(s, \theta_1, \theta_2) = \frac{-a}{s^6} + \frac{b}{s^{12}} + \frac{kQ_d Q_a}{\epsilon s} - \left[\frac{-a_0}{s^6} + \frac{b_0}{s^{12}} \right] f(\theta_1, \theta_2) \qquad (2\text{-}104)$$

C. Some Quantum Mechanical Studies of Hydrogen Bonds

A large number of quantum mechanical studies have been carried out on hydrogen bonds. We discuss here two studies by Pullman and co-workers which are sufficiently general to cover most aspects of this area of research.

1. Hydrogen Bonding in the Dimer of Formamide (41)

One way to attack the problem of the hydrogen bond would be to start with the infinitely separated components and allow them to approach each other, investigating the potential energy surface and the variations of the electronic characteristics along the path of approach. This procedure would require a large number of calculations in different conformations so as to allow for all possible deformations of angles and bond lengths inside each constituent. Another possibility is to study the characteristics (structure and energy) of the hydrogen-bonded complex and compare it to the nonbonded individual units. The first approach would seem ideally the best if all factors could be taken into account, and if among other things, no change of phase occurred in the process. If, however, a comparison with experimentally existing compounds is desired, the less ambitious second approach is probably more realistic. Thus, in the case of formamide, the hydrogen-bonded entities are well defined in the crystal where cyclic dimers occur in which two practically coplanar formamide units are linked by two hydrogen bonds, as revealed by the X-ray diffraction study.† This geometry was adopted for the calculation of the dimer. On the other hand, a calculation of the monomer, frozen in the geometry of the half-dimer was also carried out. The differences observed between the dimer and the two isolated half-dimers can be considered as representing the intrinsic effect of "bonding" through the hydrogen bond. In particular, any transfer of electrons observed under these conditions will result only from the establishment of the interactions between the monomeric units.

The quantum mechanical method adopted was a nonempirical self-consistent molecular orbital calculation including all electrons, using an atomic basis set of Gaussian-type functions (GTF) (43). The input geometry used for the dimer is given in Fig. 2-21. Since only the coordinates of the heavy atoms are known by X-ray investigation, it was assumed that the hydrogen atom was collinear with the end atoms. Reasonable values of 1.0 Å for NH and 1.1 Å for CH bonds were adopted. The monomer was treated as a half-dimer as explained above. Two rather different GTF sets were used in these calculations.

† See Pullman (42) and other papers and remarks therein.

Set A, which was initiated by Clementi and his group for calculations on large aromatics (44–46), and is a ($7^s1^p/3^s$) atomic basis contracted to ($2^s3^p/1^s$).

Set B, which was partially optimized on small molecules and is a ($4^s2^p/3^s$) basis, contracted into ($2^s1^p/2^s$) (47). This second set is less rich in GTF than set A, but it involves less contraction, in particular on the hydrogens, a feature which may have its importance in the problem investigated.

The exponents and coefficients corresponding to these two sets can be found in the original publications. For set A the dimer is found to be more stable than the monomers by 14 kcal/mole while for set B the dimer is more

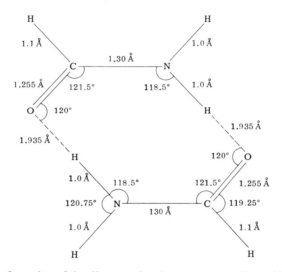

Fig. 2-21. Configuration of the dimer used as input geometry. From (41), M. Dreyfus, B. Maigret, and A. Pullman, A Non-Empirical Study of Hydrogen Bonding in the Dimer of Formamide, *Theoret. Chim. Acta (Berlin)* **17**, 109–119 (1970), Berlin-Heidelberg-New York: Springer.

stable by 19 kcal/mole indicating a hydrogen bond strength of −7 to −9½ kcal/mole. A comparison with the experimental value −7 kcal/mole of H-bond in solids (8b) shows that the difference obtained is reasonable. Table 2-15 contains a summary of the Mulliken populations in basis sets A and B, respectively, for the monomer and dimer. The hydrogen atom engaged in the hydrogen bond, and the oxygen, lose σ electrons, while the nitrogen atom and, to a lesser extent, the carbon atom gain σ electrons. In turn the oxygen gains π electrons, whereas the nitrogen and carbon atoms lose them. Globally, the hydrogen of the bridge loses electrons for the benefit of both the proton donor and the proton acceptor. It is remarkable that these conclusions are qualitatively entirely similar to those obtained by an all-valence-electrons semi-empirical method (48). Moreover, the qualitative conclusions regarding the charge shifts do not depend on the GTF basis set. This seems to be also the case with the dimer of H_2O (49).

The overlap populations are decreased in the dimer for NH and for CO, while they increase on CN. The overlap population on the H···O "bond" is 0.040 in set A and 0.055 in set B.

It is possible to get an idea of the redistribution of charge density upon the

Table 2-15

Mulliken Population Analysis[a]

	Set A		Set B	
	Monomer	Dimer[b]	Monomer	Dimer[b]
σ net charges	H +337 O −89 / −838 N—C +75 / H +321 H +194	H +370 O −59 / −882 N—C +55 / H +320 H +195	H +413 O −119 / −1112 N—C +205 / H +391 H +223	H +451 O −99 / −1160 N—C +195 / H +391 H +224
π net charges	O −294 / N—C +199 +95	O −362 / N—C +236 +126	+242 O −389 / N—C +147	+281 O −450 / N—C +169
Total net charges	H +337 O −383 / −639 N—C +170 / H +321 H +194	H +370 O −420 / −646 N—C +181 / H +320 H +195	H +413 O −508 / −870 N—C +352 / H +391 H +223	H +451 O −549 / −879 N—C +364 / H +391 H +224

[a] Unit of charge = $+10^{-3}$ e. From (41), M. Dreyfus, B. Maigret, and A. Pullman, A Non-Empirical Study of Hydrogen Bonding in the Dimer of Formamide, *Theoret. Chim. Acta (Berlin)* **17**, 109–119 (1970), Berlin-Heidelberg-New York: Springer.

[b] The oxygen and the upper hydrogen are the atoms engaged in hydrogen bonding.

formation of hydrogen bonds by constructing the electron density difference contours for the dimer and for two noninteracting monomers for the same geometry as the dimer. Such a map is shown in Fig. 2-22. Three observations can be made from this map:

(1) There is an electron density buildup across the C—N and N—H bonds.

(2) There is a loss in electron density about both the O and the H involved in the hydrogen bond.

(3) There is *no* buildup in electron density across the hydrogen bonding O···H bond as many workers feel should occur.

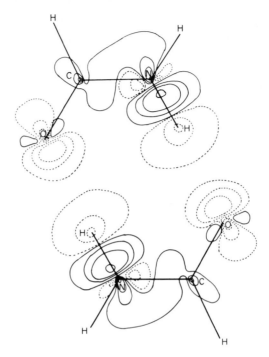

Fig. 2-22. This map represents the difference in electron density between the dimer and two monomers at the same distance, but supposed ideally not interacting, in set A. The dashed lines represent the contours inside which electron density is smaller in the dimer than in the monomer, and full lines represent the reverse situation. The values of this density difference are successively $\pm 10^{-3}$ (outer curves), $\pm 5 \times 10^{-3}$, $\pm 10^{-2}$, and $\pm 3 \times 10^{-2}$. From (41), M. Dreyfus, B. Maigret, and A. Pullman, A Non-Empirical Study of Hydrogen Bonding in the Dimer of Formamide, *Theoret. Chim. Acta (Berlin)* **17**, 109–119 (1970), Berlin-Heidelberg-New York: Springer.

2. *Hydrogen-Bonding between Peptide Units* (50)

In this study the hydrogen bond energy was taken to be the difference in total energy between two peptide units in a hydrogen-bonded configuration (the dimer) and two isolated peptide units (the monomers). The total molecular energies were calculated using the LCAO selfconsistent procedure of Roothaan (51). Because of the large number of electrons in such systems no geometrical optimization was attempted: the two interacting molecules were supposed to keep a constant geometry. Furthermore, as the dimer of formamide is a model for the NH···O=C hydrogen bond in proteins these calculations were restricted to the conformations corresponding to the structures β antiparallel (I) and parallel (II) occurring in fibrous proteins (see Fig. 2-23).

The geometry for each unit was taken from X-ray studies of formamide crystals (52). Reasonable geometrical parameters were assumed for the hydrogens: 1.0 and 1.09 Å bond lengths for NH and CH bonds respectively. Specifically, two linear structures, I and II, were examined and the O$\cdots$N distance varied. In the next calculation the O$\cdots$HN bond length was held fixed in a linear configuration and the C=O$\cdots$H angle was allowed to vary. In a final set of calculations the H-bonded proton was allowed to move along the O$\cdots$N line, in order to find the N—H bond length of minimum energy.

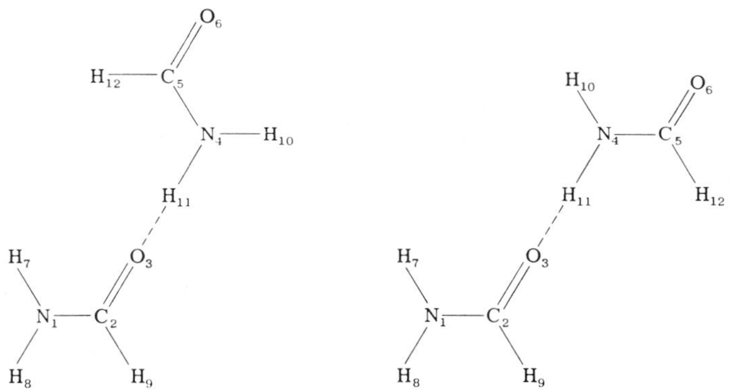

Fig. 2-23. (I) antiparallel β structure. (II) parallel β structure. From (50) M. Dreyfus and A. Pullman: A Non-Empirical Study of the Hydrogen Bond between Peptide Units, *Theoret. Chim. Acta (Berlin)* **19**, 20–37 (1970), Berlin-Heidelberg-New York: Springer.

Table 2-16 contains the values of the hydrogen-bond energy as a function of O$\cdots$N distance. The minimum in energy at 2.85 Å of -7.94 kcal/mole is in reasonable agreement with the minimum in energy for $R_0 = 1.85$ Å given in Table 2-14. The difference in hydrogen bonding energy between antiparallel dimer I and parallel dimer II is approximately zero. This suggests that any relative difference in stability between parallel and antiparallel structures is not due to variations in hydrogen bond strength.

The geometry used in the study of the energy versus O$\cdots$HN distance in the dimer is shown in Fig. 2-24. The N$\cdots$O distance was held fixed at a value of 2.85 Å, the energies and wave functions were computed for $\phi = -45°$, $0°$, $+45°$, $+75°$, and $+90°$. In fact, for $\phi > 70°$ and $\phi < -50°$, atoms other than those involved in the hydrogen bridge come within the sum of their van der Waals radii so that studying the angular variation has little meaning outside these limits where the repulsion increases dramatically. Figure 2-25 shows the energy variation in terms of the angle.

TABLE 2-16

Energy of the Antiparallel Dimer I, and H-Bond Energy as a Function of the O$\cdots$N Distance[a]

Distance O$\cdots$N (Å)	2.65	2.85	3.05
Antiparallel dimer SCF energy (a.u.)	−328.908174	−328.910607	−328.909970
H bonding energy (kcal/mole)	−6.41	−7.94	−7.54
Distance O$\cdots$N (Å)	3.25	5	∞
Antiparallel dimer SCF energy (a.u.)	−328.908339	−328900.765	−328.897952
H bonding energy (kcal/mole)	−6.52	−1.77	0

[a] The energy of the parallel dimer II has been computed only for O$\cdots$N = 2.85 Å and is found to be −328.910652 a.u. From (50) M. Dreyfus and A. Pullman: A Non-Empirical Study of the Hydrogen Bond between Peptide Units, *Theoret. Chim. Acta (Berlin)* **19**, 20–37 (1970), Berlin-Heidelberg-New York: Springer.

It is seen that the SCF curve has a very flat minimum between 45° and 75° and a slight shoulder for the linear arrangement. The decomposition of the energy into its components indicates that the electrostatic attraction E_C is quite dissymmetrical on each side of $\phi = 0°$, and is the source of the dissymmetry observed in E_{SCF}, the exchange repulsion variation as well as that of E_{P+CT} being symmetrical on each side of $\phi = 0°$. The variation of the exchange repulsion, E_E, with the angular displacement is rather large: it seems to show,

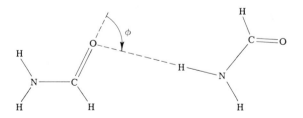

Fig. 2-24. Configuration of the dimer used for the angular variation. The arrow indicates a positive sense of rotation. From (50) M. Dreyfus and A. Pullman: A Non-Empirical Study of the Hydrogen Bond between Peptide Units, *Theoret. Chim. Acta (Berlin)* **19**, 20–37 (1970), Berlin-Heidelberg-New York: Springer.

among other things, that there is little hope to account accurately for the repulsion between nonbonded atoms with an expression taking into consideration only their distance, at least in the case of heteroatoms of nonspherical environment.

Finally, the energy corresponding to polarization + charge-transfer is minimal (in absolute value) for $\phi = 0°$ and increases on each side. An explanation of this fact is partially provided by the population analysis on the monomer: the highest occupied σ molecular orbital is practically a pure p lone-pair orbital, perpendicular to the CO bond. Consequently the s-character

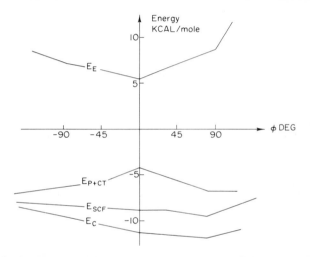

Fig. 2-25. Coulomb energy (E_C). Exchange energy (E_E), polarization + charge transfer energy (E_{P+CT}) and SCF interaction energy (E_{SCF}), as a function of the angle ϕ (kcal/mol). From (50) M. Dreyfus and A. Pullman: A Non-Empirical Study of the Hydrogen Bond between Peptide Units, *Theoret. Chim. Acta (Berlin)* **19**, 20–37 (1970), Berlin-Heidelberg-New York: Springer.

of the lone-pair electrons facing the NH bond decreases from $\phi = 0°$ to $\phi = 90°$ so that their lability increases, thus favoring lone-pair delocalization terms. Indeed the charge transfer is 30% larger for $\phi = 75°$ than for 0°.

Hydrogen bonding AH···B is known to cause important modifications of the properties of the AH bond. The AH stretching force constant is lowered (up to 10 to 20% in strong complexes); the AH equilibrium distance is lengthened, and the infrared intensity of the AH stretching band is usually dramatically increased.

In this series of calculations the O···N distance was fixed at 2.85 Å and the hydrogen-bonded proton was allowed to move along the O···N line. A search for the equilibrium NH bond length was made in both the monomer and in the dimer, the NH bond direction being kept constant (Table 2-17). Assuming the energy to be a quadratic function of the NH distance, one finds an equilibrium value of 1.0617 Å in the monomer and 1.0718 in the dimer, thus an increase in length of 10^{-2} Å. The corresponding force constants are 0.488 and 0.474 a.u., respectively. The effects calculated are thus in the right direction although the numerical values by themselves should not be given too much significance.

TABLE 2-17

Variation of the Monomer and of the Dimer I Energy with the N—H Bond Length[a]

NH (Å)	0.95	1.00	1.05	1.07	1.09
SCF energy (monomer)	−164.438603	−164.448976	−164.452768	−164.452827	−164.452189
SCF energy (dimer)	—	−328.901607	−328.915162	−328.915563	−328.915287

[a] In the dimer, O···N was fixed at 2.85 Å, and only the NH involved in the H bridge was varied. From (50) M. Dreyfus and A. Pullman: A Non-Empirical Study of the Hydrogen Bond between Peptide Units, *Theoret. Chim. Acta (Berlin)* **19**, 20–37 (1970), Berlin-Heidelberg-New York: Springer.

X. Bond Angle and Bond Length Distortion Potential Functions

A. Hookean Distortions of Valence Bonds and Bond Angles

It was mentioned earlier that empirical quantum mechanical procedures could be employed to calculate bond angle and bond length distortion potential functions. In this section we discuss how data obtained from infrared and Raman spectra can provide a basis for calculating the parameters needed to describe such strain potential functions. At the end of this section is a list of a set of force constants which can be used in parabolic strain potential functions (see Table 2-18).

The energy needed to strain valence bonds and bond angles can usually be accurately estimated from knowledge of the "force constants" of the group of atoms within the molecule. The concept of a force constant implicitly assumes that valence bonds are essentially springs which obey Hooke's law,

$$F_i = -k_i r_i \tag{2-105}$$

where F_i is the force acting on particle i, r_i the distance from equilibrium for

TABLE 2-18

A Set of Force Constants for Bond Length Stretching S (mdyn/Å), Bond Angle Bending B (mdyn-Å), and Out-of-Plane Bending W (mdyn-Å)

Group	Force constants	
	Observed	Predicted[a]
S (C—C)	4.5 ± 0.3	4.4
S (C—C(sp²))	3.1 ± 0.2	—
S (C=O)	12.1 ± 0.2	13.8
S (C(sp²)=O)	8.15 ± 0.22	—
S (C=C)	9.6 ± 0.1	8.7
S (C—N)	5.5 ± 0.35	5.7
S (C=N)	8.3 ± 0.32	8.6
S (C—H)	4.8 ± 0.42	5.2
S (C(sp²)—H)	4.0 ± 0.31	—
S (C—CH₃)	1.96 ± 0.12	—
S (N—H)	4.8 ± 0.27	4.7
S (C—H) aromatic	5.0 ± 0.08	—
B (X—CH₂—H)	0.55 ± 0.08	—
B (C=CH—H)	0.68 ± 0.06	—
B (H—C—H)	0.32 ± 0.07	—
B (C=C—H) aromatic	0.86 ± 0.05	—
B (H—N—C)	0.44 ± 0.05	0.38
B (H—C(sp²)—CH₃)	0.325 ± 0.020	—
B (H—N—C(sp²))	0.26 ± 0.03	—
B (N—C(sp²)—C)	0.33 ± 0.04	—
B (C(sp²)—N—C)	0.58 ± 0.03	—
B (N—C—C(sp²))	0.60 ± 0.03	—
B (N—C(sp²)—O)	0.56 ± 0.02	—
B (N—C—M)	0.45 ± 0.02	0.57
B (O—C(sp²)—C′)	0.54 ± 0.02	—
B (C—C(sp²)—N)	0.45 ± 0.04	—
B (H—C—C(sp²))	0.35 ± 0.03	—
B (M—C—C(sp²))	0.67 ± 0.04	—
B (CH₂—CH₂—CH₂)	—	0.62
B (C—O—C)	—	1.76
B (C—CH—F)	—	0.49
B (F—C—F)	—	0.26
W (C=O)	0.54	—
W (N—H)	0.15	—

[a] Predictions were made using the CNDO/2 quantum mechanical approximation. All force constants reported here are assumed valid for room temperature (25°C) only. The M refers to a methyl group. The recorded deviations in the value of the force constants were computed as the average spread in the data from the mean value.

the particle i, and k_i the appropriate force constant. The potential energy associated with this deviation from equilibrium is

$$V_i = - \int F_i \, dx \qquad (2\text{-}106)$$

$$V_i = \tfrac{1}{2}kr_i^2 \qquad (2\text{-}107)$$

and the frequency of oscillation for this distortion from equilibrium is

$$\nu_i = (1/2\pi)(k_i/m_i)^{1/2} \qquad (2\text{-}108)$$

where m_i is the mass of the ith particle. The expression for V_i in Eq. (2-107) provides a basis for computing the total distortion potential energy, V_t, of the valence bonds,

$$V_t = \sum_{i=1}^{N_a} \tfrac{1}{2}k(a)_i \, \theta_i^2 + \sum_{i=1}^{N_a} \tfrac{1}{2}k(b)_i \, S_i^2 \qquad (2\text{-}109)$$

where the symbol N_a refers to all bonds angles, N_b to all bond lengths, θ to the distortion angle, and S to the distortion length. In using Eq. (2-107), that is, Hooke's law, we assume that the deformations are perfectly harmonic. Since in fact Hooke's law is not strictly obeyed, separation of the independent harmonic oscillators, that is, the normal vibrations is only an approximation and may lead to serious errors. Nevertheless, this approach has proved remarkably useful, and successful, in the elucidation of molecular structure.

Bond distance and bond angle force constants may be found from experimental studies of the infrared and/or Raman spectra of various molecules. The ability to determine these force constants depends upon being able to assign a particular band in a spectrum with a specific deformation.

Much of the work that has been done, and is being done, in IR and Raman spectroscopy is in the development of methods which can be used to deduce which bands in a spectrum are caused by which distortions. From Eq. (2-108) it is clear that the vibrational frequency ν observed in a spectrum will vary inversely with the square root of the oscillating mass m (normally the reduced mass). This relationship has been put to practical use.

Isotopic substitutions are made in a molecule being investigated and the isotopic spectrum of the molecule is compared to the normal spectrum. If one knows what specific isotopic substitutions have been made, then the bands which have shifted in the isotopic spectrum may be assigned. This method, isotopic substitution, is most useful for low atomic weight atoms because the effect is most pronounced for these atoms. For example, the O—H stretch occurs near 3570 cm^{-1} and the O—D stretch occurs near 2630 cm^{-1}. The frequency shift is nearly cut in half in going from O—H to O—D.

The IR and/or Raman spectrum(a) of a molecule also contains information about the local environments of the atoms in the molecule. That is, the spectrum(a) contains information about the molecular conformation. The shape (i.e., size of half-widths of the bands) of the bands gives information about the time-relaxation processes involved with the vibrational modes of an atom in a molecule. In turn, the time-relaxation processes of an atom in a molecule are dependent upon the "stiffness" of valence bond(s), as measured by the valence bond and angle force constants, *and* the available "room" in the vicinity of an atom in a molecule to vibrate (as measured, by the nonbonded interaction potential functions). Atoms which are located in a well packed conformation are subject to damped vibrational modes which "die off" rapidly. On the other hand, atoms in a loosely packed structure can vibrate rather freely without "bumping" into other nearby atoms and thus the vibrations should be long lived. The analysis of the information contained within the shape of the absorption band is virgin research, and we anticipate seeing much done in this field.

It is instructive to demonstrate how the force constants can be computed from the vibrational frequencies found experimentally when the "valence bond" approximation is invoked. That is, we assume the forces between *bonded* atoms operate only along valence bonds in accordance with Hooke's law and a force constant may be assigned to the stretching (or compression) of each valence bond and to the bending of each valence angle in the molecule. Consider linear CO_2 which we will restrict to vibrating along the line defined by the three atoms

$$\underset{m_O}{\overset{(1)}{O}} \overset{k_s}{-----} \underset{m_C}{\overset{(2)}{C}} \overset{k_s}{-----} \underset{m_O}{\overset{(3)}{O}} ----------x$$

Assuming Hooke's law is valid the total potential energy is

$$V = \tfrac{1}{2}k_s(x_2 - x_1)^2 + \tfrac{1}{2}K_s(x_3 - x_2)^2 \tag{2-110}$$

The forces on atoms (1), (2), and (3) are, respectively,

$$F_1 = \frac{-\partial V}{\partial x_1} = k_s(x_2 - x_1)$$

$$F_2 = \frac{-\partial V}{\partial x_2} = -k_s(x_2 - x_1) + k_s(x_3 - x_2) \tag{2-111}$$

$$F_3 = \frac{-\partial V}{\partial x_3} = -k_s(x_3 - x_2)$$

But, from Newton's second law

$$F_1 = m_O \frac{d^2 x_1}{dt^2}$$

$$F_2 = m_C \frac{d^2 x_2}{dt^2} \tag{2-112}$$

$$F_3 = m_O \frac{d^2 x_3}{dt^2}$$

Therefore,

$$k_s(x_2 - x_1) = m_O \frac{d^2 x_1}{dt^2}$$

$$-k_s(x_2 - x_1) + k_s(x_3 - x_2) = m_C \frac{d^2 x_2}{dt^2} \tag{2-113}$$

$$- k_s(x_3 - x_2) = m_O \frac{d^2 x_3}{dt^2}$$

Since we assume each of the three atoms of carbon dioxide vibrates harmonically, it follows that

$$x_1 = A_1 \sin 2\pi \nu t + B \cos 2\pi \nu t \tag{2-114}$$

and

$$\frac{d^2 x_1}{dt^2} = -4\pi^2 \nu^2 x_1 \tag{2-115}$$

Since all these atoms must vibrate with some frequency ν,

$$\frac{d^2 x_2}{dt^2} = -4\pi^2 \nu^2 x_2, \qquad \frac{d^2 x_3}{dt^2} = -4\pi^2 \nu^2 x_3 \tag{2-116}$$

substitution of these expressions for the 2nd derivatives into Eq. (2-113) yields

$$-4\pi^2 m_O \nu^2 x_1 + k_s x_1 - k_s x_2 = 0$$
$$-k_s x_1 + 2k_s x_2 - 4\pi^2 m_C \nu^2 x_2 - k_s x_3 = 0 \tag{2-117}$$
$$-k_s x_2 + k_s x_3 - 4\pi^2 m_O \nu^2 x_3 = 0$$

For these three equations to yield a nontrivial solution it follows that the determinant of the set of equations vanish. That is,

$$\begin{vmatrix} k_s - 4\pi^2 m_O \nu^2 & -k_s & 0 \\ -k_s & 2k_s - 4\pi^2 m_C \nu^2 & -k_s \\ 0 & -k_s & k_s - 4\pi^2 m_O \nu^2 \end{vmatrix} = 0 \tag{2-118}$$

Expansion of the determinant leads to the equation

$$4\pi^2 \nu^2 (k_s - 4\pi^2 \nu^2 m_O)(4\pi^2 \nu^2 m_O m_C - k_s m_C - 2k_s m_O) = 0 \qquad (2\text{-}119)$$

The solutions of this equation are

$$\nu_1 = \frac{1}{2\pi} \left(\frac{k_s}{m_O} \right)^{1/2} \qquad (2\text{-}120)$$

$$\nu_3 = \frac{1}{2\pi} \left(\frac{k_s(m_C + 2m_O)}{m_C m_O} \right)^{1/2} \qquad (2\text{-}121)$$

Observed frequency for $\nu_1 = 1337$ cm^{-1} or $k_s = 16.8 \times 10^5$ dynes/cm and ν_3 has observed frequency 1349 cm^{-1} or $k_s = 14.2 \times 10^5$ dynes/cm.

B. The General Relationship between Molecular Vibrations and Molecular Structure

In the most general sense, for both Hookean and non-Hookean deformations, the corresponding force constant k is a local measure of the total molecular potential P. If a specific molecular deformation is characterized by a change in the spatial variable s, then

$$\partial^2 P / \partial s^2 = k \qquad (2\text{-}122)$$

relates the potential to the force constant. If we now assume, as is always done, that P can be written as an additive sum of potentials of the form P_s in which each potential is explicitly dependent upon only some spatial variable s, then the complete molecular structure can be represented by the set of equations

$$\partial^2 P_s / \partial s^2 = k(s) \qquad \text{for all} \quad s \qquad (2\text{-}123)$$

All the equations in (2-123) are implicitly interrelated through the primary structure of the molecule. If a complete set of $k(s)$, that is a unique $k(s)$, can be assigned to each possible equation in (2-123), and each P_s is specified, then the set of equations in (2-123) may be used to determine the values for the set of spatial variables $\{s\}$. In other words it is possible to arrive at the detailed molecular structure. Unfortunately, three factors hinder the use of this scheme in calculating molecular structure:

(1) The set of equations in (2-123) are nonlinear in $\{s\}$. In general, sets of nonlinear equations are very difficult to solve. Fortunately, however, we usually have a rather good "trial structure" for the molecular structure which very much enhances the chances of numerically solving the set of nonlinear equations.

(2) The separable and additive potential functions, $\{P_s\}$, are of limited accuracy.

(3) A sufficiently complete set of $\{k(s)\}$ cannot normally be determined from the vibrational spectra of the molecule. However, the $k(s)$ found in the analysis of any molecule are usually "almost" transferable to other molecules. This near-transferability is enough to provide a reasonable trial solution for equation set (2-123).

Because of the errors introduced in (2) and (3) above, the best that can be done in arriving at the $\{s\}$ is to determine that set of $\{s^*\}$ which yields the best linear least-square fit to equation set (2-123).

An alternate use of equation set (2-123) is to assume a molecular geometry for the molecule and a set of corresponding force constants, and then to proceed to deduce the values of the parameters in the potential functions. Usually the resulting set of equations are linear and much easier to solve than in the previous case. The geometry of the molecule might be arrived at from X-ray studies, NMR investigations, as well as other vibrational experiments. The implicit assumption made is that the set of force constants are in fact consistent with the molecular geometry.

A third and final means of using equation set (2-123) is to predict a set of force constants from a given set of $\{P_s\}$ and a given set of $\{s\}$. This use of equation set (2-123) has most often been used in the past and constitutes the basis of normal coordinate analysis.

XI. Torsional Potential Functions

If one carries out a rotation χ about the C—C bond in ethane one notes a periodic dependence in the *total* molecular energy $E(\chi)$ of the form shown in Fig. 2-26, where $\chi = 60°$ corresponds to the hydrogen atoms on one carbon eclipsing the hydrogens on the other carbon when looking down the C—C bond. The value of E^*, which is the difference between the maximum and minimum in total molecular energy, is the barrier height to the bond rotation. This type of energy curve as function of χ is called the torsional potential about the C—C bond. The actual periodic energy dependence arises from two sources: (1) the varying extent of interaction of the H's on one carbon with the H's on the other carbon. (2) the orbital–orbital interactions of the electrons centered about the two carbons. This is a purely quantum mechanical effect due to the geometry of the orbitals in the molecule. For a perfect single bond the energy contribution from such orbital–orbital interactions is zero since the orbitals of perfect single bonds are completely symmetric with respect to rotations about the bond. Thus it is deviations from the formation of single bonds which is the contributing agent to the torsional potential.

Since it is possible to calculate the nonbonded interactions separately, it is important that one specify whether or not a given torsional potential already includes nonbonded contributions. If this is not done the nonbonded interactions could be overlooked, or counted twice.

For rotations about near-single bonds the nonbonded interactions make the major contributions to the torsional potential, while for near-double and near-triple bonds the orbital–orbital interaction terms dominate. For near-single bonds in small molecules like ethane the value of E^* is not usually more than 1 kcal/mole. Obviously, bad contact nonbonded interactions, when possible, could raise E^* to very large values for some bond rotation values. Nevertheless, however, for near single bonds the values of E^* are sufficiently small so that inclusion of this energy contribution is conceptually more important than its impact upon dictating molecular structure. For double and

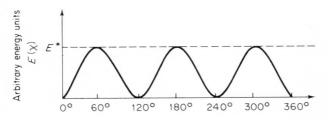

Fig. 2-26. Characteristic variation of the molecular energy of ethane with the torsional rotation χ about the C—C bond.

triple bonds the torsional potentials play major roles in dictating the structure of a molecule.

In general, a near-single bond torsional potential can be represented by

$$E(\chi) = E^*(1 \pm \cos(n\chi - b)) \tag{2-124}$$

where E^* is again the barrier height, n the periodicity of the function, and b the phase angle. This function is valid only for representing torsional potentials in which all of the atoms, or groups of atoms, bonded to each respective atom forming the bond about which rotation occurs are identical. If the atoms, or groups of atoms, are not all identical (i.e., a fluorine is substituted for a hydrogen in ethane), then the torsional barrier heights are not all identical and Eq. (2-124) is not valid unless some correction is added, normally the appropriate nonbonded interactions.

The barrier heights, E^* may often be found from microwave spectroscopy (53) or from temperature dependent NMR investigations (54, 55). In both cases the E^* are found for small model molecules and are assumed to be transferable to other larger molecules. The NMR investigations have the disadvantage of including the effects of molecule–solvent interactions. The periodicity, n, must be determined from the *orbital* geometry. The values of E^*, n, and b may be theoretically estimated by carrying out quantum mechanical calculations on model molecules.† The magnitude of the orbital interactions may be estimated by subtracting the nonbonded interaction energy, computed using the classical potential energy functions discussed in this chapter, from the total molecular energy. The orbital interaction energy may then be adopted whenever needed. By adding this orbital term to the appropriate nonbonded term it is possible to approximate reasonably the total barrier to rotation about any bond in nearly any molecule.

Lastly, the use of equations like that given in (2-124) to describe torsional potentials requires making two assumptions which must be remembered when considering the consequences of these types of calculations. These assumptions are:

(1) For a sequence of connected bonds, $1, 2, \ldots, j-1, j, \ldots, m$, the torsional potential about bond l is energetically independent of the rotations about all other bonds.

(2) the bond energy of any bond about which rotation does *not* occur is constant and independent of the effects of other bond rotations.

A listing of some proposed torsional potential functions is presented in Table 2-19.

In past conformational energy calculations, rotations about partially double and double bonds have not been allowed. The rationale for this imposed stereochemical restriction is based upon the high energy barriers to rotation about such bonds. However, small fluctuation rotations about these bonds may take place with little expenditure of energy. Very often bond rotations of up to 15° from the minimum energy state may occur about partially double and double bonds with less than 1 kcal/mole expenditure of energy. Further, the additional gain in conformational freedom resulting from these small bond rotations can result in a gain in conformational free energy, through nonbonded interactions and entropic considerations, greater than the expended bond rotation energy.

Slight deviations from planarity in the peptide unit, that is rotations about the amide, or imide, bonds, normally denoted by ω, appear to exist in most proteins. This is not yet completely certain, since the X-ray structures are not sufficiently resolved to supply very precise atomic models, but considerable evidence does point to the existence of nonplanar peptide units in proteins.

Conformational energy calculations carried out for the ordered forms of

† See Hopfinger (56) for discussion and references.

TABLE 2-19

Proposed Torsional Potential Functions for a Variety of Bonds[a]

I. Single bonds $E(\chi) = E^*(1 \pm \cos(n\chi - b))$

Bond	Experimental				Theoretical			
	E^* (kcal/mole)	sign	n	b	E^* (kcal/mole)	sign	n	b
—C—C—	1.20	+	3	0	1.25	+	3	0
	1.40	+	3	0				
⫤C—C—	0.55	−	3	0	0.40	−	3	0
	0.10	−	3	0				
—C—⬡	0.29	+	6	0	0.53	+	6	0
H₂N—C—	0.99	+	3	0	1.15	+	3	0
	0.29	+	3	0				
O₂N—C—	0.06	+	6	0	—	—	—	—
—C—O—	1.07	+	3	0	1.00	+	3	0
⫤C—C—	—	—	—	—	0.63	−	3	0

II. Double bonds $E(\chi) = E^* \sin^2(n\chi)$

Bond	Experimental		Theoretical	
	E^*(kcal/mole)	n	E^*(kcal/mole)	n
C⫤N	—	1	23.2	1
C=C	24.0–31.0	1	27.7	1
⫤C—C⫤	9.0–17.0	1	—	1

[a] The experimental and theoretical E^* are due only to orbital–orbital interactions. The extended Hückel scheme was used to evaluate the values of the parameters in the torsional potential functions under the theoretical heading. $E(\omega) = (A/2)(1 - \cos 2\omega)$ for $A = 20$ kcal/mole is also used to describe the torsional potential about the amide, or imide, bond.

poly-L-alanine and poly-L-tyrosine indicate that the right-handed α-helices of these two biopolymers are considerably stabilized by allowing ω rotations, i.e. rotation about the peptide bond C=N. Table 2-20 lists the values of $\bar{E}$, $\langle S \rangle$ and $\langle A \rangle$, for $T = 300°K$, for poly-L-alanine and poly-L-tyrosine in and out of aqueous solution when the ω rotation is fixed at 180°, the trans configuration, and when ω is allowed to vary. There is virtually no gain in free energy through polymer–solvent interactions, at least for aqueous solution, by allowing ω rotations, but for both biopolymers there is a gain in the stabilization free energy through intrachain nonbonded interactions $\bar{E}$ and entropic effects $\langle S \rangle$. Moreover, poly-L-tyrosine shows a greater gain in stabilization free energy than poly-L-alanine as a result of ω rotations. This is

TABLE 2-20

The Conformational Energy Characteristics of Poly-L-alanine and Poly-L-tyrosine, in and out of Aqueous Solution, for Disallowed and Allowed ω Bond Rotations

Molecule	Conformation	$\bar{E}$ (kcal/mole-residue)	$\langle S \rangle$ (seu/residue)	$-T\langle S \rangle$ (300°K) (kcal/mole-residue)	$\langle A \rangle$ (kcal/mole-residue)
Poly-L-alanine					
$\omega = 180°$ fixed	RHα(H$_2$O)	-7.8	5.4	-1.6	-9.4
planar unit	RHα(vacuo)	-7.8	5.5	-1.6	-9.4
ω allowed	RHα(H$_2$O)	-8.3	6.4	-1.9	-10.2
to vary	RHα(vacuo)	-8.3	6.6	-2.0	-10.3
Poly-L-tyrosine					
$\omega = 180°$ fixed	RHα(H$_2$O)	-8.7	5.9	-1.8	-10.5
planar unit	RHα(vacuo)	-6.1	7.0	-2.1	-8.2
ω allowed	RHα(H$_2$O)	-9.8	9.6	-2.9	-12.7
to vary	RHα(vacuo)	-7.1	10.2	-3.1	-10.2

attributed to the ω rotations providing considerable side chain conformational freedom in addition to backbone conformation freedom for poly-L-tyrosine. Thus, the size and shape of the peptide side chain may markedly influence the extent of the ω rotation in the peptide backbone.

Figures 2-27(a) and (b) demonstrate the gain in conformational freedom in $\phi\psi$-space near the right-handed α-helical conformation when the ω rotation is allowed for poly-L-alanine. The gain in the stabilization internal energy $\bar{E}$ is also apparent. This additional steric freedom probably is important in the folding of secondary structures in proteins to form the tertiary structure. Without this steric freedom it might prove difficult to distort segments of α-helical structure in the manner necessary for the formation of the overall tertiary structure.

In summary, fluctuation rotations about partially double and double bonds should not be neglected in conformational analysis. A high torsional potential barrier does not negate the existence of significant torsional fluctuations which can modify secondary structure.

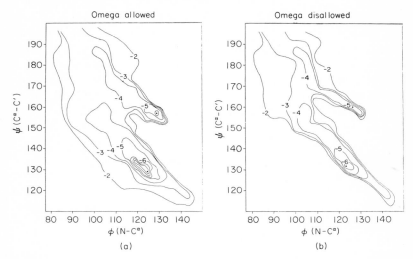

Fig. 2-27. (ϕ,ψ) conformational energy maps of poly-L-alanine in the vicinity of the right-handed α-helix. The ● denote relative minima. (a) The total conformational energy was minimized as a function of ω. S is 6.58 seu. (b) the ω rotation was fixed at 180° corresponding to a trans planar residue unit. S is 5.46 seu.

XII. Conformational Entropy

A. Introduction

Thus far we have discussed various types of energy interactions which are responsible for inducing specific preferred conformations in particular macromolecules. In principle we can construct an analytic function which yields the total conformational *internal* energy of the macromolecule as a function of the variables which are sensitive to conformation. Further, and again strictly in principle for many real situations, we can compute the set of relative minima associated with this function, and thus identify the set of stable conformations of the macromolecule. But what will be the probability of finding the macromolecule in any one particular conformation? In order to answer this question we must be able to compute the conformational entropy associated with each of the relative minima. It should be noted that knowledge of the total conformational entropy of the macromolecule is not sufficient to provide a means of calculating the probability of observing any particular stable macromolecular conformation. Further, it should be noted that the term "conformational entropy" is used to indicate specifically the entropy contribution from the distribution of atoms in space of one molecule. Similar arguments to those developed here could be given for the description of the configurational entropy, that is, entropy due to the arrangement of two or more macromolecules in time and space.

Entropy is a measure of the order in a system. The higher the order in the system, the lower the value of the entropy. In terms of a potential function, the entropy is a measure of the shape of the surface generated from the potential function when conformational energy is plotted as a function of the conformational variables. If the potential energy is constant for all conformational variables, then the entropy is maximized. If, on the other hand, the potential obeys a delta function, then the entropy is minimized. Implicitly the calculation of the entropy of a system requires complete knowledge of the internal energy of the system. It is this requirement, knowledge of the internal energy of the system at all points, which makes it difficult, if not impossible, to compute the entropic properties of most macromolecules.

From a statistical mechanical point of view, the calculation of the entropy of a system, like the calculation of any thermodynamic variable of a system, reduces to the ability or inability to accurately compute the partition function Q associated with the system. For conformational calculations on macromolecules we will assume that the ability to calculate Q for the macromolecule implies the ability to calculate the partition functions q associated with each of the relative minima (i.e., stable conformations).

B. Expansion of the Potential Function about a Minimum in Order to Compute the Entropy

First, let us see how we might calculate the conformational entropy of a macromolecule which has one degree of conformational freedom per monomer unit and where the equivalence condition has been employed. Examples of such macromolecules are polyethylene and poly-L-proline. In the vicinity of each of the minima in conformational potential energy the associated potential function can be approximated by

$$P(\xi) = P_0 + \left(\frac{\partial P}{\partial \xi}\right)_{P_0} (\xi - \xi^0) + \left(\frac{\partial^2 P}{\partial \xi^2}\right)_{P_0} (\xi - \xi^0)^2 \qquad (2\text{-}125)$$

But $(\partial P/\partial \xi)_{P_0}| = 0$, since we are centered at a relative minimum. Hence,

$$P(\xi) = P_0 + \left(\frac{\partial^2 P}{\partial \xi^2}\right)_{P_0}\Bigg|(\xi - \xi^0)^2 \tag{2-126}$$

Then the approximate partition function q_0 associated with this minimum in conformational potential energy is

$$q_0 = 2 \int\limits_{x=\xi^0}^{x=\infty} \exp\left\{\left(\left(-P_0 - \left(\frac{\partial^2 P}{\partial \xi^2}\right)_{P_0}\Bigg|(x - \xi^0)^2\right)\Bigg/RT\right\} dx \tag{2-127}$$

which when integrated yields

$$q_0 = (\pi RT)^{1/2}\left[\left(\frac{\partial^2 P}{\partial \xi^2}\right)^{-1/2}\right]_{P_0} \exp\left(-\frac{P_0}{RT}\right) \tag{2-128}$$

In general the entropy is given by

$$S = RT\left(\frac{\partial \ln q_0}{\partial t}\right)_{V,N} + R \ln q_0 \tag{2-129}$$

Using q_0 we obtain the following expression for the entropy of the minimum:

$$S_0 = \frac{-R}{2} \ln \left(\frac{\partial^2 P}{\partial \xi^2}\right)_{P_0}\Bigg| + \left[\frac{R}{2}\left(1 + \ln(\pi RT) + \frac{P_0}{T}\right)\right] \tag{2-130}$$

Since we are not dealing with absolute entropies it is possible, as suggested by the form of Eq. (2-130), to have negative values for the entropy.

Obviously the one-dimensional expansion of the conformational potential function in Eq. (2-125) is a special case of the general n-dimensional expansion (57) which has the general form

$$f(a_1, a_2, \ldots, a_n) = f_0(a_1, a_2, \ldots, a_n) + \sum_{i=1}^{n} \frac{\partial f(a_1, a_2, \ldots, a_n)}{\partial a_i}\Bigg|_{f_0}$$
$$\cdot (a_i - a_i^0) + \sum_{i=1}^{n}\sum_{j=1}^{n} \frac{\partial^2 f(a_1, a_2, \ldots, a_n)}{\partial a_i\, \partial a_j}\Bigg|_{f_0}$$
$$\cdot (a_i - a_i^0)(a_j - a_j^0) \tag{2-131}$$

Thus we can apply this expansion technique to each of the minima of a macromolecule having n degrees of conformational freedom and proceed to compute the entropy associated with each of the minima. The general expression for the entropy using the expansion given in Eq. (2-131) is

$$S = \frac{R}{2} [n + n \ln \pi R + n \ln T - \ln \det \mathsf{F}] \tag{2-132}$$

where n is the number of degrees of freedom, and the F_{ij} are given by

$$F_{ij} = \frac{\partial^2 f(a_1, a_2, \ldots, a_n)}{\partial a_i \, \partial a_j} \bigg|_{f_0} \tag{2-133}$$

The reasonableness of applying this technique to a specific relative minimum in order to compute the associated entropy is contingent upon the validity of at least one of the two following assumptions:

(1) The potential well about the relative minimum is sufficiently steep in all directions (degrees of conformational freedom) so that only the immediate neighborhood about the minimum (for which the general expansion is valid) makes a contribution to the partition function.

(2) The shape of the potential well is very similar in topology to the parabolic expansion given in Eq. (2-131).

C. Statistical Methods to Approximate the Conformational Entropy

Now what does one do if neither of the assumptions listed above is valid for the potential well associated with a relative minimum? In principle one could "tack" additional terms onto the expansion in Eq. (2-131) and then attempt the associated higher order integrations of these terms. The main drawbacks to this approach are (a) performing the higher-order integrations is usually difficult if not impossible in most situations, and (b) the time involved in such computations is excessive.

The most direct means to the calculation of the entropy associated with a relative minimum is by a digital scan technique. That is, the topology of the potential well is deduced by sampling the conformational potential energy at various points in conformational hyperspace "near" the minimum. Since a sample of the system is taken, this technique is, by definition, statistical. From the collection of sample points the approximate partition function is given by

$$q_0 = \sum_{i=1}^{N} \exp(-E_i/RT) \tag{2-134}$$

where N is the number of sample points and E_i is the conformational potential energy associated with the ith point in conformational hyperspace. If one retains the set of E_i, then the probability of observing the macromolecule at conformational point k (remember the one relative minimum and its local neighborhood constitutes the entire system in this discussion) is

$$\mathscr{P}_k = \frac{\exp(-E_k/RT)}{q_0} \tag{2-135}$$

and the entropy identified with the relative minimum is

$$S_0 = -R \sum_{i=1}^{N} \mathscr{P}_i \ln \mathscr{P}_i \tag{2-136}$$

The critical question to be asked when applying this technique is: How do we pick the set of scan points so as to most accurately approximate q_0 and hence S_0? Unfortunately there is no definite answer to this question. There are two general schemes for picking the set of scan points, and both schemes have individual merits and faults. The first scheme simply assumes the existence of a uniformly spaced grid in conformational hyperspace which is centered at the relative minimum. Then, say, the points of intersection of the imaginary lines composing the grid are chosen to represent the set of scan points and the corresponding values of S_0 and q_0 are computed. To check the sensitivity of q_0 and S_0 to the choice of the set of scan points, a new set of scan points might be chosen. For example, the set of points located at the geometric centers of the hyperdimensional cubes defined by the lines of the grid might be picked and the values of q_0 and S_0 for this set of data computed. A large difference in the values of S_0 and/or q_0 for these two choices of the set of scan points would suggest that a different technique is needed to accurately estimate S_0 and q_0. On the other hand, if S_0 and q_0 are very nearly identical for the two calculations, then the choice of a uniformly distributed set of scan points is probably pretty good.

Let us derive the expression for q_0 using the conditions imposed by assuming that a set of uniformly distributed scan points can be used in such a computation, and also list the implicit assumptions that are made in such a derivation. The general expression for q_0 is

$$q_0 = \int_v \exp(-E(v)/RT) \, dv \tag{2-137}$$

We now assume that $E(v)$ is continuous so that it is possible to partition q_0 uniformly about the relative minimum, thus giving

$$q_0 = \sum_{i=1}^{N} \int_{v_i} \exp(-E(v_i)/RT) \, dv_i \tag{2-138}$$

By insisting that some set of conformational energies $\{E_i^*\}$ which correspond to a set of uniformly distributed set of points about the minimum, can be used to estimate q_0 accurately, we assume that

$$\int_{v_i} \exp(-E(v_i)/RT)\, dv_i = \exp(-E_i^* V/NRT) \qquad \text{for all} \quad i \qquad (2\text{-}139)$$

where V is the total conformational hypervolume included in the scan. This is the critical assumption and requires that the conformational energy variation in the neighborhood of each E_i^* be small. The expression for q_0 subject to the above constraints is

$$q_0 = \sum_{i=1}^{N} \exp(-E_i^* V/NRT) \qquad (2\text{-}140)$$

The two-dimensional geometry for v_i is shown in Fig. 2-28.

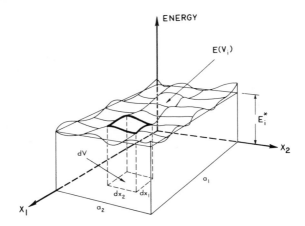

Fig. 2-28. The two-dimensional geometry of the volume element v_i used in the entropy calculations. $V/N = a_1 a_2 = V_i$.

Whenever the technique involving a uniformly distributed set of points does not provide an adequate base for calculating q_0 and S_0 it is necessary to resort to a semirandom method to select the set of $\{E_i^*\}$. Situations involving the calculation of q_0 and S_0 for macromolecules having many degrees of conformational freedom (i.e., proteins in which the number of degrees of freedom is normally 200 or more) also require the semirandom means of selecting the E_i^*. The reason this is so for such macromolecules is simply due to the fact that the number of uniformly distributed points needed to accurately estimate q_0 and S_0 about some minimum is so large as to make the calculation impractical in terms of the time needed.

An actual description of the most reliable semirandom method for picking the E_i^* used to calculate q_0 and S_0 cannot be given since little work has been done in this area. What is presented below is a description of two computer routines which are being presently used to compute q_0 and S_0 associated with the crystal-structure potential-energy minima of the proteins ribonuclease and lysozyme (58).

1. Minimum Energy Constraint

In this scheme one first defines an upper limit distance p_0 in conformational hyperspace. One then chooses a random direction ζ and a random distance p, such that $0 < p < p_0$. Using this vector centered at the point at which the energy is a minimum one can define some new point located at the tip of the vector. The energy associated with this point is determined. This procedure is repeated in order to locate a second point and the corresponding energy. The point corresponding to the lower energy is now chosen as the new origin and the procedure described in the previous few sentences is repeated for, say, m cycles. The value of m is arbitrarily selected. While it is possible to move out of the hypersphere defined by p_0 centered at the point corresponding to the energy minimum, there are two factors which make it reasonable to neglect any hypervolume contributions greater than $\frac{4}{3}\pi p_0^3$. These are:

(1) Since a minimum energy constraint is imposed and the energy decreases as one approaches the energy minimum there is a bias to stay within p_0.

(2) Since the entropic contributions depend upon Boltzmann statistics, the higher energy contributions outside the hypersphere will make small contributions to the total entropy.

In this method the entropy is approximately given by

$$S = (1/QT) \sum_{i=1}^{m} [E_i \exp(-E_i/RT)] + R \ln Q \qquad (2\text{-}141)$$

where Q and E_i are the same as defined above. The values of the entropies of some polymers in some stable conformations using this technique are presented in Table 2-10.

2. Convergence Constraint

In this technique we once again define some hypersphere about the point corresponding to the minimum in energy having a radius p_0. Next we randomly locate m_1 points within the hypersphere and calculate the energy at each point. We then proceed to compute the approximate entropy, $S(m_1)$ using Eq. (2-140). We now repeat the process using m_2 points where $m_2 > m_1$ and find $S(m_2)$ which allows us to compute the difference $\Delta S(1,2)$. We repeat the entire scheme to find $\Delta S(2,3)$, $\Delta S(3,4)$, ..., until $\Delta S(i, i+1) < \epsilon$, where ϵ is a desired criteria of convergence.

References

1. R. Hoffmann and A. Imamura, *Biopolymers* **7**, 207 (1969).
2. G. Nemethy and H. A. Scheraga, *Biopolymers* **3**, 155 (1965).
3. W. L. Bragg, *Phil. Mag.* **40**, 169 (1920).
4a. S. B. Hendricks, *Chem. Rev.* **7**, 431 (1930).
4b. C. H. Bamford, A. Elliott, and W. E. Hanby, "Synthetic Polypeptides." Academic Press, New York, 1956.
5. J. C. Slater and J. G. Kirkwood, *Phys. Rev.* **37**, 682 (1931).
6a. R. A. Scott and H. A. Scheraga, *J. Chem. Phys.* **45**, 2091 (1966).
6b. K. S. Pitzer, *in* "Advances in Chemical Physics" (J. Prigogine, ed.) Vol. 2, p. 59. Wiley (Interscience), New York, 1959.
7. G. Del Re, B. Pullman, and T. Yonezawa, *Biochim. Biophys. Acta* **75**, 153 (1963).
8a. G. W. Wheland and L. Pauling, *J. Amer. Chem. Soc.* **57**, 2086 (1935).
8b. D. Poland and H. A. Scheraga, *Biochemistry* **6**, 3791 (1967).
9. A. J. Hopfinger and A. G. Walton, *J. Macromol. Sci. Phys.* **B3**(1), 171 (1969).
10. W. A. Hiltner, M.S. Thesis, Case Western Reserve Univ., Cleveland, Ohio (1972) (has a comprehensive review and set of references).
11. H. J. C. Berendsen, *J. Chem. Phys.* **36**, 3297 (1962).
12. C. A. Swenson and R. Formanek, *J. Phys. Chem.* **71**, 4073 (1967).
13. R. F. Steiner, "The Chemical Foundations of Molecular Biology." Van Nostrand-Reinhold, Princeton, New Jersey, 1968.
14. G. Nemethy and H. A. Scheraga, *J. Chem. Phys.* **36**, 3401 (1962).
15. K. D. Gibson and H. A. Scheraga, *Proc. Nat. Acad. Sci. U.S.* **58**, 420 (1967).
16. J. A. Glasel, *J. Amer. Chem. Soc.* **92**, 375 (1970).
17. S. Krimm and C. M. Venkatachalam, *Proc. Nat. Acad. Sci. U.S.* **68**, 2468 (1971).
18. E. A. Moelwyn-Hughes, "Physical Chemistry." Pergamon, Oxford, 1957.
19. W. C. Davidson, AEC Res. and Develop. Rep. No. ANL-5090 (1959).
20. A. J. Hopfinger, *Macromolecules* **4**, 731 (1971).
21. L. Mandelkern, *in* "Poly-α-Amino Acids" (G. D. Fasman, ed.). Dekker, New York, 1967.
22. D. Aebersold and E. S. Pysh, *J. Chem. Phys.* **53**, 2156 (1970).
23. P. J. Flory, *in* "Conformation of Biopolymers" (G. N. Ramachandran, ed.), Vol. I. Academic Press, New York, 1967.
24. G. D. Fasman, *in* "Polyamino Acids, Polypeptides and Proteins" (M. A. Stahmann, ed.), p. 221. Univ. of Wisconsin Press, Madison, Wisconsin, 1962.
25. G. D. Fasman, *in* "Poly-α-Amino Acids" (G. D. Fasman, ed.), p. 499. Dekker, New York, 1967.
26. W. B. Rippon and A. G. Walton, *Bull. Biophys. Soc.* **11**, 188 (1971).
27. T. Ooi, R. A. Scott, G. Vanderkooi, and H. A. Scheraga, *J. Chem. Phys.* **46**, 4410 (1967).
28. D. A. Brant and P. R. Schimmel, *Proc. Nat. Acad. Sci. U.S.* **58**, 428 (1967).

29. J. A. Schellman, *C. R. Lab. Carlsberg Ser. Chim.* **29**, 223 (1955).
30. J. T. Edsall and J. Wyman, "Biophysical Chemistry," Vol. 1, p. 124. Academic Press, New York, 1958.
31. S. Krimm and J. E. Mark, *Proc. Nat. Acad. Sci. U.S.* **60**, 1122 (1968).
32. W. A. Hiltner, A. J. Hopfinger, and A. G. Walton, *J. Amer. Chem. Soc.* **94**, 4324 (1972).
33. S. Hanlon and I. M. Klotz, *Biochemistry* **4**, 37 (1965).
34. J. Semen, Ph.D. Thesis, Case Western Reserve Univ. Cleveland, Ohio (1972).
35. R. E. Dickerson and I. Geis, "Structure and Action of Proteins," Chapter 2. Harper, New York, 1969.
36. J. C. Slater, *Phys. Rev.* **36**, 57 (1930).
37a. E. R. Lippincott and R. Schroeder, *J. Chem. Phys.* **23**, 1099 (1955).
37b. R. Schroeder and E. R. Lippincott, *J. Phys. Chem.* **61**, 921 (1957).
38. W. G. Moulton and R. A. Kromhout, *J. Chem. Phys.* **25**, 34 (1956).
39. H. A. Scheraga, R. A. Scott, G. Vanderkooi, S. J. Leach, K. D. Gibson, T. Ooi, and G. Nemethy, *in* "Conformation of Biopolymers" (G. N. Ramachandran, ed.), Vol. I. Academic Press, New York, 1967.
40. K. D. Gibson and H. A. Scheraga, *Proc. Nat. Acad. Sci. U.S.* **58**, 1317 (1967).
41. M. Dreyfus, B. Maigret, and A. Pullman, *Theoret. Chim. Acta (Berlin)* **17**, 109 (1970).
42. A. Pullman, *in* "Quantum Aspects of Heterocyclic Compounds in Chemistry and Biochemistry," Jerusalem Symposium, number II (in press).
43. E. Clementi and D. R. Davis, *J. Comput. Phys.* **1**, 223 (1966).
44. E. Clementi, H. Clementi, and D. R. Davis, *J. Chem. Phys.* **46**, 4725 (1967).
45. E. Clementi, H. Clementi, and D. R. Davis, *J. Chem. Phys.* **46**, 4731 (1967).
46. E. Clementi, H. Clementi, and D. R. Davis, *J. Chem. Phys.* **47**, 4485 (1967).
47. B. Mely and A. Pullman, *Theoret. Chim. Acta (Berlin)* **13**, 278 (1969).
48. A. Pullman and H. Berthod, *Theoret. Chim. Acta (Berlin)* **10**, 461 (1968).
49. P. A. Kollman and L. C. Allen, *J. Chem. Phys.* **51**, 3286 (1969).
50 M. Dreyfus and A. Pullman, *Theoret. Chim. Acta (Berlin)* **19**, 20 (1970).
51. C. C. J. Roothan, *Rev. Mod. Phys.* **23**, 69 (1951).
52. J. Ladell and B. Post, *Acta Crystallogr.* **7**, 559 (1954).
53. D. R. Hershbach, Bibliography for Hindered Internal Rotation and Microwave Spectroscopy, Lawrence Radiat. Lab. Univ. of California, Berkeley, California (1962).
54. J. D. Roberts, "Nuclear Magnetic Resonance." McGraw-Hill, New York, 1959.
55. F. A. Bovey, "High Resolution NMR of Macromolecules." Academic Press, New York, 1972.
56. A. J. Hopfinger, Ph.D. Thesis, Case Western Reserve Univ. (1969).
57. E. Isaacson and H. B. Keller, "Analysis of Numerical Methods." Wiley, New York, 1966.
58. M. Levitt and S. Lifson, *J. Mol. Biol.* **46**, 269 (1969).

Chapter 3 | Accuracy and Refinement of the Potential Functions

I. Accuracy of the Classical Potential Energy Functions

A. Introduction

In view of the many inherent approximations, just how accurate are the empirical conformational potential functions discussed previously? This is a singularly important question to pose, for the reliability of any conclusions made from conformational energy calculations will be directly dependent upon the accuracy of the potential functions. The answer to this question will not be resolute, but rather depend upon what "quantity" one is interested in measuring. We present the results of several investigations in which it was possible to compare theoretical predictions to experimental findings. It is left to the reader to decide upon the reliability of the proposed potential functions for a specific type of structural calculation.

B. Investigations

1. Consistent Force Field Calculations on 2,5-Diketopiperazine and Its 3,6-Dimethyl Derivatives

The diketopiperazines (Figs. 3-1a and 3-1b) are cyclic anhydrous dipeptides for which there exists a variety of experimental data so that they are well

(a)   (b)

Fig. 3-1. (a) 2,5-Diketopiperazine (DKP) (glycylglycine anhydride); (b) 3,6-Dimethyl 2,5-diketopiperazine (DMDKP) (alanylalanine anhydride); DL, trans; LL, cis. From Karplus and Lifson (1).

suited as a means of testing the accuracy of the proposed conformational energy calculations. In the consistent force field calculation (CFF) a particular conformational energy function is chosen and the parameters on which it depends are varied to give the best simultaneous fit to the available experimental data; e.g., the molecular structures, the vibrational spectra, the strain energies and the vibrational enthalpies. Karplus and Lifson (1) chose the potential function given by Eq. (3-1) to represent the molecular energy

$$2V(s) = \sum K_b(b - b_0)^2 + \sum K_\theta(\theta - \theta_0)^2$$
$$+ \sum F_{ij}(r_{ij} - r_{ij}^0)^2 + 2 \sum F'_{ij}(r_{ij} - r_{ij}^0)$$
$$+ \sum K_\phi \cos n(\phi - \phi_0) + \sum K_\chi \chi^2$$
$$+ 2 \sum [\epsilon_{ij}^*(r_{ij}^*/r)^{12} - 2\epsilon_{ij}^*(r_{ij}^*/r)^6] + (e_i e_j/r) \qquad (3\text{-}1)$$

b and θ are the bond lengths and bond angles, respectively, r_{ij} and r are the 1,3 and higher-order interatomic distances, respectively, ϕ is a dihedral rotation angle, and χ is an out-of-plane deformation angle. The values n and ϕ_0 for the torsional potentials of rotations around the bonds N—C^α, C^α—C', and C^α—C^β were taken as $n = 3$ and $\phi_0 = 0$; for C'—N it was assumed that $n = 2$ and $\phi_0 = \pi/2$; i.e., that the torsional potential has the same minimum value for the cis and the trans configurations. The out-of-plane angle quadratic potentials in Eq. (3-1) approximate the restoring forces (for small χ) which tend to keep the bond C'=O (or, similarly, N—H) in the plane formed by the two other bonds belonging to C' (or, respectively, to N). $\chi C'$ has been defined (2) as the angle between the planes NCC^α and NC$'$O, and χN, similarly, as the angle between the planes C'NC^α and C'NH. The nonbonded steric interactions are represented by a Lennard-Jones 6-12 function and the very last term accounts for electrostatic interactions.

Values for the parameters in Eq. (3-1), after energy minimization, are given in Table 3-1. The only parameters unique to the diketopiperazines are those involved in quadratic bond angle terms and 1,3 interactions for $C'C^\alpha$N. A linear relationship has been assumed between F_{ij} and F'_{ij},

$$F'_{ij} = -0.2 F_{ij} r_{ij}^0 \qquad (3\text{-}2)$$

In investigating the shape of the conformational energy minima, a single dihedral angle, ψNC$'$$C^\alpha$N was held fixed over a range of values, while the constrained minimum energy conformation was computed as a function of all other internal coordinates. Once the potential had been determined as a function of ψNC$'$$C^\alpha$N, the absolute minimum was found by varying all internal coordinates including ψNC$'$$C^\alpha$N in the neighborhood of the low energy region.

TABLE 3-1

Parameters

Bond distance	$\frac{1}{2}K_b$	b_0 (Å)	Dihedral angle	$\frac{1}{2}K_\phi$
$C^\alpha H$	287	1.100	$C'NC^\alpha C^\beta$, $C'NC^\alpha C'$	1.500
$C^\alpha C^\beta$	112	1.457	$NC'C^\alpha C^\beta$, $NC'C^\alpha N$	1.400
NC	405	0.980	$C^\alpha NC'C^\alpha$	1.655
$C^\alpha N$	261	1.458	$C^\alpha NC'O$	2.679
C'O	595	1.200		
C'N	403	1.279	Deformation angle	$\frac{1}{2}K_x$
$C'C^\alpha$	187	1.470		
			$C^\alpha N$, C'O	4.045
			$C^\alpha C'$, NH	0.686

Bond angle	$\frac{1}{2}K_\theta$	θ_0 (rad)	$\frac{1}{2}F_{ij}$	r_{ij}^0 (Å)
$HC^\alpha H$, $HC^\beta H$	38.2	1.911	2.8	1.800
$HC^\alpha C^\beta$	27.0	1.911	43.2	2.200
$C^\alpha NH$	31.4	2.094	26.0	1.790
$C^\alpha NC'$	54.5	2.094	16.2	2.400
C'NH	26.7	2.094	27.1	2.019
$C^\beta C^\alpha N$	21.0	1.911	50.5	2.200
$NC^\alpha H$	30.1	1.911	41.0	1.900
NC'O	48.5	2.094	90.00	2.186
$C^\alpha C'O$	40.9	2.094	52.0	2.400
$C^\alpha C'N$	33.1	2.094	50.5	2.229
$C'C^\alpha H$	28.7	1.911	38.4	1.975
$C'C^\alpha C^\beta$	21.6	1.911	37.0	2.200
$C'C^\alpha N$	21.0	1.911	30.0	2.000

Lennard-Jones[a]	$(\epsilon^*)^{1/2}$	$\frac{1}{2}r^*$	Effective charge	e_i (units of e)
			C	0.144
H	0.067	1.468	H	0.272
C	0.300	1.800	N	−0.305
N	0.400	1.800	C'	0.449
O	0.480	1.482	O	−0.406

[a] Note: $\epsilon_{ij}^* = (\epsilon_i^* \epsilon_j^*)^{1/2}$. The ϵ_k are the coefficients of the attractive term in a Lennard-Jones 6-12 function for a *homo* $k \cdots k$ interaction. The r_k are the corresponding distances in the $k \cdots k$ interactions for which the function has its minimum value. From Karplus and Lifson (1).

The vibrational analysis is carried out by evaluating the second derivatives of the potential energy function at the equilibrium conformation; transforming these into a mass-weighted coordinate system; and solving the Lagrangian equation in this coordinate system to obtain the normal modes corresponding to the eigenvalues of the transformed matrix of second derivatives.

The predicted molecular structures of DKP, DL-DMDKP, and LL-DMDKP are given in Table 3-2 and compared to the crystallographic data. Agreement

TABLE 3-2

Molecular Structures[a]

Bond length	Obs (Å)	Calc (Å)	Angle	Obs (deg)	Calc (deg)
		DKP (symmetric)			
$C'N$	1.325^b	1.336	$C'NC^\alpha$	126.0	125.9
NC^α	1.449	1.464	$NC^\alpha C'$	115.1	113.4
$C^\alpha C'$	1.499	1.523	$C^\alpha C'N$	118.9	118.2
			Dihedral	Planar	Folded
		DL-DMDKP (symmetric)			
$C'N$	1.325^c	1.336	$C'NC^\alpha$	127.9	127.0
NC^α	1.462	1.463	$NC^\alpha C'$	112.8	113.7
$C^\alpha C'$	1.470	1.529	$C^\alpha C'N$	119.2	118.9
$C^\alpha C^\beta$	1.509	1.524	Dihedral	Planar	~Planar
		LL-DMDKP (asymmetric in crystal)			
$C'N$	1.344, 1.323^d	1.337	$C'NC$	127.1, 126.2	126.2
NC^α	1.457, 1.462	1.463	$NC\ C'$	109.9, 112.0	112.7
$C^\alpha C'$	1.512, 1.534	1.528	$C\ C'N$	117.4, 115.1	118.5
$C^\alpha C^\beta$	1.530, 1.537	1.521	Dihedral	Folded (↑)	Folded (↓)

[a] From Karplus and Lifson (1).
[b] R. Degeilh and R. E. Marsh, *Acta Crystallogr.* **12**, 1007 (1959).
[c] E. Benedetti *et al.*, *Proc. Nat. Acad. Sci. U.S.* **62**, 650 (1969).
[d] E. Benedetti *et al.*, *Biopolymers* **7**, 751 (1969).

for the bond lengths and bond angles is reasonable, but major differences are found for the folding of the molecules. The minimum energy configurations for DKP and DL-DMDKP correspond to "boat" conformations with the two peptide groups nearly coplanar. In the case of LL-DMDKP the two methyl groups are directed toward each other and away from the carbonyl oxygens. This disagrees with the conformation in the crystal state, where the ring is folded in the opposite direction. The observed differences in molecular folding may be due to the intermolecular forces present in the solid state. This possibility gains some support from the solution studies of Schellman and Nielsen (3) on LL-DMDKP who suggest a structure like that found for the absolute energy minimum. Further, the "boat" structure for DKP is consistent with NMR studies by Sykes *et al.* (4).

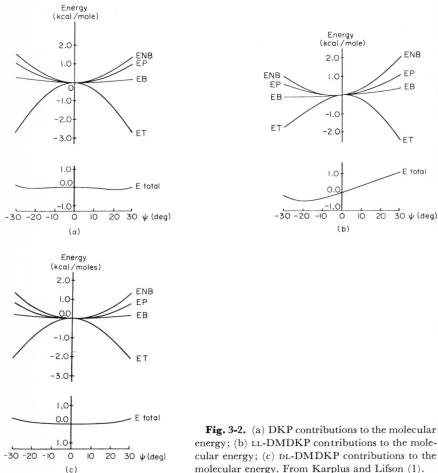

Fig. 3-2. (a) DKP contributions to the molecular energy; (b) LL-DMDKP contributions to the molecular energy; (c) DL-DMDKP contributions to the molecular energy. From Karplus and Lifson (1).

The different contributions to the potential energy function are represented over a range of dihedral angles in Fig. 3-2. The contributing energy functions can be identified as follows: EB, quadratic bond stretching terms; ET, quadratic bond bending plus 1,3 Urey–Bradley interactions; EP, torsional interactions plus out-of-plane quadratic terms; ENB, Lennard-Jones potential plus electrostatic interactions. In these figures it can be seen that for all three molecules the terms that contribute to lowering the energy as the molecule is folded away from planarity are included predominantly in the ET functions, which involve changes in the bond angles. Although the total energy represents a small difference between large terms, and dihedral angles and bond angles are coupled in the potential function, one can reach the qualitative conclusion that neglect of bond bending and Urey–Bradley 1,3 interactions would result in planar minimum energy conformations for DKP and DL-DMDKP, while LL-DMDKP would be only slightly distorted from planarity. The net tendency as $\psi NC'C^\alpha N$ increases in absolute value is for $\theta_{C'C^\alpha N}$ to decrease from the high value required in the planar ring conformation.

The results of the vibrational analysis for DKP are given in Tables 3-3 and 3-4 and compared with the experimental results. The agreement is generally

TABLE 3-3

Infrared Frequencies of DKP $(cm^{-1})^a$

Obs[b]	Obs[c]	Calc	Obs[b]	Obs[c]	Calc
1690	1640–1740	1655	1075	1070	1109
1468	$(1520)^d$	1475	998	(1000)	929
1443	1450–1490	(1443)	910	915	920
		1421	840	840	882
1340	1340	1340	806	810	742
1249	(1290)	(1270)		(780)	(696)
	(1260)		449		424
	(1220)				
	(1180)				
	(1150)				

[a] From Karplus and Lifson (1).
[b] T. Miyazawa, *J. Mol. Spectrosc.* **4**, 155 (1960).
[c] R. Newman and R. M. Badger, *J. Chem. Phys.* **19**, 1149 (1951).
[d] ( ) = weak intensity bands.

good for the values of the observed and calculated frequencies. For the observed 449 cm^{-1} frequency a C=O in-plane bending is suggested in contradiction with Miyazawa's (5) suggested assignment but in agreement with an earlier calculation by Fukishima (6). The NH in-plane bending contributes to two of the calculated frequencies, 1475 cm^{-1} and 1421 cm^{-1}, and a CH$_2$ scissors bending is assigned to an intermediate calculated frequency, 1443 cm^{-1}, in agreement with the general frequency range in Miyazawa's assignments (5). Here, as in general, it is found that the normal modes involve complicated mixtures of the various nuclear motions. The results agree with those of Miyazawa in the order of the CH_2 wag, CH_2 twist, and CN stretch assignments. However, the CH_2 rocking motion is assigned to a lower frequency and there are two, not one, NH out-of-plane bending contributions. If the calculated 742 cm^{-1} out-of-plane bending mode is to be identified with the observed 840 cm^{-1} frequency for an out-of-plane bend, this difference represents the greatest discrepancy found between calculated and experimental results. Trial changes of parameters in the potential function do not reduce this difference significantly.

For all three molecules, the inclusion in the potential function of terms involving deformations in the bond angles is decisive in determining the

TABLE 3-4

Frequency Assignments for DKP[a]

Frequency (calc, cm^{-1})	Assignment
1655	C'O, C'N stretch
1475	C'O stretch; NH in-plane bend
1443	CH_2 bend
1421	C'N stretch; NH in-plane bend; CH_2 bend
1340	CH_2 wag; C'O, C'N, C'C stretch
1270	CH_2 twist; C'O, NH out-of-plane bend
1109	CN stretch
929	C'C stretch
920	CH_2 twist; C'O, NH out-of-plane bend
882	CH_2 rock; ring out-of-plane bend
742	NH out-of-plane bend, "symmetric"
696	NH out-of-plane bend, "antisymmetric"
424	C'O in-plane bend

[a] From Karplus and Lifson (1).

minimum energy conformations. Although the difference between the final set of bond angles and a standard *cis*-peptide set may not be large, small deformations in bond angles are seen to be coupled with large changes in the internal rotation angles. These results demonstrate that in ring systems the minimum energy conformations cannot be safely estimated by assuming fixed values for the bond angles, an assumption which is commonly made in calculations on peptide conformations.

2. *Molecule–Solvent Interactions of Phenethylamines in Aqueous Solution*

A very important class of biological molecules are the phenethylamines. These molecules affect the central nervous system in varying ways depending upon the types and positions of group substitutions on the basic phenethyl-amine skeletal structure. Amphetamine, whose primary structure is shown in Fig. 3-3, is a potent stimulant. This molecule has been the subject of many structural investigations, both experimental and theoretical. The theoretical investigations of Kier (7) and Pullman *et al.* (8) involved quantum mechanical calculations of the total energy of the molecule as a function of the bond rotations χ_1, χ_2, and χ_3 which are defined in Fig. 3-3. Both workers concluded that the amphetamine molecule assumed one of two general conformations. First the ethylamine chain can be extended so that the amine group points away from the ring, or, secondly, the ethylamine chain can fold back on itself so that the amine group points toward the ring. The extended form is depicted for a phenethylamine skeletal backbone in Fig. 3-4a and the folded form is shown in Fig. 3-4b. The energy contour map calculated by Pullman *et al.* (8)

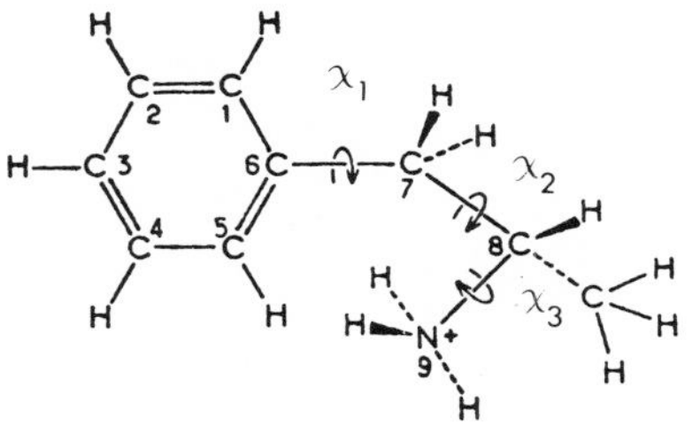

Fig. 3-3. Amphetamine.

is shown in Fig. 3-5. The minima corresponding to $\chi_2 = 0°$ correspond to extended conformations, while the minima at $\chi_2 \approx 120°$, $240°$ correspond to folded structures. As can be seen from the contour map, Pullman *et al.* (8) concluded that the extended and folded forms were essentially equally probable since their respective energies were virtually identical. Kier (9), on the other hand, found the folded structures to be of lower energy than the extended forms. The crystal structures of four other phenethylamines indicate each of the molecules is in the extended conformation. Pullman and Kier attribute the added stability of the extended form in the solid state to interchain packing forces. This certainly seems reasonable and does not contradict either experimental or theoretical findings. However, Neville *et al.* (10) have concluded from NMR calculations that amphetamine exists in the extended form in aqueous solution. Thus while both Pullman *et al.* (8) and Kier (9) have deduced stable-isolated-vacuum conformations for amphetamines, they cannot relate their theoretical finding to experimental data. Inclusion of intermolecular interactions or molecule–solvent terms is necessary in order to hope to achieve agreement between theory and experiment.

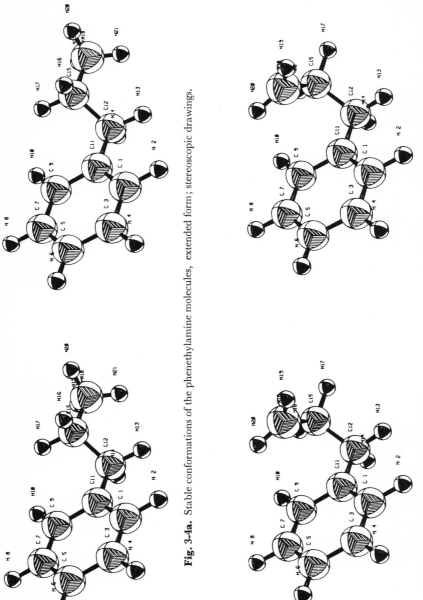

Fig. 3-4a. Stable conformations of the phenethylamine molecules, extended form; stereoscopic drawings.

Fig. 3-4b. Stable conformations of the phenethylamine molecules, folded form; stereoscopic drawings.

Weintraub and Hopfinger (11) have carried out the conformational analysis of amphetamine in vacuum and in aqueous solution. The energy contour maps are shown in Fig. 3-6. The vacuum-energy map is very similar to the map determined by Pullman *et al.* (8) (see Fig. 3-5). The vacuum calculations indicate that the folded form is preferred over the extended conformation which is in agreement with the findings of Kier (9). However, when aqueous solution–amphetamine interactions are taken into consideration the extended

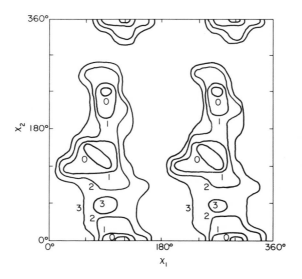

Fig. 3-5. Energy contour map of χ_1 versus χ_2 for amphetamine. The energy contours are in kcal/mole relative to the absolute minimum which is assigned a value of zero. This map is the result of calculations by Pullman *et al.* (8). Reprinted from *J. Med. Chem.* **15**(1), 17 (1972). Copyright 1972 by the American Chemical Society. Reprinted by permission of the copyright owner.

conformation becomes, far and away, the preferred conformation (see Fig. 3-6b) in agreement with experimental results. The extended form is preferred to the folded form in aqueous solution because of the interaction of the NH_3^+ group with water molecules.

The point to be made here is that molecule–solvent interactions were essential in order to achieve agreement between the predicted solution-state conformation of amphetamine and the observed conformation in solution. The molecule–solvent model proposed in Chapter 2, i.e., the hydration shell model, was successful in correctly predicting the solution-state conformation of amphetamine in aqueous solution.

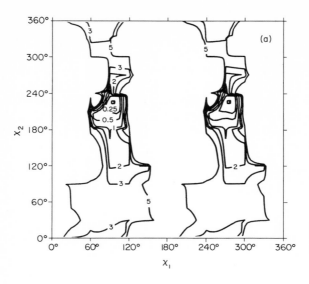

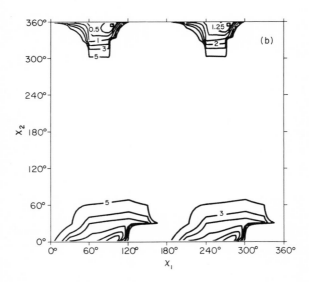

Fig. 3-6. Energy contour maps of the charged form of amphetamine. The contours are in $\frac{1}{2}$- and 1-kcal/mole intervals. These maps are the results of calculations carried out by Weintraub and Hopfinger (11) using classical potential energy functions: (a) *in vacuo*, (b) in aqueous solution.

3. Comparison of Quantum Mechanical Dipeptide Energy Maps to Classical Mechanical Dipeptide Energy Maps

Molecular orbital calculations provide an independent theoretical check on classical potential energy (CPE) calculations where the total energy has been partitioned into additive components such as nonbonded interactions, barriers to internal rotations, electrostatic interactions and hydrogen bonding. Hoffmann and Imamura (12) have used the extended Hückel theory (EHT) to calculate, as a function of conformation, the total molecular energy for diglycine and dialanine in the ground and first excited states. To be absolutely correct, the molecules studied were N-acetyl-N'-methyl-glycylamide and N-acetyl-N'-methyl-alaninamide. The fixed bond distances and angles were

$$H_3-C-\overset{O}{\overset{\|}{C}}-N-\overset{R}{\underset{H}{\overset{\phi}{\underset{|}{C}}}}-\overset{}{\underset{H}{\overset{}{C}}}-\overset{O}{\overset{\psi}{\overset{\|}{C}}}-N-CH_3 \qquad R=H, CH_3$$

those used by Scott and Scheraga (13). A grid of points at 30° increments of ϕ and ψ was scrutinized. The results for the ground states of the glycyl and alanyl residues are presented in Figs. 3-7a and 3-7b as contour maps of the energy relative to an energy zero at the most stable conformation. The first electronic transition of these residues corresponds in these calculations to an excitation of an electron from an orbital identifiable as a combination of lone pairs on both carbonyl groups (considerably delocalized to nearby atoms) to a π^*-type orbital of both amide groups (14). Figures 3-8a and 3-8b show the energy contours for these glycyl and alanyl residue excited states. It is assumed that the average energy of the excited configuration will show the same conformational dependence for both singlet and triplet states arising from that configuration. A representative example of the energy contour maps constructed using classical potential energy functions are shown in Figs. 3-9a and 3-9b. These are maps constructed by Scott and Scheraga (13).

The general resemblance of (EHT) calculated ground state curves to those of previous hard sphere and extended-interaction models is good. For the glycyl residue a large low energy basin extends over the range $\phi = 0°$ to $110°$ and $\psi = -70°$ to $+70°$ (the symmetry related region for $\phi > 180°$ will not be explicitly referred to in the discussions of the glycyl residue surface). The actual absolute minimum of the calculations lies along the line $\psi = 0°$, but the energies within the basin defined above are all within 1 kcal of the minimum. The (EHT) potential energy surface approaches mirror symmetry about the $\psi = 180°$ line more closely than other calculations. The alanyl residue surface differs in minor ways from other calculations. The minimum energy is at greater ψ and the local minimum near $\phi = \psi = 240°$ is more

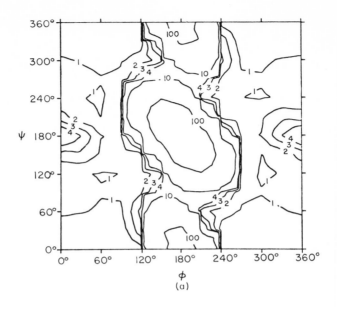

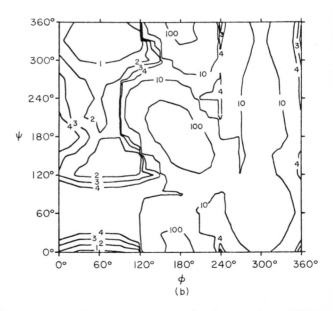

Fig. 3-7. The (EHT) calculated energy contours for the ground state. The contours are labeled in kcal/mole relative to a zero energy at the absolute minimum from the calculation: (a) diglycine, (b) dialanine. From Hoffman and Imamura (12).

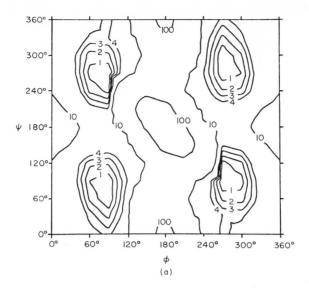

(a)

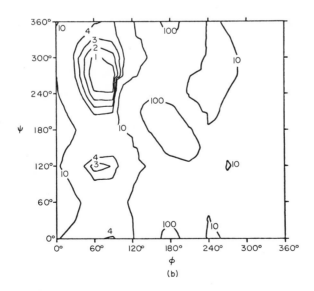

(b)

Fig. 3-8. The (EHT) calculated energies for the first excited state. The energy contours are in kcal/mole relative to a zero of energy at the absolute minimum found in the calculations: (a) diglycine, (b) dialanine. From Hoffman and Imamura (12).

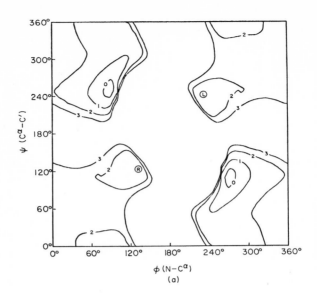

(a)

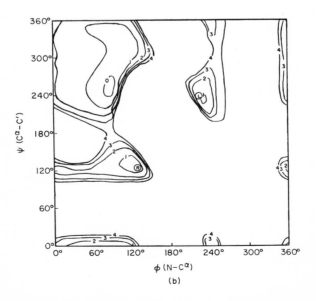

(b)

Fig. 3-9. Energy contour maps constructed by Scott and Scheraga (13) using classical potential energy functions. Contours are in kcal/mole-dimer: (a) diglycine, (b) dialanine.

unstable with respect to the absolute minimum than in the calculations of Scott and Scheraga. The sterically forbidden regions are remarkably similar to those obtained in other calculations, emphasizing once again the predominant role of nonbonded repulsions in determining those regions.

The ground state forbidden regions would be expected to remain forbidden on the excited state surfaces. This is indeed so, and the most interesting effects are in the sterically allowed regions. There one finds in the EHT calculations a general sharpening and clearer definition of the potential minima. The excited glycyl residue surface (Fig. 3-8a) has well separated "volcanic" minima at $\phi = 80°$, $\psi = 80°$, 270°. The alanyl residue retains virtually unchanged above the one minimum, at $\phi = 80°$, $\psi = 270°$. The other one is partially affected and is reduced in stability. The excited state calculations must be considered tentative until the effect of distortions corresponding to further degrees of freedom in the excited state are examined.

Pullman and co-workers (15) performed a quantum mechanical study of the model "dipeptide" N-acetyl-N'-methylalanylamide using the all-valence electrons molecular orbital method PCILO. In this calculation values of ϕ and ψ were incremented in 30° steps and, the total molecular energy was minimized for χ rotation of the methyl side chain group about the C^α—C^β bond, for each pair of (ϕ, ψ) values. This side chain energy minimization was not included in the previously discussed studies. The refined energy contour map is shown in Fig. 3-10. This map is again very similar to Fig. 3-7b for the in-common allowed regions (e.g., approximately the left-hand half of the map. However, the PCILO map indicates that a low energy region, in fact the global minimum, should be found near $\phi = 270°$, $\psi = 120°$. This is a disallowed region in both the classical and EHT potential energy calculations. Presumably, this low energy region arises because of the energy minimization with respect to χ. It is also interesting to note that this global minimum corresponds to one of the two degenerate minima found for diglycine using the (CPE) calculation scheme.

The existence, or nonexistence, of this new energy minimum found by Pullman and co-workers could have crucial implications on the predictive capability of the (CPE) calculations. Therefore, it is important to determine if this new found energy minimum has ever been observed experimentally. Poly-L-alanine has not been observed in this new conformation. The crystal structures of alanine residues in small peptide chains, e.g., dimers and trimers, again do not correspond to conformations located in the newly found low energy region. A (ϕ, ψ) map is shown in Fig. 3-11 with the conformations of alanine residues found in five different proteins, whose crystal structures have been solved, plotted as points. The proteins used in the plot were ribonuclease-S, lysozyme, chymotrypsin, α-chain of hemoglobin, and myoglobin. The reader will note that not a single alanine residue adopts a conformation

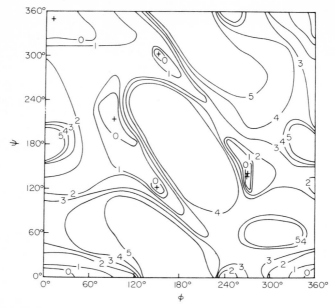

Fig. 3-10. Conformational energy map of dialanine (2 planar peptide units) obtained by a minimization procedure with respect to χ (rotation of the side chain methyl group). The contours are in kcal/mole with respect to the global minimum ($\ddagger$). Local minima are indicated by $+$. This map is the result of calculations by Pullman (15).

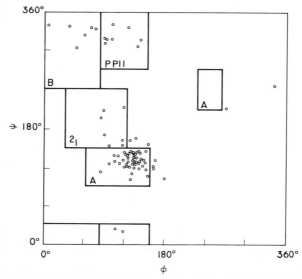

Fig. 3-11. A (ϕ, ψ) map in which the conformations of alanine residues found in a variety of proteins (see text) have been plotted. Regions associated with standard conformations are partitioned off for reference.

which is located in the new low energy region. In view of these findings one probably should conclude that the new low energy found by Pullman and co-workers either is characteristic only of an isolated *N*-acetyl-*N'*-methyl-alanylamide molecule or is a false low energy region resulting from short-comings of the PCILO method. To the degree that these conclusions are correct, the (CPE) and (EHT) calculations better describe the energetics of an alanine residue than does the PCILO calculation.

4. The Lattice Energetics of Some Polypeptide Chains

A basic means of estimating the *spatial* accuracy of the (CPE) functions is to see just how well these functions will predict crystal structures which have been determined by X-ray analysis. Such crystal structure calculations provide a direct comparison of experimental and theoretical data. We report here the results of crystal structure calculations which have been carried out for:

I. Polyglycine as a
   a. right-handed α-helix
   b. parallel β-"helix"
   c. antiparallel β-"helix"
   d. threefold polyglycine II (PG II) helix
II. Trans poly-L-proline as a
   a. threefold polyproline II (PP II) helix
   b. 3.30-fold helix
III. Poly(Gly-Pro-Pro) in the triple-helix structure proposed by Yonath and Traub (16a) as well as a left-handed threefold helix like that of PP II and PG II

The appropriate set of (τ, ϕ, ψ) are given in Table 3-5. The glycine residue geometry is that suggested by Pauling and Corey (17), the proline residue geometry is that reported by Sasisekharan (18), and the poly(Gly-Pro-Pro) triple-helix residue geometry is that reported by Yonath and Traub (16a). Very recently, Balasubramanian *et al.* have carried out a detailed analysis of the geometry of the proline residue (19). They have found that the proline residue geometry is sensitive to both the N and the C nearest neighbor residues and that, in general, the topology of the proline residue is variable. However, since the specific proline residue geometries found for both trans poly-L-proline and poly(Gly-Pro-Pro) via X-ray crystallography are used in these calculations, errors due to the variations in the proline residue geometry should be minimized. Each chain used in the study is thirty residues in length in order to minimize computer time while still including energy contributions from long range interactions. There are four parameters

TABLE 3-5

Values of the Dihedral Angles τ, ϕ, and ψ[a]

Structure	τ (deg)[b]	ϕ (deg)	ψ (deg)
1. Polyglycine			
A. Right-handed α-helix	109.5	−47.3	−57.4
B. Parallel β-helix	109.5	−117.5	+116.5
C. Antiparallel β-helix	109.5	−139.7	+147.0
D. Polyglycine II	109.3	−77.5	+145.4
2. *Trans*-poly-L-proline			
A. Poly-L-proline II	109.8	−76.5	+145.0
B. Left-handed 3.30-fold helix	109.8	−76.5	+159.0
3. Poly(Gly-Pro-Pro)			
A. Yonath and Traub triple helix			
(1) Gly	115.0	−51.0	+153.0
(2) Pro	109.0	−76.0	+113.0
(3) Pro	115.0	−45.0	+148.0
B. Left-handed threefold helix			
(1) Gly	109.3	−77.5	+145.5
(2) Pro	109.8	−76.5	+145.0
(3) Pro	109.8	−76.5	+145.0

[a] This table reports ϕ and ψ in the revised notation of the IUPAC-IUB Commission (16b) and not in the convention of Edsall *et al.* (16c).

[b] τ is the angle NC$^\alpha$C′.

($|\mathbf{S}|$, $\boldsymbol{\theta}$, $|\mathbf{a}|$, $\boldsymbol{\mu}$) which define the relative orientation of two helices whose axes are parallel or antiparallel. These parameters have been rigorously defined in Chapter 1 and will only be briefly described here with the aid of Fig. 3-12.

S A vector perpendicular to the axes of the two helices. Its magnitude is the distance between the axes of the helices.

θ The angle (always measured clockwise) which the **S** vector makes with some reference vector.

a A vector which represents the relative axial translation of one helix with respect to the other helix.

μ A rotation about one of the helical axes. This parameter measures the relative axial rotation of one helix with respect to the other. The rotation is counterclockwise.

The total interchain conformational energy for a helix–helix interaction is calculated using Eq. (3-3)

$$E = \sum_i \sum_j \left(\frac{-A_{ij}}{r_{ij}^6} + \frac{B_{ij}}{r_{ij}^{12}} + 331.81 \frac{Q_i Q_j}{\epsilon r_{ij}} \right) + H_b \tag{3-3}$$

(i and j nonbonded) where E is the total configurational energy (kcal/mole), A_{ij} the nonbonded attractive coefficient for the ijth interaction, B_{ij} the non-bonded repulsive coefficient for the ijth interaction, Q_i the partial charge (in a.u.) on atom i, Q_j the partial charge (in a.u.) on atom j, r_{ij} the interaction distance between atoms i and j, H_b the total interchain hydrogen bond energy, and ϵ the effective dielectric constant ($= 3.5$).

The A_{ij} and B_{ij} are those reported in Table 2-2 of Chapter 2. The partial charges were calculated by the method of Del Re *et al.* (20) as revised by Poland and Scheraga (21). The "corrected sum" hydrogen bond function discussed in Chapter 2 was used to calculate H_b.

An upper cutoff interaction distance of 12 Å was chosen to minimize

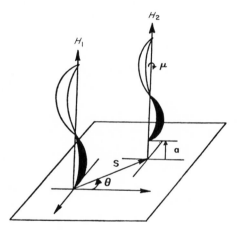

Fig. 3-12. Geometric representation of the four spatial variables ($|\mathbf{S}|$, θ, $|\mathbf{a}|$, $\mathbf{u}$) relating the orientation and relative positions of two parallel or antiparallel helices.

computer time and to correct for changes in chain–chain overlap as the $|\mathbf{a}|$ parameter is varied. Since pair interactions more than 12 Å apart have a maximum absolute value of 0.005 kcal/mole-interaction, such interactions can be validly neglected. The energy given by Eq. (3-3) is divided by the number of residues or tripeptides in a chain. Hence, energies in Table 3-6 are reported in kcal/mole-residue (or tripeptide). This allows the interchain energy to be directly compared to the intrachain energy (calculated in the usual way) which is also given in Table 3-6. There is no definite criteria for deciding how large the difference in total interaction energy between two structures must be before it is possible to meaningfully select one configuration as being more probable than another. Ooi *et al.* (22) have used intrachain energy differences as small as 0.40 kcal/mole-residue to predict successfully

TABLE 3-6

The Predicted and Observed Lattice Constants for Some Polypeptide Chains[a]

Polymer	Structure	Interchain Lattice Constants (Å) Pred a	Pred b	Pred c (FA)	Obs a	Obs b	Obs c (FA)	Packing Pred	Packing Obs	Energy (kcal/mole/residue) Intra-chain	Inter-chain	Total
	α_R	5.83	—	1.44/residue	N.O.	N.O.	N.O.	Hex	N.O.[b]	−5.3	−.6	−5.9
Polyglycine	$\beta_\parallel$	5.00	3.73	6.50	4.85	N.O.	6.50	Monocl	N.O.	−2.5	−3.0	−5.5
	β_x	9.63	3.45	7.06	9.50	3.45	7.00	Tricl	Tricl (23)	−2.6	−3.3	−5.9
	PG II	4.80	—	9.36	4.65	—	9.33	Hex	Hex (24)	−2.8	−3.4	−6.2
	PP II	6.75	—	9.36	6.68	—	9.36	Hex	Hex (18)	−3.4	−1.6	−5.0
Trans-poly-L-proline	L.H. 3.30-helix	6.23	—	3.24/residue	N.O.	N.O.	N.O.	Rect	N.O.	−3.3	−1.0	−4.3
	L.H.[d] 3.00-helix	6.67	—	9.35	N.O.	N.O.	N.O.	Hex	N.O.	−9.8[c]	−4.4[c]	−14.2[c]

Polymer	Structure	|a| (Å)	μ (deg)		|a| (Å)	μ (deg)		Packing Pred	Packing Obs	Energy (kcal/mole/tripeptide) Intra-chain	Inter-chain	Total
Poly(Gly-Pro-Pro)	YTTH	2.91	108.0		2.87	108.0		Hex	N.O.[b]	−9.7[c]	−6.1[c]	−15.8[c]

α_R right-handed α-helix   YTTH   Yonath and Traub triple helix   Rect   rectangular
$\beta_\parallel$ parallel β-sheet   FA   fiber axis
β_x antiparallel β-sheet   N.O.   not observed
PG II   polyglycine II   Hex   hexagonal
PP II   *trans*-poly-L-proline II   Tricl   triclinic
L.H.   left-handed   Monocl   monoclinic

[a] Also listed are the lattice configurational energies along with the predicted and observed lattice packing
[b] hexagonal packing has been postulated
[c] energy (kcal/mole/tripeptide)
[d] the glycine residue geometry is that suggested by Pauling and Corey and the proline geometry is that of Sasisekharan

observed helical conformations. However, this or any other criteria for choosing the existence of any structure over another lacks a solid theoretical or experimental basis. Thus, caution should be exercised in the interpretation of the results presented in Table 3-6.

The lattice maps (see Fig. 3-13) were plotted using energy contours in

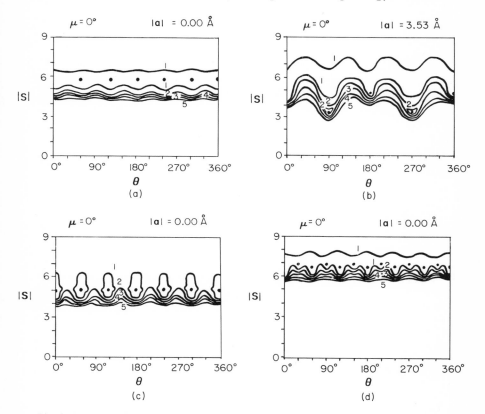

Fig. 3-13. Helix–helix lattice maps with the energy contours in kcal/mole-residue. For each point ($|\mathbf{S}|$, θ) on a map the energy is minimized with respect to ($|\mathbf{a}|$, μ). The filled circles denote relative minima and the relative minimum found for $\theta = 0$ on each map has the $|\mathbf{a}|$ and μ values listed above the figure. (a) Two right-handed α-helices of polyglycine, (b) two antiparallel β-helices of polyglycine, (c) two polyglycine II chains, (d) two poly-L-proline chains.

kcal/mole-residue relative to the absolute minimum which is assigned a value of zero. Each lattice map represents the configurational energetics of a helix–helix interaction. The maps were constructed by scanning $|\mathbf{S}| = 0.00$ to 9.00 Å at 0.25 Å intervals, $\theta = 0°$ to $360°$ at $15°$ intervals, $|\mathbf{a}| = 0.00$ Å to the length of one helix turn at 0.25 Å intervals and $\mu = 0°$ to 360° at 15° intervals. However, the lattice constants and energies reported in Table 3-6 were calculated by placing chains in each of the positions and orientations suggested by the low energy regions of the helix–helix scans and then mini-mizing the total interaction energy. Thus the lattice constants were calculated from a first nearest neighbor polymer chain model in which Eq. (3-3) was repeatedly applied to each possible helix–helix interaction in the lattice. The total lattice energy was minimized by successively minimizing the energy with respect to one variable at a time while holding the remaining variables at their previously calculated optimum values. Convergence to a local relative minimum was assumed when for all variables simultaneously the

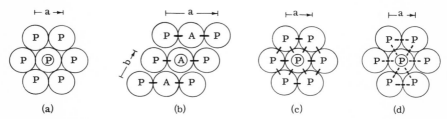

Fig. 3-14. Schematic illustration of polypeptide chain packing corresponding to the minimized nearest neighbor lattice energy. The view is looking down the chain axis, the c axis, and into the ab plane where the a and b axes are shown. The circles represent the chains, the solid lines indicate hydrogen bonds, and the broken lines indicate the direction of the proline ring planes. P stands for a parallel chain and A for an antiparallel chain. The central helix can be identified by a circle around the P or A symbol. The $\boldsymbol{\theta} = 0°$ vector would extend from the center of the central helix in the direction of the a lattice constant vector. (a) Right-handed α-helices of polyglycine, (b) antiparallel β-sheets of polyglycine, (c) polyglycine II chains, (d) poly-L-proline II chains.

absolute difference between two successively calculated optimum values for each variable was less than some tolerance level (0.02 Å for $|\mathbf{S}|$ and $|\mathbf{a}|$, and 1.5° for θ and μ).

The results of these calculations are presented in Table 3-6. Some of the lattice energy maps are presented in Fig. 3-13 and the corresponding packing schemes are illustrated in Fig. 3-14. From inspection of Table 3-6 a number of general observations may be made:

(1) The predicted and observed lattice constants are very nearly identical in all cases. The largest discrepancy is 0.15 Å.

(2) The observed lattice constants are always less than or equal to the predicted lattice constants. The differences between predicted and observed lattice constants is largely due to contact distances too large to describe the H---H interactions. The van der Waals radius of the hydrogen atom needed

to produce the correct lattice dimensions for the β structures and the PP II chain conformation varies, in a nonspecific manner, between 1.12 and 1.23 Å.

(3) The energy functions predict the correct chain packing in all cases. The sensitivity of the energy functions to chain packing is perhaps most evident from an inspection of the energy contour maps where the shapes of contours vary markedly with chain conformation.

(4) The energy differences between interchain interactions for different chain conformations is of the same magnitude as the energy uncertainty in making such calculations. Thus the energy functions do not provide a reliable base for distinguishing between the relative stabilities of different lattice packings. Of course we assume here that entropy contributions are negligible.

(5) The interchain interactions can be of the same stabilizing magnitude as intrachain interactions. In some cases there appears to be an inverse relationship between intra- and interchain interactions. Interchain stabilization takes place at the expense of intrachain interactions and *vice-versa*.

II. Refinement of Classical Potential Energy Functions

A. Refinement of the Lennard-Jones 6-12 Potential Functions

A rather novel means of refining the values of the parameters in the (CPE) functions is to make use of the conditions of static equilibrium which must. hold in crystals. If $P(r)$ is the general form of any intermolecular pair potential function, then the force between pairs of atoms described by $P(r)$ is the gradient of the potential

$$\mathbf{f} = -\nabla P(r) = -\mathbf{r}\left(\frac{1}{r}\frac{\partial P(r)}{\partial r}\right) \tag{3-4}$$

The total force acting on any molecule M in a crystal is

$$\mathbf{F} = \sum \mathbf{f} \tag{3-5}$$

where the sum extends over the interaction of all atoms in M with all other atoms in the crystal. The first condition for static equilibrium requires that the net forces acting upon M in any three orthogonal directions each be zero

$$F_x = \sum f_x = 0, \qquad F_y = \sum f_y = 0, \qquad F_z = \sum f_z = 0 \tag{3-6}$$

The second condition of static equilibrium is that the net moments of force acting on M in any three orthogonal planes each be zero

$$\sum (f_x y - f_y x) = 0, \qquad \sum (f_y z - f_z y) = 0, \qquad \sum (f_z x - f_x z) = 0 \tag{3-7}$$

Once again the sum extends over the interaction of all atoms in M with all other atoms in the crystal.

The molecule M belongs to some unit cell which defines the repetitive crystallographic unit composing the crystal. The third condition of static equilibrium requires that the stress acting upon all molecules in a unit cell have a net value of zero

$$\sum \sum f_{xx} = 0, \quad \sum \sum f_{xy} = 0, \quad \sum \sum f_{yy} = 0$$

$$\sum \sum f_{yz} = 0, \quad \sum \sum f_{zz} = 0, \quad \sum \sum f_{xz} = 0 \tag{3-8}$$

where the first sum refers to all atoms not in the unit cell and the second sum to all atoms in the unit cell.

The above equations hold for any crystal. One additional equation is needed to uniquely define the particular crystal being considered. An equation which satisfies this requirement is

$$E_{sub} = \tfrac{1}{2} \sum P(r) \tag{3-9}$$

where E_{sub} is the experimental sublimation energy, corrected for thermal energy which is obtained from translational, rotational, and vibrational heat capacities. Normally the zero-point energy correction may be neglected by assuming that it is the same in the crystalline and gas phases. The factor $\tfrac{1}{2}$ is included since the sum includes each interaction twice when summing over all pairwise interactions in the crystal.

We have a total of thirteen independent equations. The number of unique types of pair interactions which may be determined from these equations is dependent upon the functional form of $P(r)$. To date, the majority of calculations of this type have dealt with nonpolar molecules such as the hydrocarbons. For these molecules electrostatic interactions may be neglected and $P(r)$ can be represented by a Lennard-Jones 6-12 potential function

$$P(r) = (-A/r^6) + (B/r^{12}) \tag{3-10}$$

For polar molecules in which hydrogen bonds may be treated as unique angular independent functions identical to other nonbonded-polar pair interactions $P(r)$ can be represented by

$$P(r) = (-A/r^6) + (B/r^{12}) + (KQ_1 Q_2/\epsilon r) \tag{3-11}$$

where $KQ_1 Q_2/\epsilon r$ is the electrostatic pair interaction energy discussed in Chapter 2. The (A, B) for an O---H hydrogen bond interaction will be different from the (A, B) for an O---H *non*hydrogen bond interaction. If one makes the assumption that the $KQ_1 Q_2/\epsilon r$ term can be adequately calculated from the methods described in Chapter 2, then for both nonpolar and polar molecules we must compute the (A,B) for each unique pair interaction. These restrictions give rise to a set of 13 equations in $2N$ unknowns, where N is the number of unique pair interactions. One will note that had a Buckingham potential function been used for $P(r)$ each unique pair inter- action would be dependent upon three parameters *and* the set of equations would be nonlinear due to the exponential term. Hence there are distinct mathematical advantages to using the Lennard-Jones 6-12 potential function.

At this point in the development of the technique to refine the (CPE) functions a number of different routes may be taken to compute the A and B. The most straightforward means of computing the A and B for a collection of crystal structures is to solve the set of 13 linear equations in $2N$ unknowns by the method of least squares. In Table 3-7 we summarize the sparse results of carrying out these type of calculations to compute potential function coefficients.

TABLE 3-7

Values of A and B in the Lennard-Jones 6-12 Potential Function Computed from the Crystal Structures of Polyethylene, PE, and Polyoxymethylene, POM, Using the Conditions of Static Equilibrium

Source	Temperature	Interaction	A (A^6 kcal/mole)	B ($A^{12} \times 10^{-4}$ kcal/mole)
POM	300°K	C$\cdots$C	404.5	28.95
		C$\cdots$H	136.9	3.712
		C$\cdots$O	418.7	10.13
		H$\cdots$H	16.91	0.1225
		H$\cdots$O	133.7	1.411
		O$\cdots$O	313.8	14.40
PE	300°K	C$\cdots$C	394.2	26.97
		C$\cdots$H	147.5	3.650
		H$\cdots$H	28.21	0.2987

B. Temperature Dependence for Nonbonded Potential Functions

By determining the values of the unit cells and atomic parameters as a function of temperature it is possible to evaluate the effects of thermal vibra- tions upon the nonbonded potential functions. The temperature effects are reflected in a set of temperature-dependent potential coefficients $(A(T), B(T))$ for the Lennard-Jones 6-12 functions. This approach has the capability of relating molecular temperature phenomena to macroscopic temperature phenomena.

Scheraga and co-workers (25) have carried out a series of temperature-dependent crystal calculations on benzene crystals at various temperatures. The unit cell and atomic parameters of benzene, reported by Cox *et al.* (26), were used to generate the crystal space at $-3°C$, and the results of Bacon *et al.* (27) were used at $-55°$ and $-135°C$. The benzene crystal has four molecules in a unit cell with the space group $(D_{2h}^{15})P_{bca}$. C—H bond lengths were 1.09 Å, with the hydrogen atoms lying on the bisector of the CCC bond angles in the plane. This C—H bond length is somewhat larger than that obtained from X-ray data on similar systems at room temperature (28); however, it agrees with the low temperature results of Bacon *et al.* (27) and is only slightly larger than that found by electron diffraction studies of benzene vapor. The heats of sublimation were taken from two sources, because of a considerable discrepancy in the values. Wakayama (29) gives $\Delta H = -9.7$ kcal/mole, while Bondi (30) gives $\Delta H = -10.67$ kcal/mole. The results of the static solutions, and a comparison of calculated and observed lattice constants, are given in Table 3-8.

TABLE 3-8

Lennard-Jones 6-12 Potential Parameters for Benzene Crystals[a]

Temp (°C)	Atom-pair type	E_{sub}	Potential Coefficients		$\langle r_g \rangle^b$	$\langle \sigma_g \rangle^b$	ϵ^b
			$B \times 10^{-4}$ (kcal/A^{12}-mole)	$A \times 10^{-2}$ (kcal/A^6-mole)			
−3	H···H		7.2634	3.1090	2.786	2.481	0.3327
	C···H	−10.68	8.9534	2.8318	2.930	2.610	0.2239
	C···C		227.41	−3.4132	No min	—	—
−55	H···H		5.2997	2.8500	2.682	2.388	0.3832
	C···H	−11.31	6.8303	2.5448	2.851	2.539	0.2370
	C···C		219.86	−2.5122	No min	—	—
−135	H···H		4.2350	2.6780	2.610	2.325	0.4233
	C···H	−12.40	4.3810	2.1295	2.727	2.429	0.2588
	C···C		230.74	−0.6811	No min	—	—

[a] From Momany *et al.* (25).
[b] Distances in angstrom units and energies in kcal/mole.

The Lennard-Jones 6-12 potential function was used to describe intermolecular pair interactions

$$P(r) = (B/r^{12}) - (A/r^6) \tag{3-12}$$

where

$$B = -\epsilon \langle r_g \rangle^{12} = -4\epsilon \langle \sigma_g \rangle^{12}, \qquad A = -2\epsilon \langle r_g \rangle^6 = -4\epsilon \langle \sigma_g \rangle^6$$

$$B/A = \tfrac{1}{2}\langle r_g \rangle^6 = \langle \sigma_g \rangle^6 \tag{3-13}$$

and ϵ and $\langle r_g \rangle$ are defined, respectively, as the depth of the minimum of each potential curve and the average position of the potential over the various energy levels of the crystal as well as the molecule for any temperature.

Several interesting results may be seen in Table 3-8. First, it is obvious that the mean value theorem breaks down, that is, the $\langle r_g \rangle_{\text{CH}}$ value is not the mean of the H---H and C---C values. Second, the calculated and predicted lattice constants are in good agreement. Third, the force constants K_r defined by

$$K_r = \partial^2 P(r)/\partial r^2 \qquad (3\text{-}14)$$

agree with those of Harada and Shimanouchi (31) obtained from Urey–Bradley force calculation. Lastly, the magnitude of the shift of $\langle r_g \rangle_{\text{HH}}$ with temperature (anharmonicity) seems to account for the large number of empirical H---H potentials found in the literature.

C. Refinement of Polymer–Solvent Potential Functions

No actual experiments have been carried out in this area. At best, we can suggest a series of experiments which might make it possible to introduce a temperature dependence into polymer–solvent potential functions coefficients as well as refine the accuracy of these functions. These experiments appear to be applicable to the hydration shell model presented in Chapter 2.

Some polymers chain-fold in a manner which reflects the solvent from which they are crystallized. That is, the extent of polymer–solvent interaction plays a major role in the chain-folding crystallization of polymers. Lindenmeyer (32), among others, has derived a theory of chain-folding crystallization which relates the characteristic chain-fold length L to the various types of interactions which occur in such systems, including the polymer–solvent interactions.

By insisting that the predicted and observed L be equal for a number of different polymers in the same solvent it might be possible to evaluate the characteristic polymer–solvent interaction parameters. By repeating these calculations for a number of different temperatures the polymer–solvent potential functions could be made temperature dependent. Table 1-2 contains a listing of various L for various polymers in various solvents.

D. Refinement of Bond Angle and Length Distortion Potential Functions

The self-consistent field calculations of Lifson and co-workers (33) represent a sophisticated means of determining the characteristic parameters of the various types of distortion potential functions. An example of these types of calculations was presented for some diketopiperazines at the beginning of this chapter.

E. Discussion

In this section we discussed some means of calculating the parameters of the (CPE) functions. The reliability of such calculations is predicated upon being able to pick an accurate functional form for the potential functions. The transferability of the parameters calculated for any one molecular system to other molecules would be a measure, in part, of the correct functional forms of the potential functions.

References

1. S. Karplus and S. Lifson, *Biopolymers* **10**, 1973 (1971).
2. A. Warshel, M. Levitt, and S. Lifson, *J. Mol. Spectrosc.* **33**, 84 (1970).
3. J. A. Schellman and B. E. Nielsen, *in* "Conformation of Biopolymers" (G. N. Ramachandran, ed.), Vol. I. Academic Press, New York, 1967.
4. B. D. Sykes, E. B. Robertson, H. B. Dunford, and D. Konasewich, *Biochemistry* **5**, 697 (1966).
5. T. Miyazawa, *J. Mol. Spectrosc.* **4**, 155 (1960).
6. K. Fukishima, Y. Idegachi, and T. Miyazawa, *Bull. Chem. Soc. Japan* **37**, 349 (1964).
7. L. B. Kier and J. M. George, *J. Med. Chem.* **14**(1), 80 (1971).
8. B. Pullman, J. L. Coubeils, Ph. Couriere, and J. P. Gervois, *J. Med. Chem.* **15**(1), 17 (1972).
9. L. B. Kier and E. B. Truitt, Jr., *J. Pharmacol. Exp. Ther.* **174**(1), 94 (1970).
10. G. A. Neville, R. Deslauriers, B. J. Blackburn, and I. C. P. Smith, *J. Med. Chem.* **14**, 717 (1971).
11. H. J. R. Weintraub and A. J. Hopfinger, *J. Theor. Biol.* in press.
12. R. Hoffmann and A. Imamura, *Biopolymers* **9**, 207 (1969).
13. R. A. Scott and H. A. Scheraga, *J. Chem. Phys.* **45**, 2091 (1966).
14. P. Urnes and P. Doty, *Advan. Protein Chem.* **16**, 401 (1961).
15. B. Pullman, *in* "Aspects de la Chimie Quantique Contemporaine" (R. Daudel and A. Pullman, eds.). Colloque International du CNRS–Paris (1971).
16a. A. Yonath and W. Traub, *J. Mol. Biol.* **43**, 461 (1969).
16b. IUPAC-IUB Commission on Biochemical Nomenclature, *Biochemistry* **9**, 3471 (1970).
16c. J. T. Edsall, P. J. Flory, J. C. Kendrew, A. M. Liquori, G. Nemethy, G. N. Ramachandran, and H. A. Scheraga, *Biopolymers* **4**, 121 (1966).
17. R. B. Corey and L. Pauling, *Proc. Roy. Soc.* (*London*) **B141,** 10 (1953).
18. V. Sasisekharan, *Acta Crystallogr.* **12**, 897 (1959).
19. R. Balasubramanian, A. V. Lakshminarayanan, M. N. Sabesan, G. Tegoni, K. Venkatesan, and G. N. Ramachandran, *Int. J. Protein Res.* **2**, 303 (1970).
20. G. Del Re, B. Pullman and T. Yonezawa, *Biochim. Biophys. Acta* **75**, 153 (1963).

21. D. Poland and H. A. Scheraga, *Biochemistry* **6**, 3791 (1967).
22. T. Ooi, R. A. Scott, G. Vanderkooi, and H. A. Scheraga, *J. Chem. Phys.* **46**, 4410 (1967).
23. C. H. Bamford, L. Brown, A. Elliott, W. E. Hanby, and I. F. Trotter, *Nature (London)* **171**, 1149 (1953).
24. F. H. C. Crick and A. Rich, *Nature (London)* **176**, 780 (1955).
25. F. A. Momany, G. Vanderkooi, and H. A. Scheraga, *Proc. Nat. Acad. Sci. U.S.* **61**, 429 (1968).
26. E. G. Cox, D. W. J. Cruickshank, and J. A. S. Smith, *Proc. Roy. Soc. (London)* **A247**, 1 (1958).
27. G. E. Bacon, N. A. Curry, and S. A. Wilson, *Proc. Roy. Soc. (London)* **A279**, 98 (1964).
28. W. R. Busing and H. A. Levy, *Acta Crystallogr.* **17**, 142 (1964).
29. N. Wakayama and M. Inokochi, *Bull. Chem. Soc. Japan* **40**, 2267 (1967).
30. A. Bondi, *J. Chem. Eng. Data* **8**, 371 (1963).
31. I. Harada and T. Shimanouchi, *J. Chem. Phys.* **44**, 2016 (1966); **46**, 2708 (1967).
32. P. Lindenmeyer, *J. Polym. Sci. Part C* **20**, 145 (1967).
33. S. Lifson and A. Warshel, *J. Chem. Phys.* **49**, 5116 (1968).

Chapter 4 | Conformational Transitions in Macromolecules

I. Introduction

Some macromolecules adopt different conformations under different thermodynamic and/or chemical conditions. The stable spatial organization of a macromolecule for one given set of environmental conditions may be unstable in other environments. This suggests that macromolecules undergo *conformational transitions* which are dependent upon external influences. The capacity of macromolecules to undergo such conformational transitions sets each molecule apart as a unique thermodynamic entity.

The nature of conformational transitions has been, and is, of interest from both the thermodynamic and kinetic points of view. In many cases the usefulness of polymeric materials is dependent upon their conformational-transition properties. If collagen molecules, which are discussed in this chapter, were to undergo an order–disorder transition 15°C lower than they actually do, life, at least as we know it, could not exist. People who live near the equator would literally melt away. Rubber would not be very useful in most climates if the individual molecules underwent conformational transitions near room temperature which resulted in bulk crystallization of the material below room temperature.

To the macromolecular scientist, conformational transitions provide a means of studying the molecular properties of polymeric materials. Conformational transitions directly reflect the nature of the molecular forces at play. The allowed molecular motions in a macromolecule are related to the kinetic nature of a transition. Unfortunately, macromolecular transitions occur very rapidly. As a result, it has not proven possible to make a "motion picture" of a macromolecule as it goes from one stable structural state to a second stable state. Another way of saying this is that the intermediate states realized during a conformational transition are very unstable and extremely short-lived.

In order to understand what happens during a conformational transition thus requires designing a model which is consistent with experimental findings. While this approach can be considered valid only in so far as there is agreement with experiment, it does provide a basis for interpreting phenomena. A theory which has had overwhelming experimental support from a large number of different types of macromolecular transitions is the *cooperative model*. This concept, cooperativity, is the topic of the next section.

162

II. The Cooperative Phenomenon in Macromolecules

A. General Description of a Cooperative Process

All of the macromolecular transitions presented in this chapter can be considered as examples of a cooperative phenomenon. Any large change in the equilibrium molecular properties resulting from small changes in the environmental conditions is usually considered to be a cooperative phenomenon. Certain characteristics of cooperative processes can be understood in similar theoretical terms. In this section a cursory review of the theory of cooperative processes is presented. This section of the chapter is a condensation of an extensive and elegant review of the subject presented by Applequist (1).

The systems discussed here can be classified as one-dimensional, in the sense that the molecular structures are periodic in one topological dimension. While cooperative phenomena in three-dimensional systems are usually in the form of phase transitions, this is not generally the case for one dimension. Thus it is necessary to characterize a one-dimensional cooperative phenomenon in some manner that will distinguish it from ordinary chemical equilibria. This can be done in terms of a "cooperative effect," which is defined here as a tendency for residues in the same state to group together, or to "aggregate" in linear sequences. The sharpness of the transition involved is directly related to this tendency.

The models discussed in this section consist of a single chain of N residues or of two chains, each with N residues, associated in a side-by-side complex. Each residue is regarded as having two accessible states, A and B. The nature of these states for some particular systems are specified in Table 4-1. The

TABLE 4-1

Specification of Residue States for Various Systems[a]

System	State A	State B
α-Helical polypeptide	Carbonyl not hydrogen bonded	Hydrogen bonded carbonyl
One-stranded polynucleotide	No stacking interaction with (arbitrary) specified neighbor	Stacking interaction with specified neighbor
Poly-L-proline	Amide bond in trans conformation	Amide bond in cis conformation
Two-stranded polynucleotide	Base not hydrogen bonded	Base hydrogen bonded to another base in opposite strand

[a] From Applequist (1).

object of the equilibrium theory is to obtain information about the manner in which residues are distributed between the two states, expressed in terms of certain basic parameters. This information is obtained from the configurational partition function Z_N (2a) for a molecule of chain length N.

A basic parameter that is appropriate in all models is the stability constant s, which is the equilibrium constant for the conversion of a residue from state A to state B at the end of an existing sequence of one or more residues in state B. The magnitude of the stability constant is assumed to be independent of any cooperative effects, so that it is independent of chain length or the length of the B sequence to which a new B state is added. The cooperative effects are conveyed by the interruption constant ω_j, which is the equilibrium constant for the conversion of a molecular state containing a single sequence of n B states (in each strand for two-stranded cases) to a state containing two B sequences totaling n B states, separated by j repeating units (residues or residue pairs, depending on the model) in A states. The relative populations of all possible sequences of A's and B's may be readily deduced in terms of s and the ω_j's. The sum of these relative populations is Z_N.

Table 4-2 summarizes the formulas for four types of long-chain cooperative phenomena.

A quantity that can often be determined at least approximately from experiment is the fraction of residues in state B, designated f. The theory [see, for example, Zimm (3)] yields this quantity as

$$f = (1/N) \left(\partial \ln Z_N / \partial \ln s \right)_\tau \tag{4-1}$$

where τ represents variables other than s appearing in the partition function.

We may also characterize cooperative effects by means of the average length $\langle l \rangle$ of a sequence of B states, given by

$$\langle l \rangle = -\left(\partial \ln \tau / \partial \ln s \right)_\rho \tag{4-2}$$

If ω_j is proportional to τ, as is true for all cases discussed here, then it follows that

$$\langle l \rangle = 1/(1 - s\rho) \tag{4-3}$$

[see also Crothers and Zimm (4)].

A second measure of cooperativity is the equilibrium fluctuation ϕ in the number of B states in a chain, defined by $\phi = n - \langle n \rangle$, where the brackets indicate the average value. $\langle \phi^2 \rangle$ is related to the sharpness of the transition by

$$\langle \phi^2 \rangle / N = \left(\partial f / \partial \ln s \right)_\tau \tag{4-4}$$

which is exact for all N. The quantities $\langle l \rangle$, $\langle \phi^2 \rangle / N$, and $(\partial \alpha / \partial \ln s)$ are closely related characteristics of a cooperative system, since all may be increased by the same basic causes.

TABLE 4-2

Summary of Formulas for Long-Chain Cooperative Phenomena[a]

Model	ω_j	δ	f	$\langle l \rangle$ at inflection[b]	$(\partial f/\partial \ln s)_\tau$ at inflection[b]
Ising model	σ	1	$\dfrac{1}{2} + \dfrac{s-1}{2[(1-s)^2 + 4\sigma s]^{1/2}}$	$1 + \sigma^{-1/2}$	$\dfrac{1}{4\sigma^{1/2}}$
Two-strand polymer: matching case $a=2$	$\dfrac{b_1}{(j+1)^2}$	1	$\dfrac{1}{s\rho - b_1 s[\rho + \ln(1-\rho)]}$	$1 + \dfrac{1}{0.6449 b_1}$	∞
Two-strand polymer: mismatching case $a=2$	$\dfrac{b_2(j+1)}{(j+2)^2}$	$\dfrac{1}{2}$	$\dfrac{1}{s\rho + (b_2 s/2)\left[(\rho^{1/2}/1 - \rho^{1/2}) - \rho/2 + \ln(1 - \rho^{1/2})\right]}$	No simple expression found	$0.38 + \dfrac{0.1447^{\,c}}{b_2}$ $(b_2 \leqslant 0.3)$
Two-strand polymer: mismatching case $a=3$	$\dfrac{b_3(j+1)}{(j+2)^3}$	$\dfrac{1}{2}$	$\dfrac{1}{s\rho - (b_3 s/2)\left[(\rho/4) + \ln(1 - \rho^{1/2}) + \text{Li}_2(\rho^{1/2})\right]}$	$1 + \dfrac{1}{0.3179 b_3}$	∞

[a] $\text{Li}_a(x)$ is the polylogarithm function defined by $\text{Li}_a(x) = \sum_{i=1}^\infty x^i/i^a$, for $|x| \leqslant 1$ [Lewin (2b)].
[b] Where a phase transition occurs, the inflection point is the transition point.
[c] This is the maximum value for the partial derivative.

The parameters s is determined by the model under consideration. The value of ρ can be calculated by solving the convergent power series

$$s\left(\rho + \sum_{j=1}^{\infty} \omega_j \rho^{(\delta_j+1)}\right) = 1 \qquad (\delta = \tfrac{1}{2}, 1) \qquad (4\text{-}5)$$

Next we will devote a few paragraphs to discussing the nature of each of the four types of cooperative models presented in Table 4-2.

B. The Ising Model

The one-dimensional Ising model (5) is characterized by the fact that interactions between nearest neighbors only are included, with the result that a single interruption constant σ applies to interruptions of all lengths. A one-stranded polynucleotide which is able to form a helix by attractive interactions between nearest neighbor bases is an excellent example of the Ising problem, since it is reasonable in this case to neglect all but nearest neighbor interactions. The α-helical polypeptides are stabilized mainly by further neighbor interactions, and it is less obvious that the Ising model should apply. Refinements of the Ising model to take into account the effects peculiar to the α-helix (6–8) have led to the conclusion that the simple Ising model is a good approximation as long as σ is much less than unity. The reason is that those molecular states that are strongly affected by the further neighbor nature of the interactions are relatively unpopulated when σ is small. The Ising model approximation is convenient because the important expressions are relatively simple. The model will be adopted here for all of the one-stranded polymers. The transition curves are characterized by the fact that $f = \tfrac{1}{2}$ at $s = 1$ for all σ.

The noncooperative system is a special case of the Ising model with $\sigma = 1$. From the formulas in Table 4-2 it is seen that $\langle l \rangle = 2$ and $\langle \phi^2 \rangle / N = 0.25$ at the midpoint of the transition for this case. These values serve as a point of reference in considering cooperative systems. The "anticooperative" systems are examples of the Ising model with $\sigma > 1$.

1. Two-Stranded Model, Matched Case

In the two-stranded models, state B in one strand is bonded to a state B in the opposite strand. Thus an interruption between B sequences consists of a closed loop of residues in state A. The matching case (9, 10) is characterized by the fact that a loop of size j consists of j A-states in each strand. The interruption constant ω_j must contain as a factor the probability of ring closure, which we take to be $\mu/(j+l)^a$ where μ and a are constants. The value $a = 2$ was found by Wall *et al.* (11), and used by Applequist (1) in some sample calculations. (This choice departs from the usual random-flight approximation, in which $a = \frac{3}{2}$) (12). The parameter b, which is given various subscripts in Table 4-2 to distinguish it for the various models, is related to ΔF_{st}, the molar free energy of stacking of a pair of B states on an adjacent pair of B states, by

$$b = \mu \exp(\Delta F_{st}/RT) \tag{4-6}$$

where R is the gas constant and T is the absolute temperature.

2. Two-Stranded Model, Mismatched Case

The mismatching case is similar to the preceding case, except that a loop of size j is understood to consist of a total of j residues in A states distributed in arbitrary manner between the two strands, all such distributions being regarded as equally probable. This case was first treated by Hill (9), and has been further discussed by Steiner and Beers (13) and by Crothers and Zimm (4). The form of ω_j is similar to that for the matching case with the additional factor $j + 1$, which is the number of ways of forming the specified interruption.

No phase transition occurs for $a = 2$. The qualitative behavior is similar to that of the Ising model, with transition midpoints occurring in the vicinity of $s = 1$.

A phase transition occurs for $a > 2$, and is of second order for $a = 3$, being very similar to the transition described for the matching case. The critical value of s is then

$$s_t = (1 + 0.3179b_3)^{-1} \tag{4-7}$$

Formulas for the cases $a = 2$ and 3 are given in Table 4-2 to illustrate the significant differences. Thus far we have dealt with long chains which could effectively be treated as infinite. We now discuss the more difficult problem of finite chains.

Several more or less exact treatments of finite chains are available for various models (6, 7, 14, 15). Since the formulas are generally too complicated for routine use, the discussion here will be restricted to certain limiting forms which provide insight into the behavior of short chains, and which may be useful in some quantitative applications.

For sufficiently large cooperative effects, one may simplify the problem by assuming that there is at most one B sequence per chain. In addition to the stability constant, one then needs only an equilibrium constant for the formation of the first B state in a chain to describe the system.

3. One-Stranded Model, Finite Chain

The equilibrium constant for the formation of the first B state is just σs, where σ has the same significance as before. Expressions for f under the above simplifying assumption have been given by Schellman (16) and by Zimm and Bragg (6). To simplify further, we note that for $\sigma \ll 1$, we must have $s \gg 1$ in order to achieve appreciable formation of B states. Under these conditions the relative population σs^N of chains entirely in state B is much greater than that of all of the partially B chains, and one may neglect the latter. Thus the equilibrium approaches the all-or-none case. This approximation breaks down as the chains become long, due to the increasing multiplicity of ways of forming partially B molecules. Some useful formulas are given in Table 4-3. These are increasingly valid as σ and N decrease. The stability constant s_m at the midpoint of the transition serves as a fixed point in studying experimental data.

TABLE 4-3

Formulas for Short-Chain Cooperative Phenomena[a]

	One-strand model $(\sigma \ll 1)$	Two-strand models (infinite dilution)
f	$\dfrac{\sigma s^N}{1 + \sigma s^N}$	$\dfrac{1 + 4\gamma s^N - (1 + 8\gamma s^N)^{1/2}}{4\gamma s^N}$
$\left(\dfrac{\partial f}{\partial \ln s}\right)_\tau$ (max)	$\dfrac{N}{4}$ [b]	$\dfrac{N}{6}$ [c]
s_m	$\sigma^{-1/N}$	$\gamma^{-1/N}$
$\dfrac{1}{T_m}$	$\dfrac{1}{T_c} + \dfrac{R \ln \sigma}{N \Delta H_s}$	$\dfrac{1}{T_c} + \dfrac{R \ln \gamma}{N \Delta H_s}$

[a] ΔH_s is the enthalpy change per mole of residues when an A residue is converted to a B residue at the end of a sequence of B residues. T_m is the temperature for which $s = 1$. From Applequist (1).

[b] This is the maximum possible value.

[c] Valid for $f = \frac{1}{2}$.

4. Two-Stranded Model, Finite Chain

The equilibrium constant (in terms of molar concentrations) for the formation of the first pair of B states between two strands is βs, where β has units of mole^{-1}. Since dilution of the system favors dissociation of the strands, a large value of s will be required at very low concentrations to achieve appreciable association of strands. Thus, for the reason given in the one-stranded case, the residues in associated strands will be almost entirely in state B. The formulas appear in Table 4-3. We define $\gamma = \beta c$, where c is the total molar concentration of strands if the strands are identical, or c is the concentration of each type of strand if the two differ and are present in equal amounts. By this convention, βs is the equilibrium constant for formation of a species with a symmetry number of two. If the two strands differ, and if each has a symmetry number of unity, then all species have symmetry number unity, and the true equilibrium constant for formation of the first pair of B states is $2\beta s$.

ΔH_s for this model is the enthalpy change per mole of residue pairs when a pair of A residues is converted to a pair of B residues at the end of a B sequence.

The formulas of Table 4-3 are increasingly valid as γ and N decrease. Since it is possible to vary the concentration, it is feasible to approach this limit experimentally for any two-stranded system, provided the strands are not too long.

It is seen in both the one- and two-stranded models that a strong dependence of the equilibrium on chain length is expected. In the one-stranded case, this is the direct result of cooperative effects, since no chain length dependence occurs in the noncooperative case. For the highly dilute two-stranded case, however, the effect is the direct result of the chain length dependence of the association equilibrium constant, since this quantity reduces to βs^N in the transition region, regardless of the magnitude of the cooperative effects. Thus an observed dependence of the transition curve on chain length is an indicator of cooperative effects only if the equilibrium is not strongly influenced by a strand-dissociation process.

We now turn our attention to the detailed discussion of several macromolecular transitions, paying particular attention as to how conformational potential energy calculations are helpful in characterizing these phenomena. In particular we present some new results which modify our ideas concerning the helix–coil transition of poly-α-amino acids with ionic side chains as a function of pH. We conclude the chapter with an Ising model to describe protein denaturation. Appendix II of this book presents an alternative method to describing cooperative phenomena. This is the theory of detailed balancing.

III. The α-Helix to Random Chain Transition

A. Introduction

As protein chemists were able to polymerize amino acids to form homopolypeptide chains it became evident that some of these biopolymers could undergo conformational changes in solution. In particular, homopolypeptides which have ionizable side chains were found to pass from one particular conformational state to some totally different conformational state as the pH of the polymer solution was varied. Further, the transition between the two states occurred almost instantaneously for some particular pH. This suggests a high degree of cooperativity in the transition process. Thus, a "zipper" model has become a popular way of viewing this transition. When the zipper is closed the biopolymer is in one conformational state. At some specific pH the zipper is rapidly opened, and the homopolypeptide is in the second conformational state. That the transition involves two conformational states and is cooperative is discernible from a plot of, say, the intrinsic viscosity of the polymer solution as a function of pH. Such a plot for poly(L-glutamic acid) (PGA) is shown in Fig. 4.1.† Each of the two plateaus on the viscosity curve correspond to stable conformational states. The rapid change in the viscosity with a small change in pH near the middle of the horizontal axis indicates the

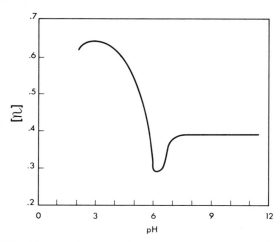

Fig. 4-1. A typical plot of viscosity, $[\eta]$, versus pH for a solution containing poly(L-glutamic acid) (PGA).

high cooperativity. If more than two conformational states (at least long-lived conformational states) were present one would expect to see other plateaus on the viscosity curve. As we shall see later, there is evidence that a right-handed to left-handed α-helical transition precedes the helix–coil transition in some polypeptides. The absence of a plateau in Fig. 4-1 corresponding to this intermediate left-handed α-helical state indicates that (a) poly(L-glutamic acid) does not undergo this intermediate transition, or (b) the intrinsic viscosity of the right- and left-handed α-helices of PGA are identical, or (c) the left-handed α-helical conformational state is so short-lived that no discernible plateau on the viscosity curve is apparent.

The helix–coil transition has also been observed to take place for some polypeptides, whose side chains are not necessarily ionizable, by simply

† See Volkenstein (17) for a discussion of this transition.

raising the temperature of the polymer solution (18).† For selective poly-peptides the helix–coil transition can also be induced by varying the polarity of the solvent medium. Transitions which are induced by heat or solvent polarity do not appear to be as highly cooperative as the ionization induced transitions. The transition curves resulting from heating or changing the polarity of the solvent are usually much broader than the corresponding ionization induced transition curves.‡

Since the helix–coil transition is, at least in the first approximation, a two-state process of high cooperativity a linear Ising lattice may be used to describe the thermodynamics of the transition. The difficulty in relating theoretical and experimental findings for the helix–coil transition is in the inability to precisely define the nature of the two states. The properties of the "helical" and "coil" states will be the topic of the next section.

B. The Conformational States

Many of the homopolypeptides that have been found to undergo a helix–coil transition have been found to be in the α-helical conformation from X-ray investigations (23). The structural situation for these biopolymers when dispersed in a solvent is known with much less certainty, owing to the fact that such knowledge must rely upon the indirect evidence which comes from solution properties such as the viscosity, light scattering, and optical rotatory dispersion (ORD). Nonetheless it is now generally agreed that, in a number of cases, the conformation of the homopolypeptides when dispersed in weakly interacting solvents or in aqueous solution of appropriate acidity is still of the helical type. Hence the rigid α-helix (normally the right-handed form) has always been considered as one of the two conformations involved in the helix–coil transition.

Macroscopic measurements such as viscosity and light-scattering experiments indicate that the other conformational state involved in the helix–coil transition has the gross properties one would expect if the polypeptide chain were folded in a *Monte Carlo* fashion. Hence this conformation came to be known as the "random coil." Without much concern, protein chemists have accepted the contention that the second conformation involved in the helix–coil transition is this random coil; a structure devoid of order at all levels. At this point we will differentiate between the coil (in contrast to random coil) resulting from the *ionization induced* helix–coil transition (for polypeptides with ionizable side chains), and the coil resulting from the *temperature induced*

---

† Go *et al.* (19, 20) gives theory and experiment, and see Urnes and Doty (21) for a classic discussion.

‡ Compare Fig. 3 in Go *et al.* (22) to Fig. 4-1 of the text.

helix–coil transition. For a lack of data and understanding, we will not discuss the coil resulting from *solvent-polarity induced* helix–coil transitions.

1. Ionization Induced Helix–Coil Transitions

The ORD/CD spectra of the so-called "random coil" conformation of PGA and poly-L-lysine (PL) when the polymer molecules are in the charged form are both remarkably similar to the ORD/CD spectra of an ordered

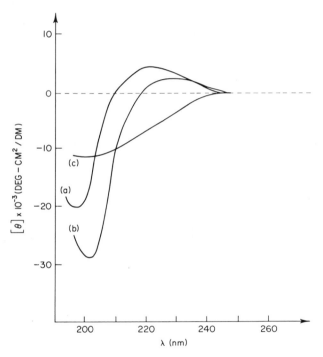

Fig. 4-2. CD spectra of (a) poly(L-glutamic acid) for pH > 8, (b) *trans*-poly-L-proline, (c) poly(L-glutamic acid) for pH < 5 in 4.5 *M* LiClO₄. From Hiltner (24).

left-handed 3_1-helix (24). Until recently, this similarity was not considered important, and the observed spectra were taken to be those of true-classical "random coils." Tiffany and Krimm (25) decided to see if they could discover the reason for the similarity between these spectra. After an intensive study of the helix–coil transition of PGA they concluded that the classical "random coil" for the charged form of PGA is, in fact, not random in its secondary structure. Instead the preferred conformation of the PGA chain is an ordered, near left-handed 3_1-helix. They further concluded that the classical "random coil" is realized only when counter ions are randomly bound (more likely a random kinetic equilibrium) to the peptide backbone and side chain groups. The CD spectra of (a) charged PGA, (b) *trans*-poly-L-proline, and (c) PGA for pH < 5 in 4.5 M LiClO$_4$ are shown in Fig. 4-2. As a side note, there appears to be a preferential type of binding between polypeptides and ions. For example, if calcium ions are bound to the charged form of PGA the

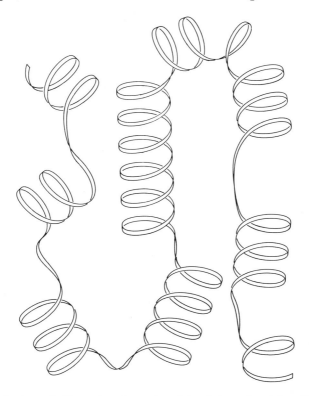

Fig. 4-3. A schematic illustration of a portion of the chain conformation of high molecular weight poly(L-glutamic acid) when in the ionized state. The helices refer to the ordered near 3$_1$-helix found by conformational energy calculations.

complex precipitates out of solution (26). Very important information might be had by investigation of polypeptide–ion complexes.

Tiffany and Krimm were able to explain the somewhat contradictory findings of the viscosity and light scattering measurements which support the classical "random coil," and the ORD/CD data which suggests an ordered structure in the following way. Since viscosity and light-scattering measurements reflect the bulk shape of the molecule the gross macroscopic structure

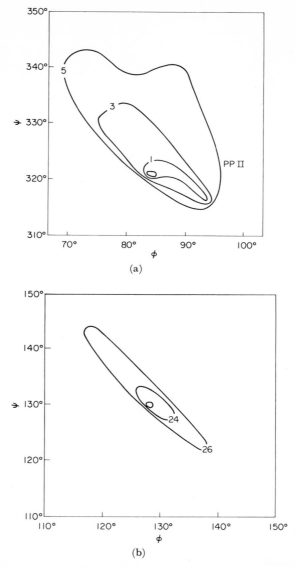

(a)

(b)

Fig. 4-4. (a) Relative energy contours in kcal/mole-residue for ionized PGA in aqueous solution, extended helix conformation. Relative energy of minimum at (84°, 321°) is 0.0 kcal/mole-residue. Position of PP II helix is indicated.

(b) Relative energy contours in kcal/mole-residue for ionized PGA in aqueous solution, α-helix conformation. Relative energy of minimum at (128°, 130°) is 21.6 kcal/mole-residue.

of the charged form of the polypeptide should be the random coil first postulated. ORD/CD measurements are, on the other hand, sensitive to ordered peptide segments. Therefore, the charged form of the polypeptide must contain regions of ordered structure. Hence the tertiary structure of the charged form of PGA is composed of segments of near 3_1-helices which are randomly oriented relative to one another. This type of molecular organization is shown in Fig. 4-3. Presumably the topological sequence of side chain ionization prevents the realization of a completely ordered structure.

When ions are brought into the vicinity of the charged-coil spheroid there is an initial interaction between ion and polypeptide which destroys a portion of one of the ordered sequences. These ion–polypeptide interactions proceed to occur until equilibrium is reached with each interaction destroying order. The result is the classical "random coil."

Hiltner *et al.* (27) performed the complete conformational analysis of PGA for (a) the neutral form, (b) the charged form, and (c) models for the ion bound form. Figure 4-4 contains the conformational maps of ionized PGA in aqueous solution, and Table 4-4 lists the relative stabilities of the various conformations in terms of the conformational free energy of each state. The findings of this investigation support the contentions of Tiffany and Krimm. These calculations predict that the most stable form of charged PGA is an ordered helix (left-handed in sense) having 2.4 residues per turn and a rise per residue along the helix axis of 3.2 Å. This helix has (ϕ, ψ) coordinates

TABLE 4-4[a]

Conformation	Environment	P (kcal/mole-residue)
Conformational energies of neutral PGA		
R.H. α-helix	*vacuo*	+14.0
	H_2O	+1.6
Extended helix	*vacuo*	+15.8
	H_2O	0.0

Conformation	Environment	P (kcal/mole-residue)	S (eu/mole-residue)	$-TS$[b] (kcal/mole-residue)
Conformational energies and entropies of ionized PGA				
R.H. α-helix	*vacuo*	+46.8	5.6	−1.7
	H_2O	+21.6	6.7	−2.0
Extended helix	*vacuo*	+31.0	7.3	−2.2
	H_2O	0.0	6.9	−2.1

[a] Reprinted from Hiltner *et al.* (27), *J. Amer. Chem. Soc.* **94**, 4324 (1972). Copyright 1972 by the American Chemical Society. Reprinted by permission of the copyright owner.

[b] $T = 300°K$.

(84°, 321°) and, as is obvious, is very similar to a left-handed 3_1-helix. The stable charged-form helix of PL may be of a slightly different conformation due to a different side chain geometry than that of PGA. Presumably the similarity of the charged-form helix and the left-handed 3_1-helix is responsible for the similarity in their ORD/CD spectra.

2. *Temperature Induced Helix–Coil Transitions*

The helix–coil transition resulting from varying the temperature of a solution containing polypeptides (including polypeptides with ionizable side chains which are in the neutral form) is probably different from the pH induced transition. The evidence for this is:

(1)   A plot of the viscosity versus temperature for a polymer solution containing a polypeptide which undergoes a thermal helix–coil transition indicates a broad range in temperature over which the transition occurs (28). This might be indicative of a less cooperative transition than the pH induced transition.

(2)   The shape of the ORD/CD spectra of the temperature induced coil are intermediate between the spectra of PGA in the charged form with and without lithium ions bound (29, 24).

It is possible to explain these observations on the molecular level in terms of changes in the stereochemical parameters as a function of temperature. When the polypeptide is in the ordered helical conformation, it is tightly packed. As the temperature of the solution rises the mean vibrational displacement of each atom becomes larger. This is equivalent, roughly speaking, to enlarging the van der Waals radii of the atoms. Actually, the shape of the atoms can no longer be treated as spheres, but rather, ellipsoids. The increased size of the van der Waals radii forces the tightly packed α-helix apart. Two possible mechanisms which might explain the kinetics of this thermally induced transition are (a) residues, or groups of residues, randomly pop out of the helical state during heating, or (b) all residues *simultaneously* reach a critical metastable state upon heating, and then, due to random fluctuations, the residues along the chain individually adopt varying conformations. For the second explanation to be correct the broadness of the transition curve must result from the out-of-phase nature of the random-fluctuation relaxations. This is equivalent to a large number of springs, all of which are coupled in some way, being subjected to random extensions and compressions and then allowed to go to equilibrium. For the first explanation to be plausible, the broadness of the transition curve is most likely due to the random, temperature-dependent "flipping" of the helical states. For both explanations, the shapes of the ORD/CD curves are probably due to the formation of segments, of varying length, of ordered structure. The chain conformations in these ordered regions very likely have (ϕ, ψ) coordinates in the upper left-hand corner of a general (ϕ, ψ) plot. This is the region where the majority of conformational freedom is found. The left-handed 3_1-helix has its (ϕ, ψ) coordinates in this region also. On a statistical basis, the majority of the (ϕ, ψ) coordinates of random-coil residues would also be found in the upper left-hand corner of the map. Hence, it is reasonable to expect the thermal coil (the tertiary structure of the polypeptide when heated) ORD/CD spectra to be intermediate to the spectra of a charged-coil helix with and without ions bound. Of course this conjecture is predicated on the assumption that similar (ϕ, ψ) coordinates implies similar conformations and similar conformations implies similar ORD/CD spectra. Figure 5-4b contains a CD dipeptide map. The sterically allowed conformations in the upper left-hand corner of the map give rise to mainly two types of CD spectra which are quite similar.

C. Application of the Linear Ising Lattice to the α-Helix–Random Coil Transition

1. Outline of the Derivation of the Transition Equations

Although the final equations describing the properties of a helix–coil transition have already been postulated in the first section of this chapter using the Ising model, it is instructive to outline the derivation of these equations as given by Applequist (30). First we define the internal partition function, Q, as

$$Q = \sum_k (\text{pop})_k \tag{4-8}$$

where $(\text{pop})_k$ is a function of the temperature and the internal energy of the kth molecular state giving the relative population of the state. Using the Zimm–Bragg (6) formulation,

$$(\text{pop})_{n,v} = g(n,v)\, s^n \sigma^v \tag{4-9}$$

where s has classically been treated as the equilibrium constant for the addition of one more hydrogen bond to the end of an already existing sequence of hydrogen bonds in an α-helix. In general, the length of this sequence is n. The basic parameter σ has been classically defined as the equilibrium constant for the formulation of an interruption in a sequence of hydrogen bonds by a process which maintains a constant number of such hydrogen bonds. The number of such hydrogen-bonded sequences is v. In view of the previous discussion concerning the nature of the conformational states these definitions should be amended to:

(1) s is the *average* equilibrium constant corresponding to the formation of one more α-helical unit from either an ordered near 3_1-helical unit, denoted by ($\sim3_1$), or from the classical unordered unit, denoted by (R), to the end of an already existing sequence of α-helical units. The average is dependent upon the relative numbers of ($\sim3_1$) to α-helix and (R) to α-helix residue linkages.

(2) σ is the *average* equilibrium constant for the formation of an interruption in an α-helical segment by a process which maintains a constant number of α-helical units. The average is over the relative number of ($\sim3_1$) and (R) residue units formed in the interruptions.

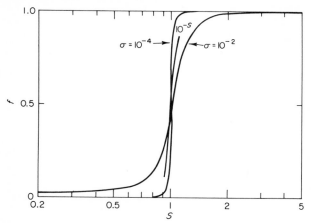

Fig. 4-5. A plot of s versus f, the average fraction of helix segments, for various values of σ. From "Structural Chemistry and Molecular Biology," edited by Alexander Rich and Norman Davidson. W. H. Freeman and Company. Copyright © 1968.

The function $g(n,v)$ gives the number of ways a given polypeptide chain can achieve n α-helical units and v interrupted sequences of α-helical segments. Upon evaluation of the partition function [see Applequist (8)] it is possible to obtain an expression, given in Table 4-2, for the average fraction, f, of helix segments in a long chain as a function of s and σ.

$$f = \frac{1}{2} + \frac{(s-1)}{2[(1-s)^2 + 4s\sigma]^{1/2}} \tag{4-10}$$

Figure 4-5 is a plot of f versus s for various values of σ.[†] From these plots it is clear that the parameter σ determines the "sharpness" of the transition.

† Taken from Giacometti (31).

Further, the parameter σ is characteristic of the particular macromolecule undergoing the helix–coil transition. Zimm and Bragg (6) assumed that the value of σ should be constant for all polypeptides. That is, they postulated that the various properties of the polypeptide side chains would not disturb the transition. Available data contradicts this assumption, as we shall see shortly. In order to characterize the helix–coil transition for a given macromolecule the value of σ must be available. Different experimental schemes are required to calculate σ, depending upon the manner in which the transition is induced.

2. Temperature Induced Transitions

For helix–coil transitions resulting from the raising or lowering of the temperature of the polymer solution the value of σ may be computed from the relationship

$$\sigma = \frac{(\Delta H)^4}{16 C_{max}^2 \, KT_{mid}^4} \tag{4-11}$$

where C_{max} is the maximum value of the heat capacity of the solution, ΔH the heat of the transition, and T_{mid} the transition temperature. The value of C_{max} may be determined from the maximum height on a heat capacity versus temperature curve for the polymer solution. The value of ΔH is the area under the heat capacity curve. T_{mid} is normally chosen to be the midpoint in the transition curve of viscosity versus temperature.

By this method values of $\Delta H = 950 \pm 30$ cal/mole and $\sigma = 1.0 \times 10^{-4}$ for poly-γ-benzyl-L-glutamate (PBG) have been determined in a mixture of dichloroacetic acid with dichloroethylene (DCA + DCE; 81 + 19) by Ackerman and Ruterjans (32). These values have been obtained by an extrapolation of calorimetric data to zero concentration of the polypeptide.

3. pH Induced Transitions

Nagasawa and Holtzer (33) have shown that

$$\int_{pH_0}^{pH} (\alpha_h - \alpha_c) \, d(pH) = 0.43 \frac{\Delta F}{KT} + \sigma^{1/2} \log \left\{ \frac{f}{1-f} \right\} \tag{4-12}$$

where α_h and α_c are the degrees of ionization of the helical and coiled polypeptide corresponding to the given pH value (these values can be obtained by extrapolation of titration curves of the helical and coiled polypeptides to the given pH) (34), pH_0 is the value of the pH corresponding to $\alpha_h = \alpha_c$ and ΔF

is the free energy difference between the helical and the statistically coiled state of the uncharged chain. Thus a plot of $\int_{\text{pH}_0}^{\text{pH}} (\alpha_h - \alpha_c) \, d(\text{pH})$ versus $\log\{f/(1-f)\}$, which can be determined from monitoring viscosity or optical dispersion as a function of pH, should be a straight line with slope $\sigma^{1/2}$ and intercept $0.43(\Delta F/KT)$.

This method has been applied to the evaluation of σ for PGA in aqueous solutions of NaCl. Various workers [Nagasawa and Holtzer (33); Nekrasova et al. (35); Snipp et al. (36)] obtained results that are relatively close to one another and are grouped near $\sigma = 2.5 \times 10^{-3}$. The shortcoming of this method is the need for extrapolating titration curves, which always leads to some uncertainty.

4. Solvent–Composition Induced Transitions

The statistical size and shape of a macromolecule is, in part, a function of the side chains of the macromolecule *and* the solvent in which the macro-molecule is located. In other words, the size and shape of the macromolecules is, to varying degrees, dependent upon the polymer–solvent interactions. Therefore, in solvent–composition induced helix–coil transitions the para-meters f and σ should be related to the size and shape of the macromolecule for both the helical and coil conformations. For polypeptides Birshtein and Ptitsyn (37), have proposed the following relationship between the geometry of the polymer and the values of f and σ:

$$\frac{\overline{h^2}}{\overline{h^2_{\text{coil}}}} = 1 - f + \frac{2L_h^2}{\sigma^{1/2} L_c^2} \frac{f^{3/2}}{(1-f)^{1/2}} \tag{4-13}$$

where $L_c = (\overline{h^2_{\text{coil}}}/n)^{1/2}$ is the effective length of one monomer unit in the coiled conformation, $L_h = (\overline{h^2}/n)^{1/2}$ is the effective length of one monomer unit in the helical chain, $\overline{h^2}$ the mean square of the end-to-end distance of the helical chain, and $\overline{h^2_{\text{coil}}}$ the mean square of the end-to-end distance of the coiled chain. Clearly this equation can be used to calculate σ only when the value of $\overline{h^2_{\text{coil}}}$ is known for some particular f. In view of the discussion of the complex nature of the conformation of the coil, reliable values of $\overline{h^2_{\text{coil}}}$ are virtually impossible to come by. Therefore, it is not fair to pass judgement on the validity of this equation until reasonable values for the constants it requires are available. Further, this equation does not take into account "end" effects. For short chains this could prove a grave shortcoming. Hence it would seem that reliable calculations involving the above equation will have to wait until $\overline{h^2_{\text{coil}}}$ can be determined accurately for high molecular weight polypeptide chains.

D. Recent Advances in the Study of the Helix–Coil Transition

Go, Go, and Scheraga† have combined theoretical conformational analysis and a modified Zimm–Bragg theory of the helix–coil transition to calculate values of σ for polyglycine and polyalanine. The results of these calculations are summarized in Table 4-5. The results indicate that poly-L-alanine

TABLE 4-5

Value of the Zimm–Bragg Parameter σ for the α-Helix–Classical Random Coil
Transition for Polyglycine and Poly-L-alanine

Parameter set[a]	Polyamino acid[b]	Regular helices[c]	Nonregular helices[d]
A	Gly	$10^{-4.4}$	$10^{-5.1}$
	Ala(R)	$10^{-3.6}$	$10^{-4.1}$
	Ala(L)	$10^{-3.4}$	$10^{-4.1}$
B	Gly	$10^{-4.5}$	$10^{-5.0}$
	Ala(R)	$10^{-3.7}$	$10^{-4.1}$
	Ala(L)	$10^{-3.4}$	$10^{-3.9}$
C	Gly	$10^{-11.0}$	$10^{-11.8}$
	Ala(R)	$10^{-9.0}$	$10^{-9.6}$
	Ala(L)	$10^{-7.3}$	$10^{-7.5}$

[a] A, dielectric constant $D = 4.0$, van der Waals radius of hydrogen, $R_h = 1.200$ Å. B, $D = 4.0$, $R_h = 1.275$ Å. C, $D = 1.0$, $R_h = 1.275$ Å.

[b] R, right-handed. L, left-handed.

[c] All ϕ, ψ, χ, respectively, kept equal in the energy minimization. The equivalence condition was invoked.

[d] Complete energy minimization with respect to the ϕ, ψ, χ. No equivalence-type relationship imposed during the energy minimization.

undergoes a transition from the right- to the left-handed α-helical form as the temperature is increased. Such transitions have been experimentally observed for several copolymers of amino acids (38a). This finding raises the conjecture that some polypeptides must adopt the left-handed α-helical conformation before the classical helix–coil transition can proceed.

E. Summary

Even with the great amount of work which has been done on the study of the helix–coil transition it appears that much exploration still needs to be carried out. While a monodimensional Ising model is an adequate phenomeno-

† This work is carried out in Go *et al.* (19, 20).

logical description of the helix–coil transition in polypeptides, the nature of the conformational change that takes place in the geometry of the polymer is not well understood. In this section more time has been spent discussing the molecular structures involved in the transition rather than deriving in complete detail the theory of the transition. The quantitative aspects of the transition are presented here in "cookbook" style since there are more than enough reports which discuss the theory in complete detail.

IV. The Poly-L-Proline I–Poly-L-Proline II Transition

The imino acid proline can be polymerized to yield a macromolecule whose chemical formula can be represented by

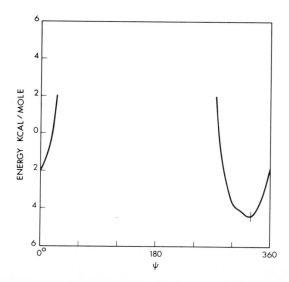

with the brackets designating the chemical repeating unit. From the conformational point of view, poly-L-proline is the simplest biopolymer. Because of the steric restriction to rotation about the N—C$^\alpha$ bond due to the pyrrolidine ring, there is only one degree of conformational freedom per

Fig. 4-6. A conformational energy plot of energy (kcal/mole-residue) versus the internal bond rotational angle ψ (rotation about the C$^\alpha$—C' bond) for *trans*-poly-L-proline.

residue—rotation about the C^α—C' bond commonly denoted by ψ. Thus, the macromolecular conformation of the ordered biopolymer is solely a function of ψ. The conformational potential energy plot of ψ versus energy (shown in Fig. 4-6) indicates only one stable conformation ($\psi \cong 325°$) for the trans form of the biopolymer. By trans it is meant that successive C^α carbons are trans to one another with respect to the (C'—N) imide bond. This stable conformation turns out to be a left-handed threefold helix. X-ray analysis also shows

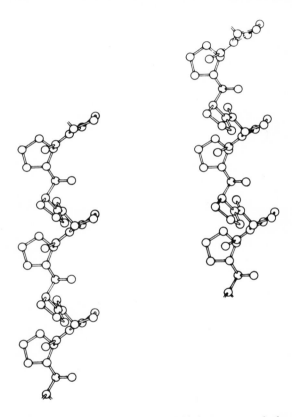

Fig. 4-7. The *trans*-poly-L-proline 3_1-helix. This is a stereoscopic drawing.

that the left-handed threefold helix is the chain conformation in the solid state for the trans form of the polymer. Figure 4-7 shows the left-handed threefold helix of poly-L-proline. This chain conformation has become known as form II (poly-L-proline II, abbreviated PP II) of the biopolymer.

By now one has guessed that there is also a form I of the biopolymer. But, as already stated, energy calculations show that form II is the only stable conformation for the sequence of trans residues. Then form I must be composed of cis residues. That is, in form I the successive C^α carbons are cis to one another with respect to the imide bond. The conformational energy versus ψ plot in Fig. 4-8 indicates that there is only one stable conformation for form I ($\psi \simeq 339°$). This energy minimum corresponds to a right-handed 3.3-helix. Table 4-6 lists the helical parameters of forms I and II. A comparison of the conformational energy versus ψ plots of forms I and II suggests

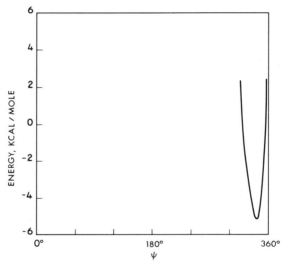

Fig. 4-8. A conformational energy plot of energy (kcal/mole-residue) versus the internal bond rotation angle ψ (rotation about the C^α—C' bond) for *cis*-poly-L-proline.

TABLE 4-6

Helical Parameters for PP I and PP II Helices

Form	Axial repeat (A) per turn	Number of residues per turn	ϕ deg[a]	ψ deg[a]
I	7.33	3.3	105	339
II	9.36	−3.0	105	325

[a] Using the convention of Edsall *et al.* (38b).

that form I is a much more rigid conformation than form II. This is verified by experimental viscosity and ultracentrifuge measurements (39). A comparison of Fig. 4-7 to Fig. 4-9, where the form I helix is shown, suggests that form II would be more soluble in polar solutions than form I. The 3_1-helix of form II readily exposes the carbonyl backbone oxygens while in the form I helix the carbonyl oxygens are partially shielded from solvent and the effective surface of the helix is largely composed of hydrophobic pyrrolidine rings. This observation will be important when next we discuss the transition between PP I and PP II.

When poly-L-proline is put into solution the biopolymer will adopt either form I or form II, depending upon the nature of the solvent. Form II prefers a polar solvent while form I prefers a less polar, slightly hydrophobic solvent.

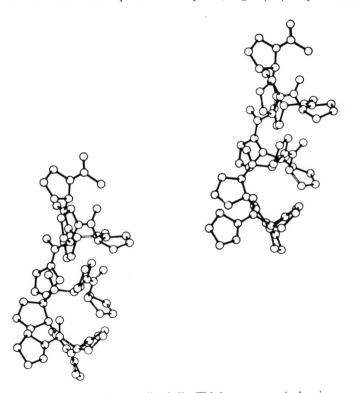

Fig. 4-9. The *cis*-poly-L-proline helix. This is a stereoscopic drawing.

This is in keeping with the nature of the helices of forms I and II discussed above. The two forms are readily interconvertible by variation of solvent composition. For example, form I will exist in pyridine. When water, aliphatic acids, benzyl alcohol, or chloroethanol are added to the solution form I slowly mutarotates to form II. Form II can be converted to form I, for example, by the dilution of water, acetic acid, or formic acid solutions with *n*-propanol or *n*-butanol. The transformation of one form to the other can be accelerated by increasing the temperature. More important, in terms of explaining the energetics of the form I–form II transition, is the fact that the transition is markedly accelerated by the addition of small amounts of strong acids. The fact that only a very small amount of acid is required to accelerate the rate in either direction has suggested that proton binding at the imide linkage play an important role (39, 40). It is postulated that protonation of the imide nitrogens leads to the temporary destruction of the double bond character of the (C′—N) bond. Rotation then occurs about the imide (C′—N) bond, making it possible for a configurational change to occur in the residue.

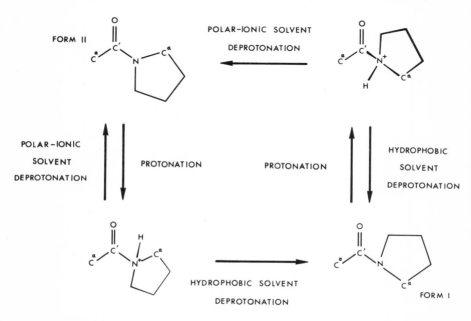

Fig. 4-10a. A schematic illustration of the possible cyclic nature of the poly-L-proline I–poly-L-proline II transition.

Once the rotation has occurred the proton is removed and the double bond character of the imide bond is restored. Whether or not a configurational change will occur depends upon the initial form of the biopolymer and the nature of the solvent. Form I in a polar solvent will tend to go to form II, while form II in a slightly hydrophobic solvent will tend to go to form I. Figure 4-10 illustrates possible transition paths between PP I and PP II.

In the mechanistic sense, the PP I ⇌ PP II transition is fairly well characterized. The proton might possibly attack the carbonyl oxygen rather than the nitrogen in the breaking down of the imide bond, but the sequence of events necessary for interconversion between form I and form II is not altered. However, the very fact that the imide bond can be weakened to the extent of allowing cis–trans interconversion does, in itself, raise some fundamental questions concerning our understanding of peptide chemistry. First

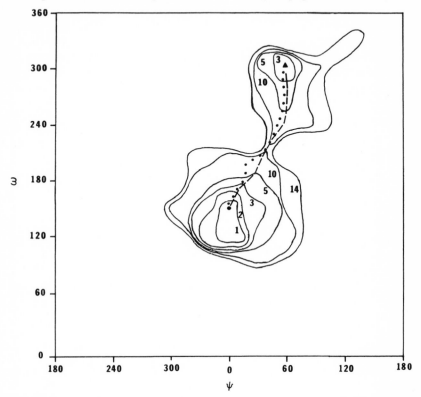

Fig. 4-10b. The conformational energy map of ω, rotation about the C′—N imide bond, versus ψ for $\phi = 120°$ and the imide nitrogen in the sp³ quaternary state. The equivalence condition holds for ω and ψ. ● defines the location of the trans form and ▲ locates the cis form. The trans → cis path of minimum conformational interconversion energy is denoted by ---, while the cis → trans path is defined by ····. The conformational interconversion energy barrier for the trans → cis transition is 13.6 kcal/mole-residue while the cis → trans transition has a barrier height of 11.6 kcal/mole-residue. Free rotation is assumed about the C′—N⁺ imide bond. Obviously, these energy barriers may not be the global minima since the equivalence condition may not hold for the transition. The energy contours are in kcal/mole-residue above the deepest minimum which is located at the trans form for the polymer.

of all, why does the cis–trans interconversion occur for the imino acid homopolymer poly-L-proline, and not for amino acid polymers? It should be noted that poly-L-hydroxyproline, which differs from poly-L-proline in that an OH group has replaced one of the protons on the C^γ atom in the proline residue, does not undergo the cis–trans transition. The reason for this, as found from conformational energy calculations (41a), is that the OH group sterically prohibits the formation of the form I helix. Thus, the non-existence of the form I $\rightleftharpoons$ form II transition in poly-L-hydroxyproline cannot be attributed to the inability to weaken the imide bond. The only apparent difference between the imide bond of imino acids and the amide bond of amino acids is the considerably lower electron density on the nitrogen atom of the imide bond. Although a number of speculative mechanisms can be invoked to explain the cis–trans behavior of the imide bond using this noted

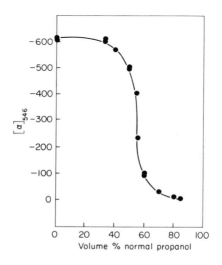

Volume % normal propanol

Fig. 4-11. Demonstration of the PP I to PP II transition. The specific rotation $[\alpha]_{546}$ versus volume percent of normal propanol yields a very sharp transition curve suggesting a highly cooperative transition. Mandelkern (41b). Reprinted from Gornick *et al., J. Amer. Chem. Soc.* **86,** 2549. Copyright 1964 by the American Chemical Society. Reprinted by permission of the copyright owner.

reduction in electron density on the imide nitrogen, none are completely satisfactory.

In addition to not being able to explain why the cis–trans isomerization occurs only in imino acid homopolymers, it is also difficult to explain the cooperative nature of this transition. The cooperativity of the poly-L-proline I–poly-L-proline II transition can be inferred from a study of the plot of equilibrium values for specific rotation $[\alpha]_{546}$ versus solvent composition (see Fig. 4-11). The sharpness of the curve indicates a highly ordered two-state cooperative transition. However, this cooperative change occurs in an isolated molecule devoid of any intramolecular hydrogen bonds. In other synthetic polypeptides composed of amino acids, hydrogen bonds are presumed to be an inherent part of the transition. The sequential formation or destruction of these hydrogen bonds (like the opening or closing of a zipper) affords a rationale to explain the basis of the cooperativity of these transitions. In the case of poly-L-proline, unfortunately, one is forced to presume that the cooperative nature of the transition must reside in the relative ease with which certain rotational angles or bond orientations will be perpetuated once initiated within the chain. That is, once one residue "flips" from the trans to the cis form, or vice versa, to initiate the transition, the nearby residues find themselves in an environment which makes it easier for them to "flip" than the other residues located a greater distance from the initial "nucleation" site.

Perhaps the nature of the poly-L-proline I–poly-L-proline II transition will be elucidated when detailed studies are made of sequential polypeptides such as $(Pro-X)_n$, $(Pro-Pro-X)_n$, $(Pro-Pro-X-Y)_n$, etc., where the X and Y refer to *amino* acid residues. From such sequential polypeptides it should be possible to evaluate both *intra-* and *inter*residue contributions to the promotion of the PP I $\rightleftharpoons$ PP II type of transition.

In the next section we will discuss the conformational properties of some polytripeptides and polyhexapeptides which contain proline.

V. The Collagen-Like Triple Helix–Random Chain Transition†

A. Introduction

Collagen (42a, 43) is one of the few proteins which is fibrous rather than globular in bulk molecular structure. It is now known that the basic tropocollagen molecule is composed of three distinct polypeptide chains, each of which is nearly a left-handed threefold helix. These three helical chains wrap around one another to form a single superhelix (the tropocollagen triple helix) whose repeat and pitch appears to be variable and dependent upon the primary structures of the polypeptide chains. The three polypeptide chains are held together by interchain hydrogen bonds which are able to come "near" to one another because glycine residues are found in every third residue position of each of the three chains. Thus the repetitive glycine residue sequencing and the mode of interchain hydrogen bonding are responsible for the manner in which the three polypeptide chains "fit together" to form the super helix. Figure 4-12 shows how three left-handed 3_1-helical polypeptide chains can be brought together to form a collagen-like triple helix.

† This section is a summary of work carried out by F. R. Brown III, E. R. Blout, and the author at Harvard Medical School. I am grateful for the permission of my co-investigators in allowing me to discuss this work.

The left-handed 3_1-helix is preferred to any other secondary structure because of the high content of the imino acid residues proline and hydroxyproline in each of the three chains. In fact, the tropocollagen molecule is unique among the proteins in having such a high percentage of proline and hydroxyproline. Because of the steric and rotational restrictions inherent in both the proline and hydroxyproline residues, the left-handed 3_1-helix is the most probable secondary structure for both these residues.

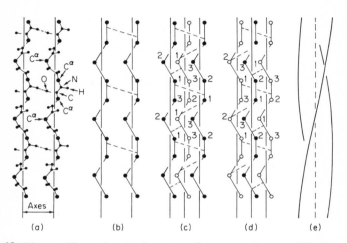

Fig. 4-12. Diagram illustrating the formation of the two collagen models from the polyglycine II lattice.

(a) Two strands of polyglycine with black dots representing various atoms and dashed lines as hydrogen bonds.

(b) Two chains of the polyglycine II lattice in which only the C^α atoms and hydrogen bonds are shown.

(c) The third chain shown with open circles lies behind the two shown in (b) to make a collagen I arrangement. The numbers indicate the phasing of the residues on the polypeptide chains.

(d) The third chain is added in front of the two in (b). The chain in front is shown by solid lines. This gives rise to the collagen II model.

(e) Solid lines represent the axes around which the polyglycine chains are coiled. These axes are gently coiled around each other in the collagen molecule. From Rich and Crick (42b).

From the brief description of the tropocollagen molecule given in the preceding paragraph it can be realized that the thermodynamic stability of this molecule should be dependent upon (a) the amount of proline and hydroxyproline present in each of the polypeptide chains and the distribution of these residues in the primary structures of the three chains, (b) the triplet sequencing of glycine residues in each of the chains, and (c) the number and mode of interchain hydrogen bonds. Although the precise amino acid sequences of the three polypeptide chains are not completely known for any particular tropo-collagen molecule, it has proved possible to evaluate the importance of each of the above factors upon the thermodynamic properties of the collagen-like triple helices by testing the properties of synthetic models. That is, various polytripeptides of the form $(X-Y-Gly)_n$ and hexapeptides of the general primary structure $((X_1-Y_1-Gly)(X_2-Y_2-Gly))_n$ have been synthesized and property tested.† The polarity of the solvent is also found to be of critical importance when discussing the stability of collagen-like triple helices. The lower the polarity of the solvent molecules the greater the stability of the triple helix.

The bulk of the experimental property testing of the tropocollagen molecules and their synthetic analogs for the purpose of determining the thermodynamic stability of collagen-like triple helices has involved following the CD spectrum of a solution containing sample molecules as a function of temperature.

The CD spectrum of a macromolecule is indicative of its conformation. Hence, once the CD spectrum of a collagen-like triple helix can be assigned to a particular tropocollagen molecule, or synthetic analog, as well as the CD spectrum of an unordered chain, then the thermal stability of a collagen-like triple helix can be monitored from the CD spectrum. There are some pitfalls in applying this experimental procedure. Unless additional experiments are done to determine if a threefold drop in macromolecular weight accompanies the "melting" of the collagen-like triple helix, it is not possible to decide if the collagen-like triple helix was formed from three unique polypeptide chains or one folded polypeptide chain. However, the thermodynamic properties associated with the triple helix–random chain(s) transition for both possible types of triple helices should be nearly identical if both triple-helix geometries are identical. This raises the unresolved question, "Are all triple helices for all the polytripeptides, polyhexapeptides and tropocollagen molecules identical?" For example, do the number of hydrogen bonds per tripeptide remain constant and independent of primary structure? Although there is no definitive way to answer this question with any kind of certainty, most workers tend to think the answer to the above question is yes. In any event, one should keep in mind that some of the experimental results might be for single-chain folded collagen-like triple helices rather than three-chain collagen-like triple helices. In Figs. 4-13a to 4-13d are shown CD melting profiles for collagen and collagen-like helices. These experimental melt curves demonstrate that:

(1) The greater the percent of imino acids residues present in the molecule the greater the stability of the collagen-like triple helix. This can be seen by

† See Brown *et al.* (43) for references.

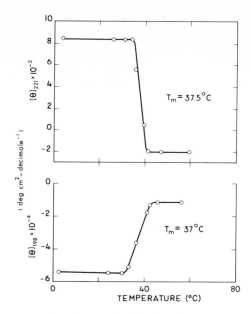

Fig. 4-13a. Temperature dependence of ordered collagen structure. Guinea pig skin collagen in water, $c = 0.5$ mg/ml. From Brown *et al.* (43).

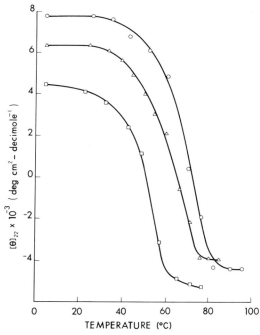

Fig. 4-13b. Temperature dependence of the collagen-like triple helix for $(Pro-Ser-Gly)_n$: $-\bigcirc-\bigcirc-$ $(Pro-Ser-Gly)_n$, fractionated (mol wt = 18,000), $T_m = 69°C$; $-\triangle-\triangle-$ $(Pro-Ser-Gly)_n$, unfractionated (mol wt = 20,000), $T_m = 62°C$; $-\square-\square-$ $(Pro-Ser-Gly)_n$, fractionated (mol wt = 10,000), $T_m = 51°C$. Solvent in all cases = 1,3-propanediol; concentration = 1.25 mg/ml. From Brown *et al.* (43).

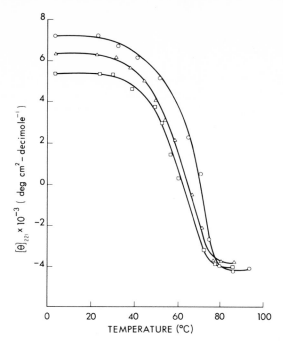

Fig. 4-13c. Temperature dependence of the collagen-like triple helix for (Pro-Ala-Gly)$_n$:
–o–o– (Pro-Ala-Gly)$_n$, fractionated (mol wt = 14,000), $T_m = 69°C$; –△–△– (Pro-Ala-Gly)$_n$,
unfractionated (mol wt = 12,400), $T_m = 63°C$; –□–□– (Pro-Ala-Gly)$_n$, fractionated (mol
wt = 7200), $T_m = 61°C$. Solvent in all cases = 1,3-propanediol; concentration = 1.25 mg/ml.
From Brown *et al.* (43).

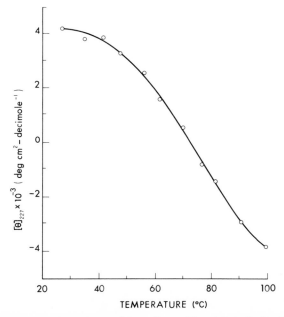

Fig. 4-13d. Temperature dependence of the collagen-like triple helix for (Pro-Gly-Pro)$_n$:
(Pro-Gly-Pro)$_n$ (mol wt = 5500), $T_m = 71°C$ (shown). Solvent = water; concentration = 0.5
mg/ml. From Brown *et al.* (43).

comparing the melt temperatures, T_m, of (Pro-Ala-Gly)$_n$ to the T_m of (Pro-Gly-Pro)$_n$ for similar molecular weight species (see legends for Figs. 4-13c and 4-13d).

(2) The higher the molecular weight of the molecule, the higher the melt temperature. This can be seen by comparing the T_m's of various molecular weight species of (Pro-Ala-Gly)$_n$ (see legend for Fig. 4-13c).

(3) The widths of each of the transition curves (ΔT) for the synthetic polytripeptide collagen models are relatively broad when compared with the corresponding transitions for collagens (approximately 15° versus 2°C). There are several possible explanations for this difference, including: (a) an incomplete attainment of thermal equilibrium, even though each point on the melting curves represents an average of several values accumulated over the course of several hours; (b) the polydispersity of the polymer preparations which yields a number of different conformations and aggregations for the various molecular weight polymers, each of which melt at slightly different temperatures; and (c) the lower molecular weight of the synthetic polymers (<20,000) as compared to single chains of collagen (100,000).

B. Theory of the Collagen Triple Helix to Random Chain Transition

Any adequate theory to describe the thermal transition of the collagen-like triple helix to unordered chain(s) must take into account the three factors just given as well as the other properties of the collagen-like triple helix mentioned in the last few pages. Since there is no experimental evidence for the existence of "long-lived" metastable intermediate states in the collagen triple helix–random chain transition, i.e. Fig. 4-13a shows only two discernible plateaus, one for the triple helix and one for the unordered chain, it is reasonable to describe this transition by a two-state model. The initial stage is characterized by the thermodynamic properties of a triple helix having the backbone geometry of the (Pro-Pro-Gly)$_n$ triple helix found by Yonath and Traub (44). An unordered chain model similar to the one proposed by Brant, Miller, and Flory (45) is chosen to represent the final stage.

1. Contributions to ΔH

The change in the conformational enthalpy ΔH_c can be expressed by the change in the internal energy plus a work term $P\Delta V$, which is assumed to be zero for reasons discussed below. Hence,

$$\Delta H_c = E^T_{x-y-g} - (E_x^R + E_y^R + E_g^R) \tag{4-14}$$

The total triple helix internal energy is given by E_{x-y-g}^{T}, where x and y are subscripts which represent any amino or imino acid residues and g denotes the glycine residue. E_{x-y-g}^{T} is further partitioned into a set of intrachain tripeptide interactions P_i, and a set of average interchain tripeptide interactions, $\langle U_i \rangle$. In this notation i refers to the ith neighbor interaction. If there are n tripeptides in each chain, then E_{x-y-g}^{T} is given by

$$E_{x-y-g}^{T} = \sum_{i=0}^{n} \left\{ \frac{n-i}{n} (P_i + \langle U_i \rangle) \right\} \tag{4-15}$$

In this work, $n = 2$ was sufficient to calculate E_{x-y-g}^{T} accurately. Since n is a function of molecular weight, E_{x-y-g}^{T} is a molecular weight-dependent term.

The internal energy E_u^{R} of a residue u in an unordered chain was calculated as the Boltzmann average energy of two peptide units and the appropriate side chain, all of which are joined at the common C^{α} atom (45). The angular increments for the backbone rotations were $10°$ and the side chain angular increments were $30°$. When the bond rotations were partitioned in this manner, the conformational partition function could be approximated by

$$Q = \sum_{(j)} \exp(-\beta E_j) \tag{4-16}$$

where the sum extends over all conformational energies, E_j. The random chain energy of residue u is

$$E_u^{R} = (1/Q) \sum_{(j)} E_j \exp(-\beta E_j) \tag{4-17}$$

In this work β was held fixed for a $T = 298°K$ distribution.

In all calculations the appropriate sets of conformation potential functions given in Chapter 2 were employed. Table 4-7 lists the values of P_i and $\langle U_i \rangle$ while Table 4-8 contains the values of E_k^{R}.

The magnitudes of the polymer–solvent enthalpy and entropy terms depend upon the extent of interaction of solvent with the backbone and side chain O, OH, and NH of the polymers. This interaction was estimated through an exposure coefficient, $m_u i$ (where u identifies a particular residue and i denotes the position of the residue in the tripeptide sequence) found by attaching solvent molecules to oxygens and hydrogens using CPK molecular models and determining the fraction of sterically allowed rotation about the oxygen or hydrogen by the solvent molecule. The values assigned to the exposure coefficients and used throughout this study are listed in Table 4-9.

The contribution to ΔH from the polymer–solvent interactions, ΔH_{p-s}, is given the form

$$\Delta H_{p-s} = \{ (m_x^{1} + m_y^{2} + m_g^{3}) - [(6 - N) + J] \} H^{p-s} \tag{4-18}$$

TABLE 4-7

Triple Helix Energy and Entropy Parameters for Various Tripeptides

Tripeptide	Neighbor (i)	P_i cal/mole-tripeptide	$\langle U_i \rangle$ cal/mole-tripeptide	S_{x-y-g}^{T} cal/mole-tripeptide-$^\circ$K
Pro-Pro-Gly	0	−7000	−4800	9.8
	1	−2400	−1300	
	2	−300	0	
Pro-Ala-Gly	0	−6400	−5200	
	1	−1400	−1500	18.5
	2	−200	0	
Ala-Pro-Gly	0	−6600	−5000	
	1	−1400	−1200	19.2
	2	−200	−100	
Pro-Ser-Gly	0	−6600	−4800	
	1	−2600	−1600	14.2
	2	−200	−100	
Ser-Pro-Gly	0	−6500	−4900	
	1	−1700	−1100	17.6
	2	−200	−100	
Ala-Ala-Gly	0	−6400	−5100	
	1	−1700	−1300	25.2
	2	−100	−100	
Gly-Pro-Gly	0	−6300	−5000	
	1	−1700	−1300	22.7
	2	0	−100	
Gly-Ala-Gly	0	−6500	−5100	
	1	−1700	−1400	27.8
	2	−200	−100	

TABLE 4-8

Random-Chain Energy and Entropy Parameters

Residue (u)	E_u^{R} (cal/mole-residue)	S_u^{R} (cal/mole-residue-$^\circ$K)
Gly	−3300	10.4
Pro(Hypro)	−2200	3.7
Ala	−2800	8.0
Ser	−2500	7.7

where the $m_u{}^i$ are the exposure coefficients, N the number of pyrrolidine ring-containing residues per tripeptide, J the number of O's, and/or OH's, and/or polar H's on the side chains in each tripeptide, and H^{p-s} is a characteristic polymer–solvent hydrogen-bond enthalpy.

The values of H^{p-s} were calculated by setting the theoretical melting temperature equal to the experimental melting temperature of one polymer. It is important to point out that once a particular solvent H^{p-s} had been determined using one polymer, this value was then used in *all* subsequent calculations. In this work H^{p-s} for water was calculated to be -2202 cal/mole and -2114 cal/mole for 1,3-propanediol. For water, the polymer used in the calibration was $(Pro-Pro-Gly)_n$ with $n = 10$, Kobayashi *et al.* (46), and for 1,3-propanediol, $(Pro-Ala-Gly)_n$ with $n = 32$, Brown *et al.* (47), was chosen as the calibration polymer. It should be noted that these values for hydrogen

TABLE 4-9

Exposure Coefficients Based on Solvent Accessibility to a Triple-Helical Structure with Coordinates of Yonath and Traub (44)[a]

$m_u{}^i$ (exposure coefficient)				
Residue one (i = 1)		Residue two (i = 2)		Residue three (i = 3)
$m_g{}^1 = 0.9$	$m_p{}^1 = 0.0$	$m_g{}^2 = 1.8$	$m_p{}^2 = 0.8$	$m_g{}^3 = 0.8$
$m_a{}^1 = 0.9$	$m_s{}^2 = 1.7$	$m_{hp}{}^2 = 1.8$	$m_a{}^2 = 1.8$	
		$m_s{}^2 = 2.5$ (water)		

[a] g = Gly, a = Ala, p = Pro, hp = Hypro, s = Ser.

bond enthalpy are within the known range of values (-1500 to -5000 cal/mole). Moreover, in an independent study (48) of the enthalpy of hydrogen bonds the value of H^{p-s} was computed to be -2180 cal/mole.

In summary ΔH is expressed by two terms:

$$\Delta H = \Delta H_c + H_{p-s} \qquad (4\text{-}19)$$

2. Contributions to ΔS

Just as ΔH was partitioned into a conformational term and a polymer–solvent interaction term, so also is ΔS. The conformational term, ΔS_c, has a form analogous to ΔH_c,

$$\Delta S_c = S^T_{x-y-g} - (S_x{}^R + S_y{}^R + S_g{}^R) \qquad (4\text{-}20)$$

where S_{x-y-g}^{T} is the conformational entropy of the ordered triple helix, which in turn is given by the formulation presented in Chapter 2:

$$S_{x-y-g}^{T} = \frac{R}{2}\ln(\det F) + C_{x-y-g} \tag{4-21}$$

with

$$F = \frac{\partial^2 U}{\partial u_i \, \partial v_j} \Bigg|_{U = E_{x-y-g}^{T}}$$

where U is a symbol used to represent the conformational potential energy, C_{x-y-g} is an additive constant entropy of integration, and u_i and v_j are any elements of the variable set $\{\phi_l, \psi_l, \chi_{l,m}, \mathbf{a}_k, \mu_k\}$ which define the conformation of the triple helix (see Chapter 1 for meaning of symbols).

The elements of F were calculated by including both first and second nearest tripeptide neighbor interactions. A paraboloid was fitted through a set of points located on the vertices and midpoints of the sides of a square centered about the points $(u_i{}^*, v_j{}^*)$ corresponding to the triple-helical configuration in u_i versus v_j conformational hyperspace. The sides of the square were 6° and/or 0.2 Å and the variables defining the triple helix conformation other than $(u_i{}^*, v_j{}^*)$ were held fixed during the fitting of the paraboloid. The general equation of a paraboloid is

$$U(u_i, v_j) = au_i{}^2 + bv_j{}^2 + cu_i + dv_j + eu_i v_j + fu_i v_j{}^2 + gu_i{}^2 v_j + hu_i{}^2 v_j{}^2 + k \tag{4-22}$$

so that

$$\frac{\partial^2 U}{\partial u_i \, \partial v_j} = e + 2fv_j + 2gu_i + 4hu_i v_j \tag{4-23}$$

The entropy of the random chain residue is calculated using the same procedure employed to calculate the internal energy of a random-chain residue.

$$S_u^{R} = E_u^{R}/T + R\ln Q + C_u \tag{4-24}$$

where C_u is an additive constant entropy of integration.

The values of S_{x-y-g}^{T} are given in Table 4-7 and the S_u^{R} are listed in Table 4-8. The partitioning used to obtain S_{x-y-g}^{T} is sufficiently similar to the random-chain partitioning so that the additive entropy constant of integration cancels,

$$C_{x-y-g} - (C_x + C_y + C_g) = 0 \tag{4-25}$$

and thus the additive entropy constants do not enter into the calculations and need not be considered.

The contribution to ΔS from polymer–solvent interactions is given a form identical to H_{p-s},

$$S_{p-s} = [\{(m_x{}^1 + m_y{}^2 + m_g{}^3) - [(6 - n) + J]\}] S^{p-s} \qquad (4\text{-}26)$$

where S^{p-s} is the characteristic entropy of hydrogen-bond formation between polymer and solvent.

Since solvents studied in this work are polar, it is assumed that all possible polymer–solvent hydrogen bonds are formed and maintained so that the polymer–solvent interaction is a fixed one-state system. Thus, S^{p-s} could be set equal to zero. However, for neutral and nonpolar solvents, it is necessary to incorporate S^{p-s} in the initial calibration calculations.

3. The Equation

Using the above formulation, the melting temperature of a collagen-like triple-helical structure T_m is given by

$$T_m = \frac{\Delta H_c + \Delta H_{p-s}}{\Delta S_c + \Delta S_{p-s}} \qquad (4\text{-}27)$$

For hexapeptide polymers composed of unique tripeptides A_1 and A_2, T_m has the form

$$T_m = \frac{\Delta H_c(A_1) + \Delta H_c(A_2) + \Delta H_{p-s}(A_1) + \Delta H_{p-s}(A_2)}{\Delta S_c(A_1) + \Delta S_c(A_2) + \Delta S_{p-s}(A_1) + \Delta S_{p-s}(A_2)} \qquad (4\text{-}28)$$

This formulation assumes that the ΔH and ΔS cross terms between the unique tripeptides can be arithmetically averaged. This assumption has been substantiated for $(Gly\text{-}Pro\text{-}Pro\text{-}Gly\text{-}Pro\text{-}Ala)_n$.

C. Limitations and Discussion of the Model

Contributions which have been explicitly neglected in this formulation are: (a) the solvent–solvent interaction free energy; (b) the polymer–polymer interaction free energy; and (c) the work term $P \Delta V$. Although solvent–solvent interactions are not explicitly included in the equation, they are considered to be contributing factors in the polymer–solvent terms. For low concentrations, ($<0.1\%$) as used in the experiments described here, polymer–polymer interactions can be neglected. There is no *a priori* reason to set the $P \Delta V$ terms equal to zero unless the experiments are done at constant volume. However, for polymers in polar solvents it is not unreasonable to suggest that the polymer chains, when in the random state, will be in an extended conformation, i.e., polyproline II, in order to maximize exposure of polar groups to solvent. Thus the change in volume in going from triple helix to a random chain under these conditions should be small and can be neglected. When the theoretical results on calculations of the melting temperatures for the poly-tripeptides and polyhexapeptides are compared with the observed melting temperatures, Table 4-10, it is apparent that these three terms simultaneously neglected do not affect the results.

TABLE 4-10

Comparison of the Observed and Predicted Melting Temperatures for Some Polytripeptide and Polyhexapeptide Sequences[a]

Polymer	Solvent	n	T_m (obs)	T_m (pred)
(Pro-Gly-Pro)$_n$[b]	H_2O	22	67°C	57°C
	H_2O	100	—	78
	1,3-Propanediol	22	89	90
	1,3-Propanediol	100	—	105
(Pro-Pro-Gly)$_n$[c]	H_2O	10	25	25[e]
		15	52	43
		20	65	53
(Pro-Ala-Gly)$_n$[b]	1,3-Propanediol	32	58	58[e]
		55	69	79
(Pro-Ser-Gly)$_n$[b]	1,3-Propanediol	42	51	56
		75	69	63
(Ala-Pro-Gly)$_n$	H_2O	40	—	−60
(Ala-Ala-Gly)$_n$	H_2O	20	—	−169
		40	—	−102
(Ser-Pro-Gly)$_n$	H_2O	40	—	−160
(Gly-Pro-Gly)$_n$	H_2O	40	—	−142
(Gly-Ala-Pro-Gly-Pro-Pro)$_n$[d]	H_2O	17	26	9
		24	32	20
(Gly-Pro-Ala-Gly-Pro-Pro)$_n$[d]	H_2O	12	32	12
		20	41	33
		26	49	40
(Gly-Ala-Ala-Gly-Pro-Pro)$_n$[d]	H_2O	16	19	6
		25	35	25

[a] From Brown *et al.* (43).   [b] Brown *et al.* (47).   [c] Kobayashi *et al.* (46).
[d] Segal (49).   [e] Calibration values for H^{p-s} (see text).

Because of end effect contributions at low molecular weights, the melting temperature equation is valid only for values of $n \geqslant 10$. Below this chain length (cf. Fig. 4-14), the formulation indicates a critical dependence on molecular weight. In this regard n must be greater than 100, in most cases, to get agreement with the melting temperature predicted for infinite chain length.

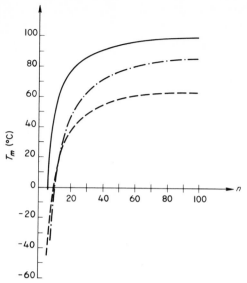

Fig. 4-14a. Melting temperature versus number of polytripeptides per chain n, for the collagen-like triple helix–random state transition. The solvent is 1,3-propanediol: ———— $(\text{Pro-Pro-Gly})_n$; —·—·— $(\text{Pro-Ala-Gly})_n$; – – – $(\text{Pro-Ser-Gly})_n$.

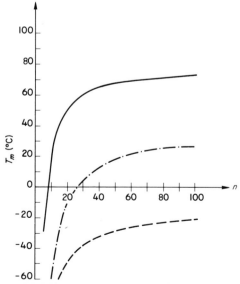

Fig. 4-14b. Melting temperature versus number of polytripeptides per chain n, for the collagen-like triple helix–random state transition. The solvent is water: ———— $(\text{Pro-Pro-Gly})_n$; —·—·— $(\text{Pro-Ala-Gly})_n$; – – – $(\text{Pro-Ser-Gly})_n$.

It should be noted that the melting temperature equation formulated here is designed to predict and partition thermodynamic variables affecting the melting temperatures of polymers known to form a collagen-like triple-helical structure in solution. No significance should be attached to the exact melting temperature predicted for nontriple helix-forming polymers, although such structures are all characterized by melting temperatures for the triple helix to random-chain transition which are much below room temperature.

The agreement of the predicted and observed T_m's in Table 4-10 suggests that the model presented here is quite realistic, at least in describing the thermodynamics of the triple helix to random-chain transition. The usefulness of this model in predicting other thermodynamic or structural properties has not been considered.

This formulation demonstrates the importance of the presence and sequencing of proline upon the stability of the triple helix. Moreover, from the predicted and observed melting temperatures in Table 4-10 for the poly-hexapeptides, it can be seen that when nonproline containing tripeptide sequences are incorporated into polyhexapeptides adjacent to Pro-X-Gly, X-Pro-Gly, or Pro-Pro-Gly trimers, the melting temperatures of the resulting polyhexapeptides are not simply the arithmetic averages of the melting temperatures of the respective tripeptides. Consideration of the (Gly-Ala-Ala-Gly-Pro-Pro)$_n$ melting temperature ($T_m = 25°C$ for $n = 25$), for example indicates that adjacent prolyl residues can stabilize the triple-helical conformation for the entire hexapeptide. As a result, the melting temperature is close to that observed for collagen, and it can be reasoned that the nontriple helix-forming regions (e.g., Ala-Ala-Gly) are stabilized in a collagen-like conformation by their proximity to the more stable proline containing regions.

Lastly, the model indicates that no single type of interaction is predominantly responsible for the maintenance of the triple helix. Rather, a number of interactions work in unison to stabilize this unique structure.

VI. Conformational Transition of Poly(Methacrylic Acid) in Aqueous Solution†

A. Introduction

Leyte and Mandel (50) have reported that isotactic and syndiotactic poly(methacrylic acid) (PMAA) exhibit a conformational transition in aqueous solution between approximately 15 and 30% neutralization with

† I would like to thank J. B. Lando and J. Semen of Case Western Reserve University for allowing me to summarize and discuss J. Semen's Ph.D. thesis on the transition properties of poly(methacrylic acid).

sodium hydroxide. Liquori (51) described this transition as a cooperative change from a relatively compact globular structure to a more extended chain conformation with increasing neutralization. The globular state should be stabilized by the hydrophobic forces associated with the methyl side chains (52). Thus, the compact conformation is characterized by a hydrophilic surface rich in carboxylic acid groups and a hydrophobic interior containing most of the methyl groups but relatively few acid groups. Mandel and coworkers (53) have suggested that the stabilization of the compact state is due

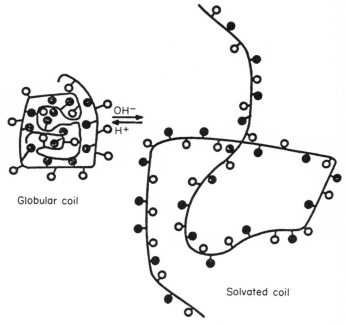

Globular coil

Solvated coil

Fig. 4-15. A schematic illustration of the PMAA transition in aqueous solution due to variation in the pH.

to van der Waals interactions between the methyl side chains. As base is added to the polymer solution, the carboxylic acid groups on the surface of the globular structure are preferentially neutralized, since the groups in the interior are much less accessible to base; thus, the hydrophobic interactions of the methyl groups continue to stabilize the globular structure even at low degrees of neutralization. However, between approximately 15 and 30% neutralization, a sufficient number of highly hydrophilic salt groups are present in the interior to overcome the interactions of the methyl groups, and therefore the structure transforms to a more open and more extended chain conformation. The conformational change is schematically shown in Fig. 4-15.

The descriptions of the conformational transition of poly(methacrylic acid) presented by Liquori and Mandel suggests that localized forces between the substituents of the polymer are responsible, at least in part, for the conformational transition. Since these localized forces are governed by the local conformation, it is important to characterize the local conformation of poly(methacrylic acid) in aqueous solution. Although the solution behavior of poly(methacrylic acid) has been studied by numerous workers (54) very little is known about the local structure in solution. Nagasawa (55) attempted to characterize the local conformations of isotactic and syndiotactic poly-(methacrylic acid) in aqueous solution with electrostatic potential calculations based on potentiometric titration data. However, the relative insensitivity of the electrostatic potential function to changes in the local chain conformation (see Chapter II) makes a definite assignment of conformations impossible with this technique. In a study by Semen and Lando (56), crystalline samples of isotactic and syndiotactic poly(methacrylic acid) were obtained; although the molecules in solution probably do not exist as rigid, helical rods, the local conformation in solution should be characterized by internal rotations that are rather similar to those of the helical chains in the crystalline polymer. Unfortunately, oriented fibers could not be obtained, and therefore the helical conformations of the crystalline polymers could not be determined.

B. Local Chain Conformations of PMAA by Conformational Analysis

1. Geometry

The conformational energy calculations described in this section were performed for isolated chains of isotactic and syndiotactic PMAA with regular helical conformations. Polymer–solvent and interchain interaction energy contributions were not considered in the calculations, although these energy contributions have some influence on the tertiary structure in the solution and crystalline phases. Therefore, conformations within a couple of kcal/mole/monomer unit of the absolute minimum must be considered as possible conformations of the polymer.

The backbone C—C—C bond angles of substituted vinyl polymers are generally much larger than the normal tetrahedral angle (57). In the initial energy calculations, the backbone bond angles were assumed to be 115°. In the subsequent calculations, the optimum bond angles were determined by including the bond deformation energy contribution in the calculated conformational energies.

A section of an isotactic poly(methacrylic acid) (PMAA) chain in the all-trans conformation is shown in Fig. 4-16a. The internal rotation angles are

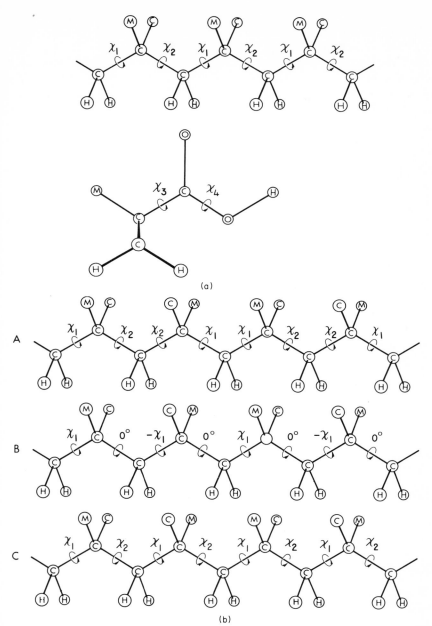

Fig. 4-16. (a) Definitions of the internal rotations and the atomic numbering scheme for isotactic PMAA. The backbone and side chain rotations are shown in A and B, respectively. "M" represents the methyl group.

(b) Definitions of equivalence cases A, B, and C for syndiotactic PMAA. The side chain rotations are as shown in (a). The direction of the side chain rotations alternates in successive monomer units for cases A and B.

designated by χ_K. Positive rotation is performed by holding atom i stationary and rotating about bond i—j such that the groups bonded to atom j move counterclockwise when viewing along the bond with atom i closer to the eye. There are four internal rotations per monomer that need to be considered in the conformational energy calculations. The backbone internal rotation angles are defined as $0°$ for the trans position. χ_4, which defines the position of the acidic hydrogen atom with respect to the O—C=O plane, is $0°$ when the hydrogen lies in this plane in the orientation shown in Fig. 4-16a. χ_3, which is equal to the angle between the plane of the O—C=O group and the plane bisecting the backbone C—C—C bond angle, is defined as $0°$ when bonds C=O and C—M are planar and trans to one another relative to the C—C bond. The values of the bond lengths and bond angles for the isotactic chain are given in Table 4-11.

TABLE 4-11

Bond Angles and Bond Lengths of PMAA

Bond	Length (Å)	Bond angle[a]	Angle (deg)
C—C	1.54	C_1—C_4—C_{10}	115
C—H	1.10	C_4—C_{10}—C_{13}	115
C=O	1.23	H_2—C_1—H_3	110
C—O	1.36	M_5—C_4—C_6	110
O—H	1.04	C_4—C_6=O_7	120
		C_4—C_6—O_8	120
		C_6—O_8—H_9	110

[a] Refer to Fig. 4-16a for numbering system used to identify the atoms.

Since the chemical repeat unit of the syndiotactic polymer contains two monomer units, there are eight independent internal rotation angles, which is an unwieldy number of parameters to consider in a digital scan over all conformational hyperspace. Fortunately, the number of independent parameters is reduced substantially by invoking the equivalence postulate of Natta (58) for syndiotactic polymers. The equivalence postulate yields two simplified coordinate systems for consideration as shown in Fig. 4-16b; in case A, there are two independent backbone rotations, χ_1 and χ_2, and two independent side chain rotations, χ_3 and χ_4, where the side chain rotations are identical to those of the isotactic chain shown in Fig. 4-16a; in case B, the number of independent rotations is reduced to three, χ_1, χ_3, and χ_4. In both case A and case B, the direction of the side chain rotations alternates in successive monomer units. A third possibility, which is not derived from the equivalence postulate of Natta, is also considered; in case C (see Fig. 4-16) there are four independent rotations, and the direction of the side chain rotations is the same in all monomer units. In all cases, positive rotation is counterclockwise as defined for the isotactic chain; the convention of reversing the direction of positive rotation in every other monomer unit (as used by some workers) is not employed. The bond angles and bond lengths in the syndiotactic chain are identical to those used for the isotactic PMAA.

2. Results: Isotactic PMAA

The conformational analysis of isotactic PMAA indicates the existence of four stable isolated chain conformations. Conformational potential energies were calculated for 5° increments in the backbone internal rotation angles. Figure 4-17 contains a potential energy map of isotactic PMAA which is

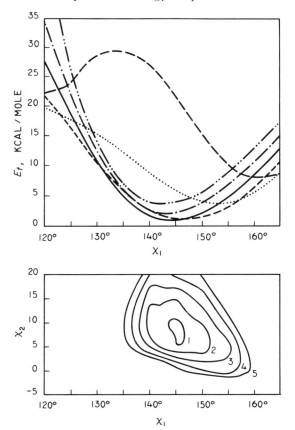

Fig. 4-17. Conformational energy maps of isotactic PMAA with side chain rotations of 0°. In the top energy map χ_2 has the value: ———, $-5°$; $\cdots\cdots$, $0°$; ---, $5°$; ———, $10°$; —·—·—, $15°$; ··—··—··, $20°$.

representative of all the isotactic PMAA calculations reported in this section. The conformational potential energies, internal rotation angles, and helical parameters of chain conformations corresponding to the four relative minima are listed in Table 4-12. The conformations given in Table 4-12 were obtained by minimizing the conformational energy as a function of all four independent internal rotation angles.

TABLE 4-12

Internal Bond Rotation Angles, Conformational Energies, and
Helical Parameters of the Four Stable Intrachain Conformations
of Isotactic PMAA

Conformation I	Conformation II
$\chi_1 = 45°$	$\chi_1 = 140°$
$\chi_2 = 0°$	$\chi_2 = 10°$
$\chi_3 = -20°, +160°$	$\chi_3 = -50°$
$\chi_4 = 0°$	$\chi_4 = -50°$
$E_t = 8.2$ kcal/mole	$E_t = -2.61$ kcal/mole
Axial repeat $= 17.5$ Å	Axial repeat $= 11.10$ Å
Type of helix: 8_1	Type of helix: 5_2
	or
	Axial repeat $= 17.79$ Å
	Type of helix: 8_3

Conformation III	Conformation IV
$\chi_1 = -135°$	$\chi_1 = 160°$
$\chi_2 = -15°$	$\chi_2 = 0°$
$\chi_3 = 180°$	$\chi_3 = 160°$
$\chi_4 = 60°$	$\chi_4 = 60°$
$E_t = -2.32$ kcal/mole	$E_t = 3.48$ kcal/mole
Axial repeat $= 11.19$ Å	Axial repeat $= 19.72$ Å
Type of helix: 5_2	Type of helix: 9_4
or	or
Axial repeat $= 17.93$ Å	Axial repeat $= 15.34$ Å
Type of helix: 8_3	Type of helix: 7_3

Conformation I is an 8_1-helix in which the side chain can have either of the two equienergy conformations.

The energy of conformation I is relatively high primarily because of the steric interactions between the oxygens of each monomer and the methyl and carbonyl-carbon atoms of the nearest neighbor monomer units. The small out of plane rotation of the carboxylic acid group (χ_3 is $-20°$ or $+160°$) optimizes the nonbonded interaction energy.

The internal bond rotation angle χ_4 can be varied over a wide range without changing the total conformational energy very much. A very weak intramolecular hydrogen bond is formed with the hydroxyl oxygen of the neighboring monomer unit when χ_3 is about $-150°$ and χ_4 is $-20°$. However, the total conformational energy of this conformation is several kcal/mole higher than the conformation with χ_4 equal to $0°$, because the hydrogen bond is far from linear; therefore, the conformation with intramolecular hydrogen bonds is energetically improbable. Hydrogen bonds, if any, must be intermolecular in the crystalline state if the chains have an 8_1-helical conformation.

Conformation II is in good agreement with both a 5_2- and an 8_3-helix. The difference in χ_1 between the 5_2- and 8_3-helices is less than $10°$ when χ_2 is $10°$. Since the energy surface is relatively flat near χ_1 equal to $140°$ both of these helices are of essentially equal stability. The conformational potential energy functions are not sufficiently accurate to warrant excluding one or the other of the two conformations. Intramolecular hydrogen bonding is not possible in conformation II. Formation of intermolecular hydrogen bonds in the crystalline state is more favorable for chains of conformation II than for the 8_1-helices of conformation I. In conformation II, the acid groups lie on the surface of the helix and are thus well positioned to form intermolecular hydrogen bonds; in the 8_1-helix, however, the methyl groups lie on the surface while the acid groups are buried in the interior. That the position of the acidic hydrogen can be varied widely with little change in the conformational energy is also conducive to the formation of strong intermolecular hydrogen bonds between chains of conformation II.

Conformation III has a backbone conformation corresponding to either a 5_2- or an 8_3-helix. It differs from conformation II only in the sense of the helical rotation; conformation III is the right-handed helix, while conformation II is the left-handed helix. As observed for conformation II, both the 5_2- and 8_3-helices must be considered to be equiprobable conformations for III. The energy is a minimum when χ_3 is $0°$ or $180°$; unlike the left-handed helical conformation, even a small out-of-plane rotation of the acid group is energetically prohibited in the right-handed conformation. No intramolecular hydrogen bonds are formed in conformation III.

Conformation IV is a 9_4-helix. The backbone internal rotation angles are also in good agreement with a 7_3-helix; the 7_3-helix is generated when either of the backbone internal rotations is about $6°$ less than the values given in Table 4-12. As was observed for conformations I–III, intramolecular hydrogen bonding is insignificantly weak, and the optimum side chain conformation produces only a small reduction in the energy compared to the conformation with both side chain rotations equal to $0°$.

Of the four conformations presented for isotactic PMAA, the 5_2- and 8_3-helices have, by far, the lowest conformational energy, and thus represent the most probable intrachain conformations for isolated chains. However, the chains in the crystalline state may not have conformations II or III since the interchain, or lattice energy, may be important in determining the conformation of lowest energy. Since the acid group of PMAA is highly polar, one can speculate that formation of interchain hydrogen bonds might alter the isolated-ordered intrachain conformation of PMAA in a crystal.

3. Results: Syndiotactic PMAA

The techniques used for successively excluding the regions of high energy conformations in the calculations for isotactic PMAA have also been used for the syndiotactic polymer.

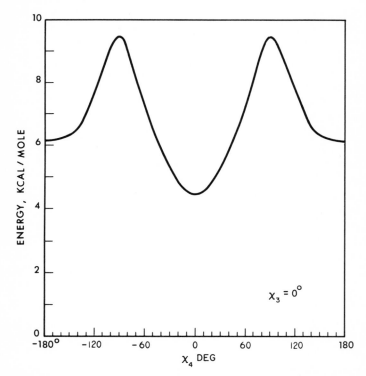

Fig. 4-18. Conformational energy of the all-trans syndiotactic chain as a function of the side chain conformation.

The backbone C—C—C bond angles have been held constant at 115°. Conformations satisfying the sets of equivalence relations, which are designated cases A, B, and C (see Fig. 4-16b) have been determined. For all three equivalence relations, the backbone conformations with reasonably low energies fall in regions where at least one of the backbone rotation angles is close to 0°.

For equivalence relation A, the conformational energy is prohibitively high for all conformations, except for those that are very close to the all-trans conformation. The trans-trans-gauche-gauche (where the gauche internal rotation angle is 120°) backbone conformation, which has been reported for syndiotactic polypropylene is energetically excluded for syndiotactic PMAA; the energy of the conformation is over 200 kcal/mole while the energy of the all-trans conformation is 4.4 kcal/mole.

The relation between the conformational energy and the side chain rotation angles is summarized in Fig. 4-18 for the all-trans conformations.

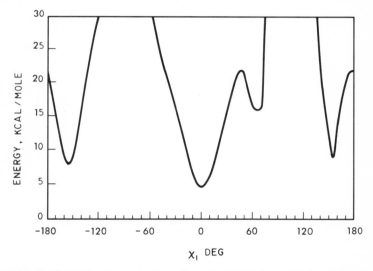

Fig. 4-19. Conformational energy of syndiotactic PMAA for equivalence case B. χ_2 and the side chain rotations are 0°.

Out of plane rotation of the carboxylic acid groups is highly restricted by the neighboring methyl groups. The acidic hydrogen atoms are confined to the plane of the O—C=O group by the neighboring methyl groups also. The all-trans conformation of syndiotactic PMAA is about 23 kcal/mole lower in energy than the corresponding conformation of the isotactic chain. The high energy of the isotactic conformation is due, principally, to the steric overlap between neighboring methyl groups; in the syndiotactic chain, however, the planar acid groups are sandwiched between the methyl groups, and the steric overlaps are substantially reduced.

Conformational potential energy plots for conformations fulfilling the equivalence cases B and C are shown in Figs. 4-19 and 4-20. In each of these figures there are conformational regions for which each of the conformational energies are not much larger than that of the all-trans chain. Therefore, the corresponding conformations may be highly probable secondary structures of the syndiotactic polymer.

The helical conformations corresponding to the relative minima near $\pm 155°$ are designated the B-helix and C-helix for equivalence cases B and C, respectively (the B-helix, although not strictly a helical conformation, is treated as a special type of helix). The two helical conformations are best described by a list of the four backbone rotation angles in one chemical

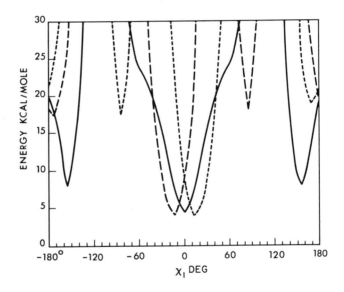

Fig. 4-20. Conformational energy of syndiotactic PMAA for equivalence case C with side chain rotations of 0°. χ_2: — — — —, $-15°$; ———, 0°; – – –, 15°.

repeat unit (two monomer units); the B-helix has backbone rotations of $\pm 155°$, 0°, $\mp 155°$, 0° (note that the direction of rotation alternates in successive monomer units), and the C-helix has backbone rotations of $\pm 155°$, 0°, $\pm 155°$, 0° (note that the direction of rotation is the same in all monomer units). The conformation of the B-helix is a nearly planar 2_1-helix (2 monomer units per turn of the helix), and the C-helix is approximately a 7_6-helix (14 monomer units in 6 turns). The conformation of the carbon backbone chain in the C-helix is very similar to the lowest energy conformations of the isotactic polymer.

C. Summary of the Conformational Analysis of PMAA

Isotactic PMAA chains very likely distribute themselves between two very nearly equally probable, locally uniform conformations. These are a 5_2-helix and an 8_3-helix. Syndiotactic PMAA chains may assume any of three nearly equally probable chain conformations. These are the all-trans conformation, a 2_1-helix and an approximate 7_6-helix. The 7_6-helix is similar to the isotactic chain conformations. No calculations have been carried out to determine the stable local chain conformations for the polysalt. However, this calculation may not be critical in order to characterize the PMAA conformational transition. Since the transition takes place for only 15 to 30% neutralization with sodium hydroxide there should, on a statistical basis, be relatively large numbers of long chain segments which remain uncharged and, therefore, in the chain conformation found for the acidic form of the polymer. Presumably at 15 to 30% neutralization electrostatic repulsions between charged groups which are in close proximity to one another force open the tertiary structure. Whether the charged groups are in close proximity because adjacent monomer units in a chain are charged, or because of intrachain folding, cannot be deduced from available data. Most likely a distribution of both types of interactions is present in the macromolecule.

Any theory to describe the conformational transition of PMAA should account for large numbers of chain segments which are locally uniform in conformation in both transition states of the macromolecule. Precisely how uniform such local conformations actually are is a matter of contention. Small variations from exactly uniform structures can be realized with little, if any, expenditure of conformational potential energy. The cumulative effect of these small perturbations can be very large and lead to globular tertiary structures provided the polymer chain is sufficiently long. We will consider this possibility next when we discuss the properties of the actual PMAA conformational transition.

D. Investigations of the PMAA Conformational Transition by Laser Raman Spectroscopy

1. The Secondary Structure

Laser Raman spectroscopy can be utilized to characterize the local conformation of PMAA in aqueous solution. Most polymers assume a regular helical conformation in the crystalline state, because this rigid, rodlike conformation contains the linear symmetry required for efficient molecular packing. In solution, however, polymer molecules often do not exist as rigid rods (except when strong intramolecular forces, such as hydrogen bonding, are present) due to the unfavorable entropy contribution to the free energy. On the other hand, the local chain conformation cannot be totally random in solution because the enthalpy is low for only a few relatively narrow ranges of backbone internal rotations, where one of these ranges obviously must include the conformation observed in the crystalline state. Two models are considered for the local conformation of PMAA. In the first model, the polymer chains are assumed to be locally helical; that is, the backbone internal rotations are nearly identical in all but a very small number of the chemical repeat units. In this model, the chains are composed of relatively long segments of helical polymer that are separated by helical defects; the defects, which are internal rotations that differ from those in the helical segments, are required in order that the molecules can form coils. In the second model, the local conformation is assumed to be defined by a small number of different internal rotations which are distributed more or less at random along the polymer chains; in this model, long segments of regular helical conformations do not exist, although there may be some helical structure over very short segments.

If model one is valid for the local conformation of PMAA, then the spectroscopic data can be used to determine the pitch of the helical segments. It has been shown that the selection rules and optical activity in the Raman and IR spectra depend on both the stereosequencing and the helical pitch (59). The selection rules for 1,1-disubstituted vinyl polymers are summarized in Table 4-13. If the optical activity of a given vibrational mode is designated by a, b, where a and b indicate the optical activity in Raman (p, d, and 0 are used to designate that the mode is polarized, depolarized, or inactive, respectively) and IR (π, σ, or 0 for parallel or perpendicular dichroism or inactive), respectively, then there are eight possible mode groupings a, b as shown in Table 4-13, and the types of mode groupings present may be used to determine the helical conformation (except for the atactic polymer, in which all helical conformations are characterized by the same four mode groupings). The optical activities of the Raman lines can be obtained from the depolarization ratios of the aqueous solutions. However, since oriented fibers cannot be obtained, the dichroic behavior of the IR bands cannot be determined. Therefore, for the data available for PMAA the observed spectroscopic modes are classifiable into one or more of the following 5 mode groupings: (p, IR), (p, 0), (d, IR), (d, 0), and (0, IR), where "IR" indicates that the mode is active in the infrared. All of the helices in Table 4-13 can be distinguished by this simplified set of mode groupings, except for the all-trans and 3_1-conformations of isotactic polymers, where both conformations have modes in the groupings p, IR and d, IR.

The frequencies of the modes observed in the Raman and IR are summarized in Table 4-14 for syndiotactic PMAA. For these samples, the modes fall into the same four mode groupings: $(p,0)$, $(0,IR)$, $(d,0)$, and (d,IR). For syndiotactic polymers, these four mode groupings are consistent with the selection rules for a helical polymer having more than 3 chemical repeat units (that is, more than 6 monomer units) in the axial repeat unit. The all-trans conformation, which has been suggested (55) for syndiotactic PMAA based on electrostatic energy calculations, must be excluded based on this spectroscopic evidence. As indicated previously, the selection rules characterizing the pitch of the helix differ for the stereoregular forms. For isotactic

TABLE 4-13

Selection Rules and Optical Activity of Vibrational Modes for
1,1-Unsymmetrically Substituted Vinyl Polymers in Helical Conformations[a]

		Optical activity							
	Raman:	p	p	d	d	p	d	0	0
Conformation	IR:	π	σ	π	σ	0	0	π	σ
Atactic									
all		x	x	x	x	—	—	—	—
Syndiotactic									
all-trans		—	x	x	x	—	x	—	—
2_1-helix		—	—	x	x	x	—	—	—
3_1-helix		—	—	—	x	x	—	x	—
$>3_1$-helix		—	—	—	x	x	x	x	—
Isotactic									
all-trans		—	x	x	—	—	—	—	—
3_1-helix		x	—	—	x	—	—	—	—
$>3_1$-helix		x	—	—	x	—	x	—	—

[a] From Koenig and Angood (60).

PMAA there is no helical pitch consistent with the four observed mode groupings. Apparently model one does not describe the local conformations of this stereoregular form of PMAA; that is, there are few significantly long segments with helical conformations in aqueous solution.

It is to be emphasized that the selection rules of Table 4-13 apply only to polymers containing long helical segments. However, it is not known precisely how long a local helical segment must be before it can be considered "long." Thus, it is possible that the majority of the monomer units in PMAA may be in locally helical conformations even when the observed modes do not fit the selection rules. Unfortunately, one cannot predict what the selection rules are for such a short helical model, since this model is not subject to symmetry analysis.

Since the atactic PMAA samples contain about 40% *mr* triads, one cannot, even to a first approximation, assume that the selection rules for syndiotactic helical polymers apply to the atactic samples. Although the observed vibrational modes for syndiotactic PMAA are in agreement with a set of selection

TABLE 4-14

Raman and Infrared Frequencies of the Various Stereoregular Forms of
Poly(Methacrylic Acid)

Raman (aqueous solution)			Infrared (solid)	
Frequency (cm^{-1})	Relative intensity	Polarized	Frequency (cm^{-1})	Relative intensity
1697	strong	—	1700	very strong
1490	shoulder	—	1487	weak
			1475	medium
1450	strong	—	1449	medium
			1413	shoulder
1397	very weak	—	1392	strong
1333	weak	—		
			1270	strong
1255	weak	√		
1210	strong	√		
			1175	very strong
1126	medium	—		
			1005	very weak
972	strong	—	970	weak
943	medium	—		
			930	medium
884	weak	—		
			835	weak
820	medium	√		
			800	very weak
780	very strong	√		
728	shoulder	—		
			635	weak
			630	weak
605	strong	√		
530	medium	—	533	medium
			515	medium

rules for a helical conformation, the fact that atactic and syndiotactic PMAA have virtually identical Raman and IR spectra strongly suggests that the local conformation of syndiotactic PMAA is also not consistent with the helical model. If syndiotactic PMAA had a regular helical conformation in solution, then one would anticipate fewer active modes in the spectra of syndiotactic PMAA than for the atactic polymer due to the additional symmetry introduced by the helical structure.

The principal Raman line of interest in evaluating the applicability of the second model to PMAA in solution is the 780 cm^{-1} line (carbon–carbon stretching mode most probably associated with the carboxylic acid side chain). Since this mode is expected to be highly coupled with the backbone stretching modes, the 780 cm^{-1} line should be highly sensitive to the backbone internal rotation angles; for example, in the normal coordinate analysis of polybutene-1 (60, 61) the calculated frequency of the analogous stretching mode was 774 cm^{-1} for the 3_1-helix (backbone internal rotations of 120° and 0°, where 0° is the trans conformation) and 790 cm^{-1} for the 4_1-helix (rotations of 90° and 0°). In the Raman spectra of solid samples of isotactic and syndiotactic PMAA, the frequencies of this mode are identical. Hence, the spectroscopic evidence indicated that the backbone internal rotations of PMAA are not very dependent on stereosequencing, a conclusion which is also obtained from the conformational energy calculations for isolated isotactic and syndiotactic helices. Further, since the spectroscopic data indicate that syndiotactic PMAA does not contain long helical segments in solution, it is concluded that the backbone internal rotations in all monomer units are nearly equal in magnitude, but the direction of these rotations varies in an approximately unsystematic manner along the polymer chains.

2. The Tertiary Structure

Since the carbon–carbon stretching mode associated with the acid side chain (which shall be referred to as the carbon–carbon stretching mode, hereafter) is sensitive to the local conformation, this mode is utilized to characterize the local conformation as a function of the degree of neutralization (α) with sodium hydroxide. This mode may also be affected by simple ionization. Therefore, it is necessary to examine model compounds which do not undergo a conformational transition in order to identify ionization effects. The Raman spectra of acetic acid and atactic poly(acrylic acid) (PAA) has been obtained in aqueous solution over the entire range of α. In the spectra of acetic acid, the carbon–carbon stretching line appears at 890 and 924 cm^{-1} for the unionized and ionized groups, respectively; at all degrees of neutralization, the two lines are very narrow and the relative intensities vary linearly with α. Raman spectra of poly(acrylic acid) are shown in Fig. 4-21. At 0% neutralization, the carbon–carbon stretching line is centered at about 840 cm^{-1}. There are two prominent shoulders present, and the line is considerably broader than the corresponding lines in atactic and syndiotactic PMAA. As α increases from 0 to 100%, the 840 and 895 lines behave in a manner similar to acetic acid; the breadth of the lines are independent of α, while the relative intensities change linearly with α.

Raman spectra of syndiotactic PMAA are shown in Fig. 4-22 for 10% solutions at various degrees of neutralization; all of the spectra have been obtained under identical experimental conditions. At 10% neutralization,

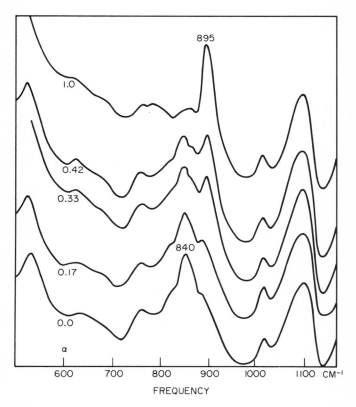

Fig. 4-21. Raman spectra of atactic poly(acrylic acid) in aqueous solution as a function of the degree of neutralization with sodium hydroxide (10% solutions).

the carbon–carbon stretching mode appears as a sharp line at 780 cm^{-1}; the line differs from that at α equal 0% only in that the intensity is slightly reduced. At 20% neutralization, there is a dramatic reduction in the intensity and considerable broadening of the line envelope, although the peak frequently is still close to 780 cm^{-1}. As α increases from 20 to about 50%, the line continues to broaden and simultaneously shifts gradually to higher frequency. In this region of α, shoulders are observed on the line. Beyond

50% neutralization, the line envelope sharpens until about 70% neutralization and subsequently remains sharp up to 100%. Although the difference in frequencies at 0 and 100% neutralization may be partially caused by simple ionization, the observed changes in the intensity and breadth cannot be explained by ionization. Since this vibrational mode is sensitive to the local chain conformation, the line broadening is indicative of a randomization of the local conformation; that is, the breadth of the line is a measure of the distribution of backbone internal rotation angles. Thus, the local conformation of syndiotactic PMAA remains highly regular up to about 20% neutralization. Between 20 and 50% neutralization, the local conformation gradually becomes more distorted with increasing α, and beyond 50% the conformation quickly becomes more regularized. Since the conformational transition of PMAA has been reported to occur between about 15 and 30% neutralization, it is concluded that the observed randomization of the local conformation is associated with the conformational transition. The Raman data suggest that the transition of syndiotactic PMAA is not cooperative, as reported previously (62), but occurs gradually over a relatively wide range of α.

The conformational transition region of syndiotactic PMAA was also studied for 6 and 2.5% aqueous solutions. The behavior of the carbon–carbon stretching mode is identical to that observed in 10% solutions.

The behavior of the carbon–carbon stretching region as a function of α is shown in Fig. 4-23 for isotactic PMAA. At 30% neutralization (the lowest α for which the solubility was high enough to obtain spectra), there are two well-defined lines at 733 and 765 cm^{-1}. As α increases, the two lines remain sharp and do not change in frequency; only the relative peak heights change with α, and the relative intensities are linear in α. This behavior is different in two respects from that of syndiotactic PMAA; (1) the two lines of the isotactic polymer do not exhibit the gradual broadening and frequency shifting observed in the transition region for the syndiotactic polymer; (2) the carbon–carbon stretching line of the ionized carboxyl group has a lower frequency than the unionized group in the isotactic polymer, while the frequency shift was in the opposite direction in the syndiotactic polymer. The failure of the isotactic polymer lines to change in breadth and frequency, especially between 30 and 50% neutralization, suggests that the conformational transition either is complete at 30% neutralization or does not involve a substantial change in the local conformation. The latter explanation is not very likely, since the direction of the shift in frequency of the stretching mode upon complete neutralization of the isotactic polymer is in the opposite direction observed for both the syndiotactic polymer and the atactic PMAA sample. In addition, the carbon–carbon stretching lines of the isotactic polymer exhibit the same behavior between 30 and 100% neutralization as was observed for the syndiotactic polymer at degrees of neutralization beyond the conformational transition region. Therefore, it must be concluded that the conformational transition of isotactic PMAA is complete at 30% neutralization. The transition of the isotactic polymer must occur over a much narrower range of α than the transition of syndiotactic PMAA and it may be cooperative in nature.

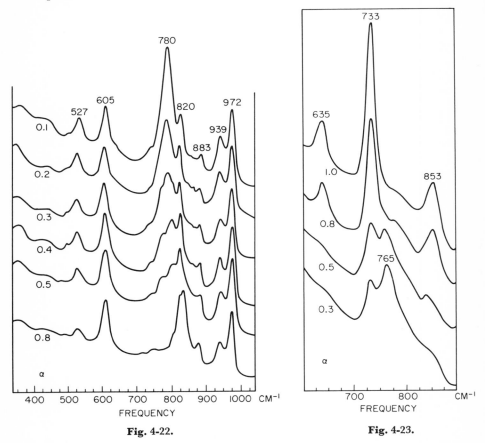

Fig. 4-22.   **Fig. 4-23.**

Fig. 4-22. Raman spectra of syndiotactic PMAA in aqueous solution at various degrees of neutralization with sodium hydroxide (10% solutions).

Fig. 4-23. Raman spectra of isotactic PMAA in aqueous solution at various degrees of neutralization with sodium hydroxide (3% solutions).

Raman spectra of atactic polymer (PMM $= 0.28$, PMR $= 0.40$, and PRR $= 0.32$) at various degrees of neutralization are shown in Fig. 4-24. The strong line at about 780 cm^{-1} exhibits the same behavior as the line in syndiotactic PMAA up to about 30% neutralization. At 30% neutralization, the line envelope shifts to about 770 cm^{-1}, and the shape of the envelope suggests that there are two strong lines present, where the frequencies of the two lines are slightly lower and slightly greater than 770 cm^{-1}. Between about 45 and 100% neutralization, the carbon–carbon stretching region has approximately the

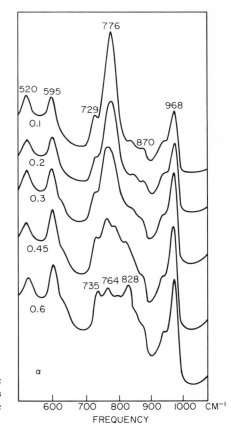

Fig. 4-24. Raman spectra of atactic PMAA in aqueous solution at various degrees of neutralization with sodium hydroxide (10% solutions).

appearance of superimposed spectra of the isotactic and syndiotactic polymers. Although the Raman spectra of atactic PMAA may contain some of the transitional features of isotactic PMAA, the structural mechanics of the conformational transition of atactic PMAA are largely identical to those of syndiotactic PMAA.

VII. The Polymorphic Phase Transition of Poly(Vinylidene Fluoride)[†]

A. Introduction

Poly(vinylidene fluoride) (PVF_2) undergoes a polymorphic phase transition which is strongly influenced by the number of head-to-head units in the homopolymer (63). The backward addition of a vinylidene fluoride molecule gives a head-to-head unit followed by a tail-to-tail unit in the polymer chain. It is worthwhile to obtain a detailed look at how the positional isomerism in the head-to-head defects might alter the conformation and packing of a polymer, and specifically, what role isomorphous replacement has in the polymorphism of poly(vinylidene fluoride).

The crystal structures for both principal phases have been determined: phase I by Lando *et al.* (64), phase II by Doll and Lando (65). The characterization of the transition from phase I to phase II, or vice versa, is limited by the constancy of the percentage of head-to-head units (HHTT) under various polymerization conditions and methods. There is no direct method of looking at the problem at hand. The close similarity between tetrafluoroethylene units and head-to-head units allows the use of copolymer studies, as done by Doll.[‡] However, this method does not allow for decreasing the defect content below the number occurring randomly during the polymerization. It does not give access to information on the defect-free polymer. It is this information that is vital to determining the effects these defects cause. Conformational potential energy calculations provide a simulation technique to calculate packing energy as a function of chain defects.

B. Effects of Head-to-Head Linkages of the Single Chain Conformation

Theoretical potential energy calculations afford the opportunity to investigate the effects of varying the amount of head-to-head units without resorting to pseudo head-to-head units such as tetra- and trifluoroethylene. At the same time, the calculations allow inclusion of comonomers so that the calculated and experimental results can be compared.

[†] I am grateful to J. B. Lando, and also to B. Farmer of CWRU, for giving me the opportunity of discussing B. Farmer's M.S. thesis concerning the transition properties of poly-(vinylidene fluoride).

[‡] See Doll (66) for a complete discussion.

The calculations were restricted to a bond rotation sequence of the type trans-A–trans-B, where $A + B = 360°$. Where A and B are both $0°$, the resulting conformation is the all-trans, planar zigzag, phase I form. Similarly, when A and B are gauche and gauche', respectively, the conformation is the phase II form, trans-gauche-trans-gauche' (TGTG'). Figure 4-25 illustrates

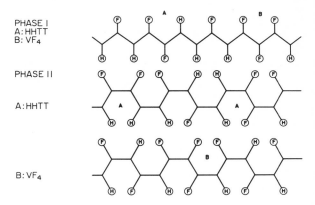

PHASE I
A: HHTT
B: VF₄

PHASE II

A: HHTT

B: VF₄

Fig. 4-25. A schematic diagram of polymer chains in the phase I and phase II conformations, with HHTT and VF₄ units.

the phase I and phase II conformations of poly(vinylidene fluoride).

A fixed set of bond angles and bond distances were used in all calculations. The values of these bond angles and lengths are listed in Table 4-15.

TABLE 4-15

Spatial Parameters Used in the PVF₂ Calculations

Bond lengths		Bond angles	
Bond	Length (Å)	Bond	Angle (deg)
C—C	1.54	H—C—H	110
C—H	1.10	F—C—F	110
C—F	1.34	C—C—C	114 (see text)

The value of the backbone C—C—C bond angle chosen for subsequent work was 114°, a value intermediate between the 112.3° found experimentally for phase I and the 115.5° found for phase II. This value was chosen to avoid weighting the subsequent calculations in favor of a particular conformation when considering the results of HHTT defects on the chain conformation.

The inclusion of head-to-head units in the polymer chain is now considered.

The potential energy curves for the TATB type of chain, as a function of the percentage of HHTT units present, is shown in Fig. 4-26. As the number of defects increases, the minimum at 150° is seen to form a plateau at approximately 120° for 5% head-to-head units, and then to split into two minima, both of which become higher in energy. At the same time, the all-trans minimum remains approximately the same in energy. Between 10 and 15% head-to-head additions, the phase I and phase II forms are equal in energy. A greater percentage of HHTT units makes the phase I form the low energy conformation. While the energy for the minimum at 150° is lower than the energy at 120°, it also increases more rapidly, so that both exceed the 0° minimum at about the same concentration of HHTT defects.

The reason for this behavior of the phase I and II minima upon addition of head-to-head units can be seen in Fig. 4-25. In the planar zigzag form, the placement of fluorine atoms in positions normally occupied by hydrogen atoms does not cause any great problem, since the sum of the van der Waals radii of a hydrogen and a fluorine is less than the distance between these atoms on the chain (2.45 Å compared to 2.56 Å). Thus, there is no strain introduced by a head-to-head unit. In fact, there will be some relief of the strain along the fluorine side of the chain. The fluorine atoms, separated by only 2.56 Å, are given somewhat more room by the placement of two hydrogen atoms along that side of the backbone.

In the phase II form, on the other hand, the interference caused by a head-to-head unit is considerable, placing two fluorine atoms only 2.3 Å apart, 0.4 Å less than the sum of their van der Waals radii. It is thus obvious why the energy increases so dramatically for the TGTG′ type chain, as illustrated in the potential energy curves.

The effects of adding tetrafluoroethylene units can also be seen in the diagram. Obviously, the tetrafluoroethylene unit represents a pseudo head-to-head defect in that the head-to-head linkage $(CF_2—CF_2)$ is present, although there is no tail-to-tail unit $(CH_2—CH_2)$ immediately following it. Thus there is a $CF_2—CF_2—CF_2$ sequence in the chain, not found in a real head-to-head unit. In the planar zigzag form, the VF_4 unit may cause a minor increase in energy due to the slightly greater bulk of the fluorine atoms on the hydrogen side of the chain. There will be no relief of strain along the fluorine side, as there is with a true head-to-head unit. In the TGTG′ conformation, the effect of a VF_4 unit will also be about the same as a true HHTT unit: similar severe fluorine–fluorine contacts occur in both cases. Since the VF_4 unit introduces two extra fluorine atoms into the chain, as opposed to simply switching the positions of fluorine and hydrogen atoms, as occurs in a true HHTT unit, a difference in the calculated energy is expected.

The effect of including a trifluoroethylene (VF_3) unit should be in the same direction as adding a VF_4 unit, although the magnitude of the change should

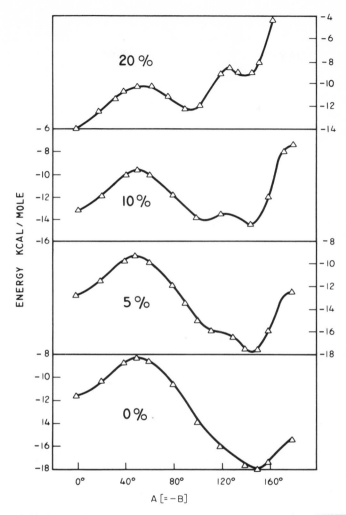

Fig. 4-26. Potential energy curves for various defect concentrations (HHTT).

be about half as large. The added complication of considering the stereo-regularity of the defects make these calculations somewhat less definitive than the symmetric VF_4 case.

The effect of head-to-head units and tetrafluoroethylene units may be determined by plotting head-to-head defects versus conformational potential energy. These plots turn out to be linear. The energy is in agreement with the intuitive analysis outlined above. The linearity may be surprising at first. However, when one considers that each defect contributes precisely the same defect interactions as all other defects, linearity is to be expected. This presumes that the defects in the PVF_2 chains are isolated from one another, as in the conformational calculations. According to the NMR study of Wilson (67) the defects are randomly distributed along the polymer chain. Since the defect concentration is low, a random distribution suggests such isolated HHTT units are present in the polymer.

Figures 4-27 and 4-28 show the conformational potential energy versus percent of defects for a particular kind of defect in one of the two conformations of interest.

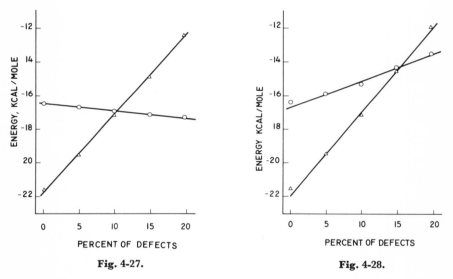

Fig. 4-27.   Fig. 4-28.

Fig. 4-27. Potential energy of the all-trans ($\circ$) and TGTG' ($\triangle$) conformations versus HHTT concentration.

Fig. 4-28. Potential energy of the all-trans ($\circ$) and TGTG' ($\triangle$) conformations versus VF_4 concentration.

The crossover point of the lines indicates at what concentration of that defect the conformations have equal energies. Above this concentration, the phase I form will be the low energy conformation. For head-to-head units, this occurs at 11% concentration, and for tetrafluoroethylene units, at 15% concentration. The higher value for VF_4 units is a result of the increase in energy of the phase I form as VF_4 units are added, as opposed to the behavior of the phase I energy upon adding HHTT units.

It is necessary to place 5% head-to-head units in the chain and then to increase the amount of VF_4 units present in order to compare calculated and experimental results. This is analogous to the various copolymer ratios used experimentally. The results of such calculations are shown in Fig. 4-29. The conformational potential energy for each conformation is plotted against the amount of tetrafluoroethylene added to the 5% HHTT units in the chain. The concentration of VF_4 which causes the two conformations to become equal in energy is 8%. This is in very good qualitative agreement with the experimental results which show the phase I form to be preferred at 7% comonomer concentration (63).

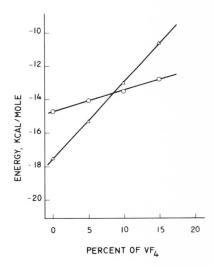

Fig. 4-29. Potential energy of the all-trans (○) and TGTG' (△) conformations versus the concentration of VF_4 added to 5% HHTT defects.

C. Effects of Head-to-Head Units on Chain Packing

The calculations have been carried out by looking first at the structure found by X-ray analysis and then perturbing the chain positions, rotations, and shifts, until the lowest energy was obtained. Such calculations, therefore, yield lattices related to the X-ray structure.

Two specific types of calculations are reported in this section. First, for determining unit cell dimensions, the system considered is one unit cell of the polymer with the chains placed so that all interaction types that occur in the crystal of PVF_2 are represented in their correct proportions. In these calculations the chains are allowed to slide along and rotate about their respective chain axes. Second, a crystal type calculation is used to obtain the total energy of a structure and to look at the effect of randomly placed HHTT defects upon the crystal packing of the chains. The interactions of all the atoms in a monomer unit located at the center of the crystal are calculated in a pairwise manner with all the other atoms on other chains located within a sphere centered at this central monomer unit. A diagram of the crystal and the interaction sphere are shown in Fig. 4-30.

The nature of the van der Waals interactions is such that above a separation distance of 10 Å, the energy for a pairwise interaction is negligible. Therefore, by taking interactions with atoms lying within a sphere with a radius of 10 Å, the energy calculated is effectively that encountered by an atom in an infinite crystal. By placing a monomer unit in the center of the sphere, the calculated energy is on a "per monomer unit" basis.

Consideration of coulombic terms necessitates expanding the interaction

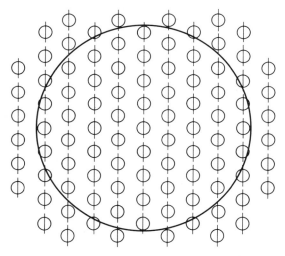

Fig. 4-30. Schematic diagram of a crystal and the interaction sphere used in these calculations.

sphere since the energy contribution between, say, two carbon atoms with two fluorine atoms attached to each, is not negligible at a distance of 10 Å. In fact, the contribution can be significant even out to 50 Å. The number of calculations involved for a sphere of this size is impractical. However, calculations show the total energy changes very little in going from spheres having radii of 10, 15, and 20 Å. Since a monomer unit in the chain is electrically neutral, and since in most cases an increase in sphere size will take in whole monomer units, the net change in energy will be approximately zero due to a balance of charge interactions.

Both the unit cell chain calculation and the crystal-type calculation have been used to investigate the packing of chains with and without head-to-head defects. The defects are handled differently in the two cases. In the crystal calculations, the defects are distributed randomly throughout the system, with some care being taken to assure that there is no buildup of defects near the center, leading to spurious results. Different sets of random defects are also used in order to avoid some fortuitous result. Calculations involving the unit cell inherently requires placing the defects in specific positions and observing their effects; then changing their locations and seeing the new effects. For example, placing a head-to-head unit in one location might cause the unit cell to expand in the a direction, while placing it in another position might cause an increase in b. Obviously, the actual situation will be a combination of both effects, since a defect in one orientation in one unit cell will be in an alternate position in a unit cell having a different origin. The overall effect of including a defect will therefore be to increase the size of the unit cell in both the a and b directions.

The starting structure used in the packing calculations is the structure determined by Lando, Olf, and Peterlin for phase I of poly(vinylidene fluoride). This structure has an orthorhombic unit cell with space group C_{m2m} and with dimensions $a = 8.47$ Å, $b = 4.90$ Å, and $c = 2.56$ Å (chain axis) (64).

Conformational potential energy calculations have been used to determine the unit cell dimensions, the placement of the center chain—that is, whether it is shifted slightly along one of the axes in the ab plane—and the rotation and shift of the chains with respect to each other. With the exception of the unit cell dimensions, each perturbation yields a higher energy than the structure found by X-ray analysis. The unit cell dimensions giving the lowest energy are somewhat smaller than those found by X ray: $a = 8.05$ Å, $b = 4.25$ Å (c was not varied).

When HHTT units are added to the phase I form, giving a simulated polymer similar to the actual polymer, the unit cell dimensions increases to $a = 8.20$ Å and $b = 4.40$ Å, but the chain shift in the c direction, the center chain position, and the chain rotations do not change from the structure of the completely regular polymer. The overall energy for the chains with defects is higher than for the defect-free chains. The potential energy curves for the calculations with and without HHTT units for the phase I form of PVF_2 are shown in Fig. 4-31.

From the results of the calculations of phase I it can be readily seen that the conformational potential energies do reflect the correct orientation between chains, even though the unit cell dimensions differ significantly from the experimentally determined values. With this in mind, we next look at the structure of phase II of PVF_2.

The starting point for these calculations is the structure for phase II found by Doll and Lando (65), except taken with the origin shifted to place chains at the corners of the cell, with the center chain located at $(\frac{1}{2}, \frac{1}{2})$, and with the chains not rotated about their respective axes. Doll and Lando were unable to distinguish between parallel or antiparallel chains, corresponding to triclinic, P_1, or monoclinic, P_{2_1}, symmetry respectively. Packing in both ways has been examined by theoretical conformational analysis.

Using the conformation of Doll and Lando with no head-to-head units, the

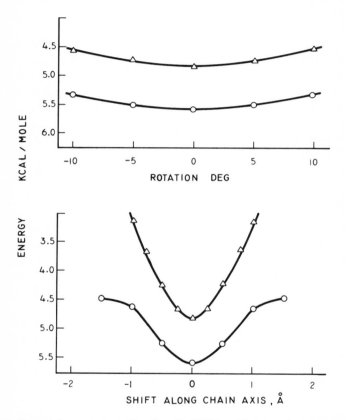

Fig. 4-31. Potential energy curves for phase I of PVF$_2$, with ($\triangle$) and without ($\bigcirc$) HHTT.

unit cell dimensions for phase II which gives the lowest energy is smaller than the experimental values.

Monoclinic:	$a = 4.65$ Å	$b = 8.75$ Å
Triclinic:	$a = 4.65$ Å	$b = 8.85$ Å
Experimental:	$a = 5.02$ Å	$b = 9.63$ Å

The calculations on chain rotation do not agree with the X-ray structure, the minimum energy corresponding to the unrotated chains. The shift of the center chain along its axis with respect to the equivalent corner chains (five chain calculation) differs from the experimental values by approximately 1 Å for both symmetries. The position of the center chain along the *a* axis gives a minimum energy when the chain is located at 0.54 of the *a* axis (0.50 being the center of the unit cell) for both parallel and antiparallel chains. This agrees well with the experimental value of 0.524 found by Doll and Lando. This shift along *a* allows the fluorine atoms on adjacent chains to move further apart.

The results of these calculations on phase II chains without head-to-head defects, while not agreeing with the structure found by Doll and Lando, do agree somewhat better with the results of Tadokoro (68), who worked with a polymer having *less* than 5% head-to-head units. This better agreement would be the expected result since the simulated polymer is more like that used by Tadokoro than that used by Doll and Lando. Although the calculated unit cell dimensions are smaller than his experimental values, the nonrotated chains, the shift of the center chain along its axis, and the shift away from $a = 0.50$ for the center chain are in agreement with Tadokoro's structure for phase II. To put it concisely, chains containing no HHTT units and in the conformation found by Doll and Lando, pack in the manner found by Tadokoro.

When 5–6% head-to-head units are included in the simulated polymer, still using Doll and Lando's conformation, the unit cell is once again found to be smaller than the experimental cell, but larger than the cell for defect-free chains:

$$\text{Monoclinic:} \quad a = 4.85 \text{ Å} \quad b = 9.00 \text{ Å}$$
$$\text{Triclinic:} \quad a = 4.85 \text{ Å} \quad b = 8.95 \text{ Å}$$

The shift of the center chain along the *a* axis for the monoclinic cell remains the same as when no HHTT units are present, while for the triclinic cell, the *a* distance between the chains is 0.52 of the *a* axis repeat. The conformational potential energy curve for shifting the center chain along its axis changes dramatically. Instead of a sharp, definite minimum, a broad energy trough occurs, allowing a shift of 2 Å along the chain axis without changing the energy significantly. The chain position found by Doll and Lando now falls within the low energy region. It is clear that the low energy region is bounded by positions where fluorine atoms on adjacent chains are at the same level.

The rotations of the chains about their axes are also changes from the results found for defect-free chains. The rotations found by Doll and Lando correspond to rotating one chain clockwise and the other counterclockwise, both by 7°. The rotations found to give the lowest energy for chains with HHTT units correspond to this same type of rotation of both chains. Rotations of 4 to 5° gives the lowest conformational potential energy for parallel and antiparallel chains.

The calculations on the polymer with head-to-head units agrees well with the experimental results of Doll and Lando. The calculations indicate the change in position along the *c* axis, the rotation of the chains, and the shift of the chains to a position where they are more than half the *a* axis repeat distance apart. When considered with the results of Tadokoro and the agreement of the calculations on defect-free chains with his experimental results, it is apparent that the presence of HHTT units is the source of the differences in the structures of phase II as proposed by Doll and Lando and that advanced by Tadokoro. In Table 4-16 is reported the total conformational potential

TABLE 4-16

Conformational Potential Energies for
Phase I and Phase II of Poly(vinylidene Fluoride)

		Without HHTT	With 5% HHTT
Phase I			
	Intramolecular	−16.4 kcal/mole	−16.6 kcal/mole
	Intermolecular	−69.5	−58.6
	Total	−85.9	−75.2
Phase II			
	Intramolecular	−21.6	−19.5
Monoclinic	Intermolecular	−57.7	−56.5
	Total	−79.3	−76.0
	Intramolecular	−21.6	−19.5
Triclinic	Intermolecular	−63.9	−56.9
	Total	−85.5	−76.4

energies for phase I and phase II of poly(vinylidene fluoride) with and without HHTT defects.

For chains without head-to-head units, the calculations indicate that for phase II, the triclinic cell is lower in energy than the monoclinic cell. The difference is due largely to the difference in the coulombic energies for the two forms. Upon adding HHTT defects to the simulated chains, the monoclinic and triclinic cells have virtually the same energy. While addition of defects to the monoclinic form causes only a small change in the packing energy, the change in the triclinic packing energy is quite pronounced. (A similar change occurs on adding head-to-head units to phase I.)

D. Summary

We arrive at virtually the same results using the conformational potential energy calculations as we do from the X-ray investigations. It is not possible to distinguish between the two possible structures for phase II of poly(vinylidene fluoride). This is a reasonable result. In both approaches, the atomic positions are the basis for the determination of the structure. Since both monoclinic and triclinic packing of the chains result in very nearly the same atomic positions, neither X-ray nor potential energy calculations could reasonably be expected to distinguish between the two possibilities. The experimental difficulties involved in X-ray analysis and the theoretical difficulties involved in potential energy computation rob the respective techniques of the exactitude necessary to probe and detect the subtle differences between the parallel and antiparallel packing of poly(vinylidene fluoride) chains. It is, of course, possible that the actual structure consists of a random array of chain directions.

VIII. Solvent-Dependent Conformational Transitions in Proteins

A. Introduction

Denaturation studies of proteins using solvents as the denaturing agents indicate the high sensitivity of the tertiary (in some cases even the quaternary) structure of the protein to interaction with solvents. It appears that different solvents will "unfold" and "refold" a protein to a degree characteristic of the macromolecule and the solvent (69). Like all conformational transitions involving macromolecules, the change in the shape of the protein, that is the "unfolding" or "refolding," is a result of a free energy transfer between protein and solvent.

While the process of denaturation provides information concerning the nature of the forces which hold proteins together, it is very difficult to determine the contribution from any single force to the overall stability of the protein. For homopolypeptides, Schellman (70) was able to estimate the enthalpy, ΔH_0, and entropy ΔS_0 changes, per residue for an infinitely long chain during the helix–coil transition. Privalov (71) obtained values of ΔH_0 and ΔS_0 very similar to Schellman using an exact calorimetric method. Nemethy, Steinberg, and Scheraga (72), as part of their detailed analysis of hydrogen and hydrophobic bonds in water–protein systems, estimated the values of the free energy of interaction of such bonds. The molecular surface tensions reported here have been calculated using a polymer–solvent model similar to that proposed by Gibson and Scheraga (73). Outside of the above-mentioned investigations little definitive information is available on the values of the various types of free energy changes accompanying protein denaturation. In this section the untested and rather bold assumption is made that conformational potential energy functions can be used in estimating these various free energy contributions. Such potential energy functions have been used extensively in the last few years† in predicting and/or verifying conformations of regularly sequenced macromolecules. Conformational potential energy calculations have also been used to predict the melting temperatures of various collagen triple-helix models as reported in Section V of this chapter.

Still, we have no reason to believe that the molecular energetics associated with such calculations are quantitatively accurate on an absolute scale. The calculations carried out for the collagen triple helices do suggest, however, that the molecular energetics are reliable on a relative basis. Although we are quite aware of the many considerations which are overlooked in the computation of the energy parameters reported here, we feel that the results from these types of studies are useful for two reasons. First, the values of the free energy changes calculated from the conformational potential energy functions should be qualitatively correct. Therefore, we should be able to pick out those types of interactions which dominate in the denaturation process. Secondly, there is a need to know which types of thermodynamic and kinetic processes can be studied with the aid of conformational energy calculations.

Once we had decided to attempt to use conformational potential functions in the study of protein denaturation we then had to decide which statistical mechanical model describing this phenomenon to adopt. After studying a few denaturation models it became apparent that the models are based upon two-state, nearest neighbor interaction constraints. Formulations differ with respect to which types of free energy interactions are retained or neglected. While no present theory of protein denaturation is quantitatively accurate, we decided to adopt a model recently proposed by Shindo (77) which accounts for nearly all the qualitative features observed in experimental protein denaturation studies. In its most basic form this theory is equivalent to the classical Langmuir adsorption model used to describe the behavior of fluid molecules above and on a surface.

B. Summary of the Shindo Protein Denaturation Model

In this model each of the residue units in the protein can exist in one of two states—a folded state (f) and an unfolded state (u). There is a characteristic equilibrium between these two states which is a function of the structural and

---

† Hopfinger (74, 75) and Flory (76) contain an extensive composite listing of work done with conformational potential energy functions.

chemical properties of the residues. For the present, one distinguishes between residues by categorizing them as either polar or nonpolar. Thus each of the residues composing the protein fits one of the four following assignments:

(1) folded—polar, (f, P)
(2) folded—nonpolar, (f, N)
(3) unfolded—polar, (u, P)
(4) unfolded—nonpolar, (u, N)

The protein is restricted to oscillating between two limited states. First is the "native" state in which the protein is folded in such a way that its interior is completely devoid of, and inaccessible to, solvent. The other limiting conformation, the "random" state, is assumed to be a flexible coil which is sufficiently "open" so that all parts of the protein are in contact with solvent. The residues have one adsorption site for binding solvent molecules. These adsorption sites for binding are identical, in each case, for both f and u states. However, actual adsorption is allowed only for u states because f states are inaccessible to solvent. To keep the model unbiased, we allow conversion from f states to u states and vice versa. At any given instant there are L_n clusters of nonpolar solvent molecules adsorbed on nonpolar units in the u state and L_p clusters of polar solvent molecules adsorbed on polar units in the u state. τ_P and τ_N are the surface tensions on the polar and nonpolar areas in the u state, respectively.

Only six types of interresidue and/or solvent energy interactions are allowed; four involving residue–residue interactions and two involving solvent cluster–solvent cluster interactions. The residue–residue interactions can be represented by $E(a, b, 1)$ where a denotes an f or a u state and b denotes a polar (P) or nonpolar (N) residue. The one (1) indicates only first nearest neighbor interactions are considered. The solvent cluster–solvent cluster interactions can be represented by $E(c/c, b/b, 1)$ where the c denotes a solvent cluster and the b and 1 are the same as before. The quasipartition functions associated with the allowed interactions are

$$Z(f, P) = \int \exp[(-E(f, P, 1))/RT] \, d\mathcal{V}$$

$$Z(f, N) = \int \exp[(-E(f, N, 1))/RT] \, d\mathcal{V}$$

$$Z(u, P) = \int \exp[(-E(u, P, 1))/RT] \, d\mathcal{V}$$

$$Z(u, N) = \int \exp[(-E(u, N, 1))/RT] \, d\mathcal{V} \tag{4-29}$$

$$Z(c/c, P/P, 1) = \int \exp[(-E(c/c, P/P, 1))/RT] \, d\mathcal{V}$$

$$Z(c/c \ N/N \ 1) = \int \exp[(-E(c/c, N/N, 1))/RT] \, d\mathcal{V}$$

Also considered as contributing to the total energy of the system are the internal residue interaction energies. These interactions give rise to four types of internal residue partition functions, $K(a,b)$, where the a and b are defined the same as above.

Shindo has shown that these assumptions lead to two denaturation equations, one for polar residues, the other for nonpolar residues, each of which have the form

$$A \ln \left[\frac{a_i}{1 + b_i \lambda_i} \right] - \left[\frac{c_i \lambda_i}{1 + b_i \lambda_i} - B \right] = 0 \qquad (4\text{-}30)$$

Each of the terms in Eq. (4-30) is defined in Table 4-17. The λ are, by the way they are defined, proportional to the effect of adsorbate concentration.

The θ denote the fraction of polar or nonpolar residues which remain in the f state, relative to the total number of polar or nonpolar residues in the

TABLE 4-17

Definition of the Terms in the General Form of the Denaturation Equation Given
by Eq. (4-30)

B	
Polar	Nonpolar
$\tau_P A(u/P)/RT$, where	$\tau_N A(u/N)/RT$, where
$\quad A(u/P) = A_P N(u/P)$	$\quad A(u/N) = A_N N(u/N)$
$\quad N(u/P) =$ number of polar residues in the u state for the native protein conformation	$\quad N(u/N) =$ number of nonpolar residues in the u state for the native protein conformation
$\quad A_P =$ area of a polar residue in the u state	$\quad A_N =$ area of a nonpolar residue in the u state

A	
Polar	Nonpolar
$\dfrac{Z(c/c, P/P, 1)}{Z(u, P)}$	$\dfrac{Z(c/c, N/N, 1)}{Z(u, N)}$

b_i	
Polar	Nonpolar
$A^{\theta_i(p)/\mathscr{P}_P}$, where	$A^{\theta_i(n)/\mathscr{P}_N}$, where
$\quad \theta_i(p) =$ fractional number of polar residues in the f state	$\quad \theta_i(n) =$ fractional number of nonpolar residues in the f state
$\quad \mathscr{P}_P = 1 - \mathscr{P}_N$	$\quad \mathscr{P}_N =$ fractional number of nonpolar residues in the protein

TABLE 4-17 (continued)

$$a_i$$

Polar

$$\left(\frac{\theta_i(\mathrm{P})}{\mathscr{P}_\mathrm{P} - \theta_i(\mathrm{P})}\right)\left(\frac{K(\mathrm{f},\mathrm{P})}{K(\mathrm{u},\mathrm{P})}\right)\left(\frac{Z(\mathrm{f},\mathrm{P})^{2(1-\theta_i(\mathrm{P})/\mathscr{P}_\mathrm{P})}}{Z(\mathrm{u},\mathrm{P})^{2\theta_i(\mathrm{P})/\mathscr{P}_\mathrm{P}}}\right)$$

Nonpolar

$$\left(\frac{\theta_i(\mathrm{N})}{\mathscr{P}_\mathrm{N} - \theta_i(\mathrm{N})}\right)\left(\frac{K(\mathrm{f},\mathrm{N})}{K(\mathrm{u},\mathrm{N})}\right)\left(\frac{Z(\mathrm{f},\mathrm{N})^{2(1-\theta_i(\mathrm{N})/\mathscr{P}_\mathrm{N})}}{Z(\mathrm{u},\mathrm{N})^{2\theta_i(\mathrm{N})/\mathscr{P}_\mathrm{N}}}\right)$$

C_i	
Polar	Nonpolar
$\left(\dfrac{\theta_i(\mathrm{P})}{\mathscr{P}_\mathrm{P}}\right) b_i$	$\left(\dfrac{\theta_i(\mathrm{N})}{\mathscr{P}_\mathrm{N}}\right) b_i$

λ_i	
Polar	Nonpolar
$\pi(\mathrm{P})\exp(\mu_\mathrm{P}/RT)$, where	$\pi(\mathrm{N})\exp(\mu_\mathrm{N}/RT)$, where
$\pi(\mathrm{P})$ = internal partition function of an adsorbed polar solvent molecule cluster	$\pi(\mathrm{N})$ = internal partition function of an adsorbed nonpolar solvent molecule cluster
μ_P = chemical potential of a solvent molecule cluster relative to a polar residue	μ_N = chemical potential of a solvent molecule cluster relative to a nonpolar residue

protein for a particular value of λ. Thus Eq. (4-30) incorporates a functionality between θ and λ. Unfortunately, Eq. (4-30) is transcendental and must be solved numerically. For a transcendental function $G(\theta_i, \lambda_i) = 0$ the Newton–Rhapheson method (78) may be used to compute λ_i as a function of θ_i. Specifically,

$$\lambda_i{}^k = \lambda_i^{k-1} - \frac{G(\theta_i, \lambda_i^{k-1})}{G'_{\lambda_i}(\theta_i, \lambda_i^{k-1})} \tag{4-31}$$

where $\lambda_i{}^m$ represents the mth estimate of λ_i, and $G'_{\lambda_i}(\theta_i, \lambda_i^{k-1})$ is the first derivative of $G(\theta_i, \lambda_i^{k-1})$ with respect to λ_i. For $G(\theta_i, \lambda_i^{k-1})$ being the function given by Eq. (4-30), $G'_{\lambda_i}(\theta_i, \lambda_i^{k-1})$ is

$$G'_{\lambda_i}(\theta_i, \lambda_i^{k-1}) = \frac{[(b_i c_i \lambda_i^{k-1})/(1 + b_i \lambda_i^{k-1})] - (Ab_i + c_i)}{(1 + b_i \lambda_i^{k-1})} \tag{4-32}$$

The major task which now remains is to evaluate the values of the A, B, a_i, b_i, and c_i for different proteins in different solvent mixutres.

C. Evaluation of the A, B, a_i, b_i, c_i

The values of $\mathscr{P}_P$ and $\mathscr{P}_N$ are calculated by simply summing up the number of polar and nonpolar side chains in a protein. The assignment of polar and nonpolar side chains is given in Table 4-18. Clearly some of these assignments

TABLE 4-18

Assignment of Polar and Nonpolar Polypeptide Side Chains

Polar		Nonpolar	
Aspartic acid	Glutamine	Tryptophan	Glycine
Glutamic acid	Histidine	Phenylalanine	Alanine
Tyrosine	Lysine	Proline	Cysteine
Serine	Arginine	Methionine	Threonine
		Valine	Asparagine
		Leucine	

can be questioned. However, a change in assignment of one or two side chains does not seriously change the θ versus λ plots. Table 4-19 contains the values of $\mathscr{P}_P$ and $\mathscr{P}_N$ for four proteins. In order to calculate the role of the K in the

TABLE 4-19

Ratio of Polar and Nonpolar Side Chains in Four Globular Proteins

Protein	Polar fraction ($\mathscr{P}_P$)	Nonpolar fractions ($\mathscr{P}_N$)
Lysozyme (human)	0.488	0.512
Ribonuclease S (bovine)	0.524	0.476
Myoglobin (Sperm Whale)	0.510	0.490
α-Chymotrypsin	0.329	0.671

G functions we make the assumption that the contributions to the K from interactions involving side chains is the same for polar and nonpolar side chains in each of the two respective conformational states f and u. Then

$$\frac{K(f, N)}{K(u, N)} = \frac{K(f, P)}{K(u, P)} = \frac{Q(f)}{Q(u)} \tag{4-33}$$

where $Q(\text{f})$ is the partition function of a "typical protein unit" constrained to the f state. We approximate $Q(\text{f})$ by the following:

$$W(\text{f}) = [(M(\text{N}) + (M(\text{P}))]^{-1} [B(\alpha, \text{H}^\beta) \, E(\alpha, \text{H}^\beta) + B(\alpha, \text{C}^\beta) \, E(\alpha, \text{C}^\beta)$$
$$+ B(\beta, \text{H}^\beta) \, E(\beta, \text{H}^\beta) + B(\beta, \text{C}^\beta) \, E(\beta, \text{C}^\beta)$$
$$+ B(3_1, \text{P}_2) \, E(3_1, \text{P}_2) + B(3_1 \, \text{H}^\beta) \, E(3_1, \text{H}^\beta)$$
$$+ B(3_1, \text{C}^\beta) \, E(3_1, \text{C}^\beta)] \tag{4-34}$$

where $M(\text{N})$ and $M(\text{P})$ are the total numbers of nonpolar and polar residues in the protein;

$$Q(\text{f}) = \exp[-W(\text{f})/RT] \tag{4-35}$$

and $B(\text{v}, \text{w})$ refers to the number of peptide residues having a w unit as a side chain and being in the v conformation when the protein is in the native state. The average internal energy of the (v, w) residue is denoted by $E(\text{v}, \text{w})$. We distinguish between three types of residues; glycyl-H^β side chain, alanyl-C^β side chain, P_2-pyrrolidine ring side chain; and three types of conformations; α-helix, β-"helix," 3_1-polyproline II helix. In other words, the native protein conformation is considered to be composed of glycyl, alanyl, and proline residues which are in the α-helical, β-helical, or PP II-helical conformations. The alanyl residue is assumed to serve as a model, with respect to the K partition function, for all amino acid residues other than glycine. The values of the partition functions are estimated using the conformational potential functions given in Chapter II and discussed by Ooi *et al.* (79). The partition function $Q(\text{u})$ is assumed to be given by

$$W(\text{u}) = \frac{[B(\text{R}, \text{H}^\beta) \, E(\text{R}, \text{H}^\beta) + B(\text{R}, \text{C}^\beta) \, E(\text{R}, \text{C}^\beta) + B(\text{R}, \text{P}_2) \, E(\text{R}, \text{P}_2)]}{(M(\text{N}) + M(\text{P}))}$$
$$\tag{4-36}$$

$$Q(\text{u}) = \exp[-W(\text{u})/RT] \tag{4-37}$$

where R refers to a "random coil" residue. The random coil residue is assumed to be thermodynamically equivalent to the thermodynamic properties of two planar peptide units and the appropriate side chain, all of which are joined at a common C^α atom. Table 4-20 contains the values of the various $E(\text{v}, \text{w})$. The justification for allowing the alanyl residue K partition function to represent all amino acids other than glycine follows from the fact that all the (ϕ, ψ) energy contour maps involving isolated residues other than glycine are very similar (80). Therefore, the corresponding partition functions should be nearly equal. The $B(\text{v}, \text{w})$ can be calculated only when the tertiary structure of the protein is known. Even when the tertiary structure is known it is sometimes a value judgement, based upon model building, to decide which assignment, α, β, or P_2 should be made for a residue which has a structure intermediate between these three fixed conformations. In cases where the tertiary structure of the native conformation of the protein is not known one is forced to choose "reasonable" sets of distributions of the $B(v,w)$ based upon the primary structure of the protein. More will be said about the $B(v,w)$ at the end of this section.

TABLE 4-20

Values for the E (v, w), the Intraresidue Energies[a]

v	w	$E(v, w)$ (kcal/mole-residue)
α	H^β	-1.83
α	C^β	-1.93
β	H^β	-2.64
β	C^β	-2.34
3_1	H^β	-2.85
3_1	C^β	-2.31
3_1	P_2	-1.14
R	H^β	-3.36
R	C^β	-2.90
R	P_2	-2.25

[a] These are, in fact, average internal energies based upon a random ensemble of conformational points located within ten degrees of the (ϕ, ψ) assigned to the appropriate "standard conformation" (80).

In order to calculate the Z partition functions we invoke the same assumptions used to calculate the K partition functions. This allows us to express

$$
\begin{aligned}
W(f,N) = W(f,P) = [&C(\alpha, H^\beta/H^\beta) \cdot E(\alpha, H^\beta/H^\beta) + C(\alpha, H^\beta/C^\beta) \\
&\cdot E(\alpha, H^\beta/C^\beta) + C(\alpha, C^\beta/H^\beta) \cdot E(\alpha, C^\beta/H^\beta) + C(\alpha, C^\beta/C^\beta) \\
&\cdot E(\alpha, C^\beta/C^\beta) + C(\beta, H^\beta/H^\beta) \cdot E(\beta, H^\beta/H^\beta) + C(\beta, H^\beta/C^\beta) \\
&\cdot E(\beta, H^\beta/C^\beta) + C(\beta, C^\beta/H^\beta) \cdot E(\beta, C^\beta/H^\beta) + C(\beta, C^\beta/C^\beta) \\
&\cdot E(\beta, C^\beta/C^\beta) + C(3_1, P_2/P_2) \cdot E(3_1, P_2/P_2) + C(3_1, H^\beta/H^\beta) \\
&\cdot E(3_1, H^\beta/H^\beta) + C(3_1, H^\beta/C^\beta) \cdot E(3_1, H^\beta/C^\beta) + C(3_1, H^\beta/P_2) \\
&\cdot E(3_1, H^\beta/P_2) + C(3_1, C^\beta/C^\beta) \cdot E(3_1, C^\beta/C^\beta) + C(3_1, C^\beta/H) \\
&\cdot E(3_1, C^\beta/H^\beta) + C(3_1, C^\beta/P_2) \cdot E(3_1, C^\beta/P_2) + C(3_1, P_2/H^\beta) \\
&\cdot E(3_1, P_2/H^\beta) + C(3_1, P_2/C^\beta) \cdot E(3_1, P_2/C^\beta)] C_T^{-1}
\end{aligned} \tag{4-38}
$$

where $C_T = \sum C(v, w_1/w_2)$ and

$$
Z(f,N) = Z(f,P) = \exp[-W(f,N)/RT] \tag{4-39}
$$

Let us define $E(\alpha, w_1/w_2)$ as (81)

$$E(\alpha, w_1/w_2) = E(\alpha, i/i + 1) + E_H \qquad \text{where} \quad E_H = -4.6 \quad \text{kcal/mole} \qquad (4\text{-}40)$$

where the first term on the right identifies the usual first nearest neighbor interaction energy and the second term accounts for the stabilizing hydrogen bond energy due to the fourth nearest neighbor. If we were to neglect the effect of the hydrogen bonds there would be no reason to choose the α-helix as a stable folded f state. In the β structure we assume that, on the average, one of the two possible hydrogen bonding groups, i.e., the C=O or the N—H, in the backbone is involved in a hydrogen bond. That is

$$E(\beta, w_1/w_2) = E(\beta, i/i + 1) + \tfrac{1}{2}E_H \qquad (4\text{-}41)$$

The first nearest neighbor partition function for a pair of residues in the u state is, by the nature of the model

$$Z(u, N) = Z(u, P) = 1 \qquad (4\text{-}42)$$

By doing this we are assuming that the unfolded state is, in fact, a random conformation. Table 4-21 contains the values of the $E(v, w_1/w_2)$. To estimate

TABLE 4-21

Values of the $E(v, w_1/w_2)$, Nearest Neighbor Interaction Energies[a]

v	w_1	w_2	$E(v, w_1/w_2)$ (kcal/mole-residue)
α	H^β	H^β	−3.54
α	H^β	C^β	−3.54
β	H^β	H^β	−2.40
β	H^β	C^β	−2.48
3_1	H^β	H^β	−1.74
3_1	H^β	C^β	−1.74
3_1	H^β	P_2	−1.90
α	C^β	H^β	−3.54
α	C^β	C^β	−3.54
β	C^β	H^β	−2.47
β	C^β	C^β	−2.55
3_1	C^β	H^β	−1.70
3_1	C^β	C^β	−2.07
3_1	C^β	P_2	−2.01
3_1	P_2	H^β	−1.73
3_1	P_2	C^β	−2.00
3_1	P_2	P_2	−2.16

[a] All random first nearest neighbor interactions are assumed to be zero.

Note: The $E(v, w_1/w_2)$ are, as the $E(v, w)$, average internal energies based upon ensemble averaging identical to that used in the computation of the $E(v, w)$.

the $Z(c/c, P/P, 1)$ and the $Z(c/c, N/N, 1)$ we computed the partition functions associated with the geometry shown in Fig. 4-32. The values of the d ranged from 2.0 to 5.0 Å and the angles Γ varied between $-45°$ and $45°$. The solvent molecules were also allowed to rotate about their own internal axes. Thus $Z(c/c, P/P, 1)$ and $Z(c/c, N/N, 1)$ were computed as a function ϕ, ψ, d_1, d_2, Γ_1, Γ_2, Δ_1, Δ_2, Δ_3, Δ_4. Only single solvent molecules–single solvent molecule interactions were considered. Normally computing the partition function of

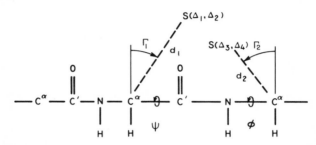

Fig. 4-32. Geometry used to calculate the $Z(c/c, P/P, 1)$ and $Z(c/c, N/N, 1)$ partition functions. The S denote the solvent molecules, the Δ are the internal rotations of the solvent molecules about their own fixed axes. The d define the distance between C^α atoms of the peptide backbone and the origins of the internal solvent molecule coordinate frames. The Γ define the dihedral angles between the d and the axes projecting from the C^α atoms which would be in common with the C^α—C^β bonds. The peptide backbone is held fixed in the nearly extended β conformation in order to simulate the u state.

TABLE 4-22

Values of the $Z(c/c, P/P, 1)$ and the $Z(c/c, N/N, 1)$
for Some Selected Solvents at $T = 298°K$

Solvent	Z
Water	41.7
Formic acid	445.9
Acetic acid	174.2
Methanol	14.4
Ethanol	7.4

a function of ten variables would be out of question. However, since the number of atoms is small for all solvents considered, and rather large incremental jumps could be employed in the hyperspace scan, it was possible to evaluate $Z(c/c, P/P, 1)$ and $Z(c/c, N/N, 1)$ for water, formic acid, acetic acid, methanol, and ethanol. The values of these partition functions are given in Table 4-22. Clearly one solvent might be considered the polar component in one solvent mixture and the nonpolar component in another mixture. For example, water is the polar component with methanol or ethanol, but the nonpolar component with formic acid and acetic acid. There are obvious pitfalls in making such assignments. Water, again as an example, would not want to *preferentially* bind to nonpolar residues in an acetic acid–water solvent mixture, and the assignment of such preferential binding, as built into this model, introduces errors. Under such conditions it would be better to decrease the value of $Z(c/c, N/N, 1)$ in some way to account for the limited amount of preferential bonding between water and nonpolar residues. The surface tensions τ_N and τ_P for nonpolar and polar residues in various solvents were estimated from the values of the polymer–solvent interaction energies of homopolypeptides in the β conformation (74)

$$\tau_N = [V_1 \tau_{v_1} + V_2 \tau_{v_2} + V_3 \tau_{v_3} + V_4 \tau_{v_4}]/(V_1 + V_2 + V_3 + V_4) \quad (4\text{-}43)$$

$$\tau_P = [S_1 \tau_{s_1} + S_2 \tau_{s_2} + S_3 \tau_{s_3} + S_4 \tau_{s_4}]/(S_1 + S_2 + S_3 + S_4) \quad (4\text{-}44)$$

where τ_{v_1} is the surface tension for valine, isoleucine, and leucine residues; τ_{v_2} the surface tension for alanine, proline, glycine, asparagine, and threonine residues; τ_{v_3} the surface tension for tryptophan and phenylalanine residues; τ_{v_4} the surface tension for methionine and cysteine residues; τ_{s_1} the surface tension for aspartic acid and glutamic acid residues; τ_{s_2} the surface tension for tyrosine and histidine residues; τ_{s_3} the surface tension for lysine and arginine and glutamine residues; and τ_{s_4} the surface tnesion for a serine residue. The V_i and the S_i denote the number of each type of residue present in the protein. These values of V_i and S_i have *not* been corrected for the residues located on the surface of the protein, which are assumed in our calculations to have identical interactions with solvent in both the native and denatured state. The β conformation was chosen in order to maximize the exposed "area" of a residue, as would usually be the case for a residue in the unfolded state. Table 4-23 contains the values of A_N and A_P as well as the $\tau_{v_i} A_N$ and $\tau_{s_i} A_P$ which are the total residue surface tension energies required in the denaturation equations. Table 4-24 contains the values of S_i and V_i for four proteins. In computing the values of τ_N and τ_P we have assumed that:

(1) Two or more "similar" residues may have the same characteristic surface tension.

(2) We may arithmetically average over a set of characteristic residue surface tensions to obtain τ_N and τ_P.

The first of these two assumptions is not too bad since residues with similar side chains (i.e., identical chemical groups, similar steric bulk, and/or similar degrees of conformational freedom) do, in fact, usually have similar polymer–solvent interactions, i.e., surface tensions, as reflected by nearly identical solubilities (74). However, the second assumption, concerning surface tension averaging, is probably a major oversimplification. This assumption destroys the surface tension–polymer geometry interrelationship which may be very important as a protein unfolds.

Lastly, in order to quantitatively apply the model presented in this section, it is necessary to determine the values of $B(v,w)$ and $C(v,w_1/w_2)$ for various proteins. This can be easily done by examining the backbone (ϕ,ψ) values

TABLE 4-23

Total Surface Tensions $\tau_{v_i} A_N$ and $\tau_{s_i} A_P$ for Nonpolar and Polar Residues, Respectively, in Various Solvents[a]

Solvent	$A_N = 20.5$ Å²				$A_P = 26.5$ Å²			
	$\tau_{v_1} A_N$	$\tau_{v_2} A_N$	$\tau_{v_3} A_N$	$\tau_{v_4} A_N$	$\tau_{s_1} A_P$	$\tau_{s_2} A_P$	$\tau_{s_3} A_P$	$\tau_{s_4} A_P$
Water	+0.3	−0.3	−0.1	+0.2	−24.4	−2.0	−8.1	−3.6
Formic acid	−1.9	−6.2	−1.7	−2.3	−36.3	−6.8	−9.6	−6.8
Acetic acid	−1.9	−8.6	−2.4	−1.9	−40.7	−9.3	−8.8	−7.2
Methanol	−1.1	−1.6	−1.7	−0.6	−3.9	−1.9	−2.0	−2.2
Ethanol	−1.3	−1.5	−1.4	−0.7	−3.4	−1.8	−2.2	−1.9

[a] The temperature is $T = 298°$K. See text for residue identification of the τ_{v_i} and τ_{s_i}.

TABLE 4-24

Values of the V_i and the S_i for Four Proteins[a]

Protein	V_1	V_2	V_3	V_4	S_1	S_2	S_3	S_4
Lysozyme (human)	19	44	6	9	6	9	23	6
Ribonuclease S (bovine)	13	38	3	12	12	10	21	15
Myoglobin (sperm whale)	35	39	11	2	19	14	27	6
α-Chymotrypsin	32	90	12	14	14	4	26	28

[a] These values have not been corrected for the residues on the surface of the protein which are assumed to have identical interactions with the solvent both in the native and denatured states.

of proteins when the native tertiary structure is known. The $B(R,w)$ are simply equal to the number of w type "units" present in the protein primary structure. If the tertiary structure of a protein is not available and one wishes to apply this model, then it is necessary to *assume* some distribution of secondary residue structures. Table 4-25 contains sets of $B(v,w)$ and $C(v,w_1/w_2)$ of four proteins. The criteria of classifying a given value of a set of (ϕ,ψ) as α, β, 3_1, or R is to use a (ϕ,ψ) map divided into four sections as shown in Fig. 4-33.

TABLE 4-25

(a) Various Values of the $B(v, w)$ in Four Proteins

Protein	$B(\alpha, H^\beta)$	$B(\alpha, C^\beta)$	$B(\beta, H^\beta)$	$B(\beta, C^\beta)$	$B(3_1, P_2)$	$B(3_1, H^\beta)$	$B(3_1, C^\beta)$	$B(R, H^\beta)$	$B(R, C^\beta)$	$B(R, P_2)$
Lysozyme (human)	5	68	7	29	1	1	13	11	109	2
Ribonuclease S (bovine)	1	53	2	52	3	0	13	3	117	4
Myoglobin	10	127	1	10	1	0	4	11	138	4
α-Chymotrypsin	13	74	5	98	9	3	39	24	209	8

(b) Various Values of the $C(v, w_1/w_2)$ in Some Proteins

Protein	$C(\alpha, H^\beta/H^\beta)$	$C(\alpha, H^\beta/C^\beta)$	$C(\alpha, C^\beta/H^\beta)$	$C(\alpha, C^\beta/C^\beta)$	$C(\beta, H^\beta/H^\beta)$	$C(\beta, H^\beta/C^\beta)$	$C(\beta, C^\beta/H^\beta)$	$C(\beta, C^\beta/C^\beta)$	$C(3_1, P_2/P_2)$
Lysozyme (human)	0	2	2	42	0	0	0	7	0
Ribonuclease S (bovine)	0	0	0	23	0	0	1	31	0
Myoglobin	0	6	7	76	0	0	0	4	0
α-Chymotrypsin	0	0	1	22	1	1	1	38	0

Protein	$C(3_1, H^\beta/H^\beta)$	$C(3_1, H^\beta/C^\beta)$	$C(3_1, H^\beta/P_2)$	$C(3_1, C^\beta/H^\beta)$	$C(3_1, C^\beta/C^\beta)$	$C(3_1, C^\beta/P_2)$	$C(3_1, P_2/H^\beta)$	$C(3_1, P_2/C^\beta)$
Lysozyme (human)	0	0	0	0	2	0	0	0
Ribonuclease S (bovine)	0	0	0	0	3	1	0	1
Myoglobin	0	0	1	0	2	0	0	1
α-Chymotrypsin	0	1	0	2	20	2	1	1

Clearly, such arbitrary partitioning of conformational states cannot be justified on first principles. However, the construction of molecular models indicate that conformations within each set of prescribed boundaries "looks" most like the standard conformation defined as being associated with the (ϕ, ψ) region.

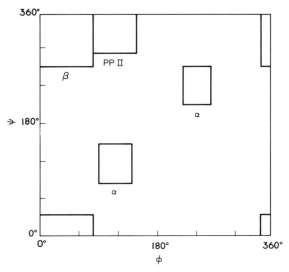

Fig. 4-33. The partitioned (ϕ, ψ) map used to classify residue conformations as α-helical, β-"helical," or 3_1 left-handed PP II helical in order to determine the $B(\mathrm{v}, \mathrm{w})$ and $C(\mathrm{v}, \mathrm{w}_1/\mathrm{w}_2)$ of globular proteins. For a residue whose (ϕ, ψ) values fall outside the specified regions it is assigned to that region to which it is closest.

D. Results and Discussion

1. Comparison of the Calculated and Observed Free Energy Changes Accompanying a Macromolecular Transition

It is possible to make direct quantitative comparisons of some predicted and observed free energy changes in macromolecular transitions using the model presented here and limited available data. This is the only *direct* means of assessing the meaningfulness of the energies reported in this paper. As mentioned in the introduction, Schellman has determined ΔH_0 and ΔS_0 for the helix–coil transition in aqueous solution. For $T = 298°\mathrm{K}$ the free energy change, $\Delta F_{\mathrm{H-C}}$ for the helix–coil transition using Schellman's values of ΔH_0 and ΔS_0 is

$$\Delta F_{\mathrm{H-C}} = \Delta H_0 - T \Delta S_0$$
$$= +1500 - 298 \times 4.2 = 0.25 \quad \text{kcal/mole-residue} \quad (4\text{-}45)$$

It is also possible to estimate ΔF_{H-C} using the values of $E(\alpha, C^\beta)$, $E(\alpha, C^\beta/C^\beta)$, and $E(R, C^\beta)$ reported in Tables 4-20 and 4-21, along with the corresponding entropy change ΔS_{H-C} and the change in polymer–solvent free energy ΔF_{P-S}. This approximation is valid at $T = 298°K$ *and* for a change in the work term, i.e. $P\Delta V$, of zero:

$$\Delta F_{H-C} = (E(R, C^\beta) - E(\alpha, C^\beta) - E(\alpha, C^\beta/C^\beta)) - T\Delta S_{H-C} + \Delta F_{P-S}$$
$$= (-2.90) - (-3.54) - (-1.93) - 298(2.9) - 1.34$$
$$= 0.37 \quad \text{kcal/mole-residue} \tag{4-46}$$

By including only intra- and first nearest neighbor residue interactions we have eliminated additional stabilizing free energy contributions which, presumably, would bring our value of ΔF_{H-C} even closer to that determined from Schellman's data. Nevertheless, the two values of ΔF_{H-C} are remarkably close, thus adding confidence to the results presented here.

Volkenstein (17) reports that a $\Delta F = -16$ kcal/mole has been determined in a *reversible* denaturation of myoglobin at pH = 9 and 298°K. It is possible to compute a value of ΔF for myoglobin in water at a high pH (>7.0) using the values of the energy parameters reported in the previous section.

$$\Delta F = \Delta F_{\text{conformational}} + \Delta F_{\text{polymer-solvent}} + \Delta F_{\text{adsorbed solvent}}$$
$$= -187.7 + 269.2 - 98.5 = -17.0 \quad \text{kcal/mole} \tag{4-47}$$

Once again there is excellent agreement with other independent data. However, this agreement may be fortuitous since we are, in fact, calculating the differences of relatively large numbers which are close in value, i.e., the difference is an order of magnitude smaller than numbers used in the computations.

2. Calculation of Denaturation Curves

The fraction of unfolded-nonpolar residues $(1 - \theta(n))$ and the fraction of unfolded-polar residues $(1 - \theta(p))$ were determined for α-chymotrypsin, ribonuclease S, myoglobin, and lysozyme as a function of binary solvent composition and solvent adsorbate concentration λ. Three binary solvent mixtures were considered in the calculations:

(1) Acetic acid and methanol which can be considered a solvent mixture having a strongly polar component and a moderately nonpolar component.

(2) Water and ethanol which can be considered a solvent mixture having a polar component and a moderately nonpolar component.

(3) Formic acid and water which can be considered a solvent mixture having a very polar component and a polar component.

In Figs. 4-34–4-37 are plotted the unfolding curves of nonpolar and polar residues as a function of solvent adsorbate concentration for the four proteins

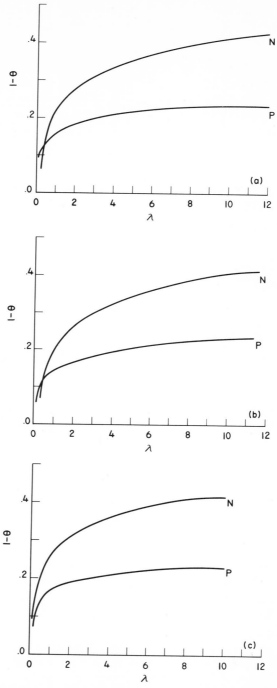

Fig. 4-34. $1 - \theta(n)$ versus λ_n and $1 - \theta(p)$ versus λ_p for α-chymotrypsin in (a) acetic acid and methanol, (b) water and ethanol, and (c) water and formic acid.

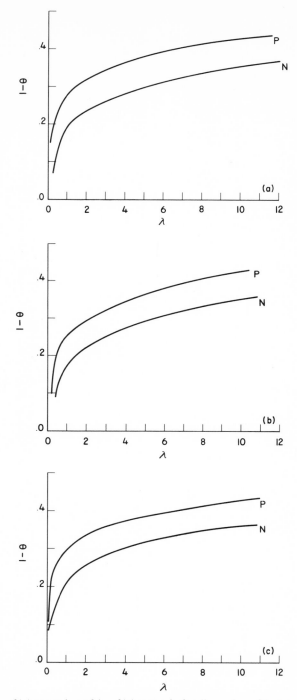

Fig. 4-35. $1 - \theta(n)$ versus λ_n and $1 - \theta(p)$ versus λ_p for ribonuclease S in (a) acetic acid and methanol, (b) water and ethanol, and (c) water and formic acid.

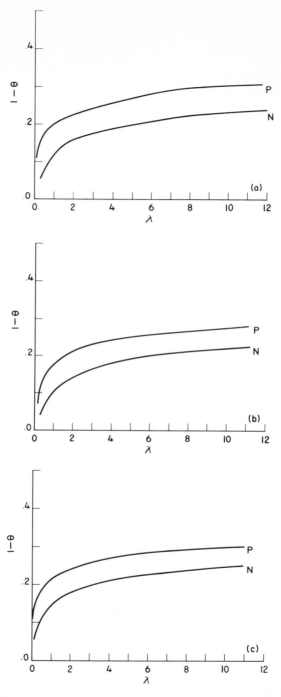

Fig. 4-36. $1 - \theta(n)$ versus λ_n and $1 - \theta(p)$ versus λ_p for myoglobin in (a) acetic acid and methanol, (b) water and ethanol, and (c) water and formic acid.

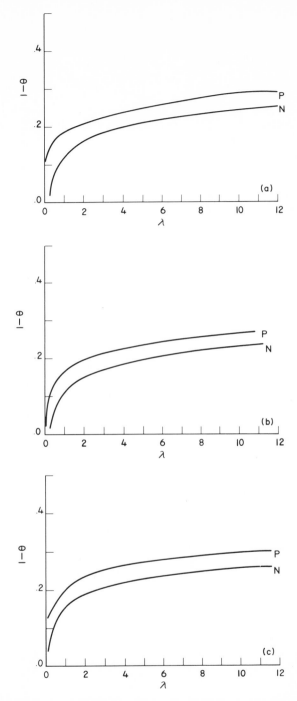

Fig. 4-37. $1 - \theta(n)$ versus λ_n and $1 - \theta(p)$ versus λ_p for lysozyme in (a) acetic acid and methanol, (b) water and ethanol, and (c) water and formic acid.

mentioned above. Since the specific relationship between λ_p and λ_n is not known for any of the solvent mixtures, the curves in Figs. 4-34 to 4-37 are plots of absolute solvent adsorbate concentration versus absolute unfolding. That is, plots of λ_p versus $(1 - \theta(p))$ are completely independent of λ_n versus $(1 - \theta(n))$.

The following general observations can be made from an inspection of the respective polar and nonpolar denaturation curves.

(1) Each protein studied has a considerably different denaturation curve, for both polar and nonpolar residues, from each of the denaturation curves of the other proteins.

(2) For a particular protein, the polar (nonpolar) denaturation curves in all three solvents are very similar. This may be an indication that the Shindo model and/or the values of the necessary parameters may not be properly sensitive to solvent.

(3) Every denaturation curve, both polar and nonpolar, is logarithmic in shape. No indication of a highly cooperative process is apparent except possibly in the neighborhood of λ_p or $\lambda_n = 0$.

(4) For each protein the absolute values of λ_p and λ_n must be large (>50) in order to achieve at least 90% total unfolding.

(5) Roughly speaking, each of the proteins studied begins to rapidly unfold for small increases in adsorbate concentrations, and then remains partially unfolded for large increasing adsorbate concentrations.

These general observations can be refined by analyzing the data in Table 4-26. The values of $(1 - \theta(n))$, $(1 - \theta(p))$ and their respective sums for $\lambda = 1$ reflects the "immediate" unfolding power of the various solvent mixtures. These same quantities for $\lambda = 10$ reflect the "overall" denaturing power of the various solvents. From the data in Table 4-26 we can make the following statements:

(1) Lysozyme and myoglobin are approximately equally difficult to unfold in all the solvents considered. Both these proteins are more difficult to unfold, in all three solvents, than α-chymotrypsin. α-Chymotrypsin, in turn, is more difficult to unfold than ribonuclease S. These findings are in correlative agreement with those of Herskovits and Mescanti (82) who concluded ribonuclease S unfolds more easily than lysozyme in a variety of solvents.

(2) The higher the polarity of the polar component of the binary solvent mixture, the greater the extent of the unfolding of the protein. This is most evident in the range of low solvent adsorbate concentrations, i.e., inspect the "*sum*" columns for $\lambda = 1$. We conclude that formic acid and water solvent mixtures have greater "unfolding power" than acetic acid and methanol solvent mixtures which, in turn, are better denaturing agents than the water and ethanol solvent mixtures.

(3) For λ_n and λ_p both equal to one, approximately $\frac{1}{3}$ to $\frac{1}{2}$ of the residues

are unfolded in the proteins. By the time both λ_p and λ_n are equal to ten only an additional $\frac{1}{4}$ to $\frac{1}{3}$ of the remaining folded residues have unfolded.

The major drawback in carrying out the type of calculations presented here is in the inability to determine the functional relationship between λ_p and λ_n for a particular binary solvent mixture and a particular protein. Thus

TABLE 4-26

Fraction of Unfolded Residues in Various Proteins as a
Function of Solvent Composition and Adsorbate Concentration λ

| | Acetic acid and Methanol | | | | | |
| | $\lambda = 1$ | | | $\lambda = 10$ | | |
Protein	$1 - \theta_n$	$1 - \theta_p$	Sum	$1 - \theta_n$	$1 - \theta_p$	Sum
α-Chymotrypsin	0.220	0.155	0.375	0.413	0.230	0.643
Ribonuclease S	0.190	0.280	0.470	0.356	0.430	0.786
Myoglobin	0.120	0.200	0.320	0.230	0.304	0.534
Lysozyme	0.118	0.185	0.303	0.243	0.290	0.533

| | Water and Ethanol | | | | | |
| | $\lambda = 1$ | | | $\lambda = 10$ | | |
Protein	$1 - \theta_n$	$1 - \theta_p$	Sum	$1 - \theta_n$	$1 - \theta_p$	Sum
α-Chymotrypsin	0.200	0.143	0.343	0.408	0.230	0.638
Ribonuclease S	0.170	0.250	0.420	0.352	0.438	0.790
Myoglobin	0.104	0.175	0.279	0.280	0.280	0.500
Lysozyme	0.110	0.160	0.270	0.230	0.255	0.485

| | Formic acid and Water | | | | | |
| | $\lambda = 1$ | | | $\lambda = 10$ | | |
Protein	$1 - \theta_n$	$1 - \theta_p$	Sum	$1 - \theta_n$	$1 - \theta_p$	Sum
α-Chymotrypsin	0.255	0.170	0.425	0.417	0.230	0.647
Ribonuclease S	0.208	0.295	0.503	0.360	0.438	0.808
Myoglobin	0.142	0.210	0.352	0.248	0.300	0.548
Lysozyme	0.155	0.200	0.355	0.260	0.295	0.555

it is not possible to relate the total residue unfolding, $1 - (\theta(n) + \theta(p))$, to total adsorbate concentration, $\lambda_p + \lambda_n$. However, it is possible to glean the probable linear relationship necessary to generate a $\lambda_p + \lambda_n$ versus $1 - (\theta(n) + \theta(p))$ curve with a particular shape. An analysis of the shape of the total denaturation curve in terms of the functional relationship between λ_p and λ_n has been carried out for lysozyme in a water–ethanol solvent mixture. The choice of this system was arbitrary. The first set of calculations were designed to determine if either the $1 - \theta(n)$ versus λ_n or the $1 - \theta(p)$ versus λ_p curves possessed any change from logarithmic behavior for small values of λ_p and λ_n. These curves are shown in Fig. 4-38a. They are logarithmic

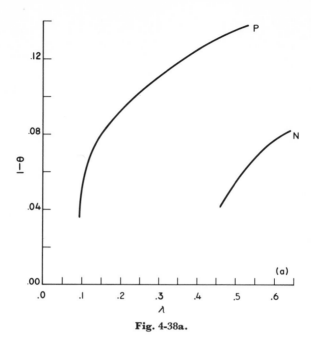

Fig. 4-38a.

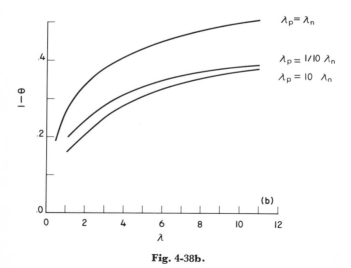

Fig. 4-38b.

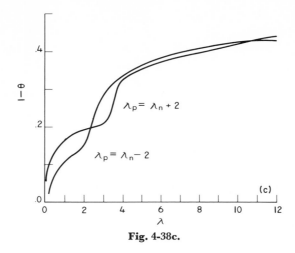

Fig. 4-38c.

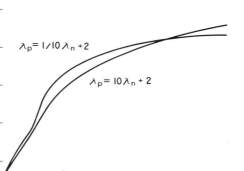

Fig. 4-38d.

Fig. 4-38. (a) An examination of the shape of λ versus $1 - \theta$ for polar and nonpolar residues at extremely low values of λ. The solvent mixture is water and ethanol, and the protein is lysozyme.

Plots of the total denaturation curves of lysozyme in water and ethanol for various inter-relationships between λ_p and λ_n: (b) $\lambda_p = a\lambda_n$; (c) $\lambda_p = \lambda_n + b$; (d) $\lambda_p = a\lambda_n + b$.

in nature at all points. This dismisses the possibility of "mini" phase transitions at extremely small values of λ_p and λ_n.

Next a series of calculations were carried out to deduce the shape of the total denaturation curve when $\lambda_p = a\lambda_n$ where $a > 0$. The results of these calculations are shown in Fig. 4-38b. Maximum unfolding is observed for $\lambda_p = \lambda_n$. For $a \neq 1$ the rate of unfolding is reduced and the curves contain slight plateaus (that is, deviations from logarithmic behavior). The next series of computations were considered for the relationship $\lambda_p = \lambda_n + b$, where $b \neq 0$. Representative results for this relationship are shown in Fig. 4-38c. These curves are reminiscent of the transition curves associated with co-operative processes (1, p. 163). However, cooperativity need not be present. There are a couple of other explanations which could account for the shape of the denaturation curves for $\lambda_p = \lambda_n + b$. These are:

(1) The difference in the chemical potentials for polar residues interacting with the polar solvent component and the nonpolar residues interacting with the nonpolar solvent component could be so large as to cause one class of residues to begin unfolding while the other class of residues remain folded in the native state.

(2) The change in the nearest neighbor interaction free energy for adsorbed solvent species between folded and nonfolded states could be more stabilizing for nonpolar (polar) solvent molecules compared to polar (nonpolar) solvent molecules so that the nonpolar (polar) residues unfold while the polar (non-polar) residues remain folded.

Lastly, a relationship of the form $\lambda_p = a\lambda_n + b$ for $a > 0$ and $b \neq 0$ was considered. Two examples of the types of denaturation curves possible from such a relationship are shown in Fig. 4-38d. Depending upon the choice of a and b one can go from a logarithmic curve to an "S"-shaped curve similar to those in Fig. 4-38c.

The single most important observation made during this series of calcula-tions is that an "S"-shaped transition curve is possible only when $\lambda_p = a\lambda_n + b$, and $a \cong 1$. Further, an "S"-shaped transition curve need not necessarily be indicative of a cooperative unfolding process for a protein in a two-component solvent.

E. Summary

The parameters necessary to quantitatively use the Shindo model for protein denaturation have been evaluated for four proteins and five solvents. There is a marked difference in the respective nonpolar and polar denatura-tion curves for the four proteins. Various linear relationships between λ_p and

λ_n were tested to determine the effect on the shape of the *total* denaturation curve. For $\lambda_p = a\lambda_n + b$ an "S"-shaped curve, indicative of a cooperative process is possible. Such a curve could also be the result of interactions which, in themselves, are not related to cooperativity.

From the results found in this series of calculations the model does not seem as sensitive to changes in solvent composition as suggested from experimental data (17). Hopefully experiment and theory will have more in common when the solvent parameters for solvents such as chloroethanol and methoxyethanol, which have been widely used in denaturation experiments, have been determined.

The evaluation of the molecular parameters has involved many assumptions of varying justification. Certainly there is much room for further refinement of the values of the necessary parameters. However, these calculations do represent an initial effort to apply conformational energy calculations to a difficult research topic—protein denaturation.

References

1. J. Applequist, *in* "Conformation of Biopolymers" (G. N. Ramachandran, ed.), Vol. I. Academic Press, New York, 1967.
2a. T. L. Hill, "Introduction to Statistical Thermodynamics." Addison-Wesley, Reading, Massachusetts, 1962.
2b. L. Lewin, "Dilogarithms and Associated Functions." Macdonald, London, 1958.
3. B. H. Zimm, *J. Chem. Phys.* **33**, 1349 (1960).
4. D. M. Crothers and B. H. Zimm, *J. Mol. Biol.* **9**, 1 (1964).
5. E. Ising, *Z. Phys.* **31**, 253 (1925).
6. B. H. Zimm and J. K. Bragg, *J. Chem. Phys.* **31**, 526 (1959).
7. S. Lifson and A. Roig, *J. Chem. Phys.* **34**, 1963 (1961).
8. J. Applequist, *J. Chem. Phys.* **38**, 934 (1963).
9. T. L. Hill, *J. Chem. Phys.* **30**, 383 (1959).
10. B. H. Zimm and S. A. Rice, *Mol. Phys.* **3**, 391 (1960).
11. F. T. Wall, L. A. Hiller, and W. F. Atchinson, *J. Chem. Phys.* **23**, 2314 (1955).
12. H. Eyring, *Phys. Rev.* **39**, 746 (1932).
13. R. F. Steiner and R. F. Beers, "Polynucleotides," Chapter 9. Elsevier, Amsterdam, 1961.
14. J. Applequist and V. Damle, *J. Amer. Chem. Soc.* **87**, 1450 (1965).
15. J. Applequist and V. Damle, *J. Amer. Chem. Soc.* **88**, 3895 (1966).
16. J. A. Schellman, *J. Phys. Chem.* **62**, 1485 (1958).
17. M. V. Volkenstein, "Molecules and Life." Plenum Press, New York, 1970.
18. R. T. Ingwall, H. A. Scheraga, N. Lotan, A. Berger, and E. Katchalski, *Biopolymers* **6**, 331 (1968).
19. N. Go, M. Go, and H. A. Scheraga, *Proc. Nat. Acad. Sci. U.S.* **59**, 1030 (1968).
20. M. Go, N. Go, and H. A. Scheraga, *J. Chem. Phys.* **31**, 526 (1959).
21. P. Urnes and P. Doty, *Advan. Protein Chem.* **16**, 401 (1961).
22. M. Go, N. Go, and H. A. Scheraga, *J. Chem. Phys.* **54**, 4489 (1971).

23. J. A. Schellman and C. Schellman, *in* "The Proteins" (H. Neurath, ed.), Vol. 2, p. 1. Academic Press, New York, 1964.
24. W. A. Hiltner, M.S. Thesis, Case Western Reserve Univ. Cleveland, Ohio (1972).
25. M. L. Tiffany and S. Krimm, *Biopolymers* **8**, 347 (1969).
26. S. Krimm, J. E. Mark, and M. L. Tiffany, *Biopolymers* **8**, 695 (1969).
27. W. A. Hiltner, A. J. Hopfinger, and A. G. Walton, *J. Amer. Chem. Soc.* **94**, 4324 (1972).
28. O. B. Ptitsyn, *in* "Conformation of Biopolymers" (G. N. Ramachandran, ed.), Vol. I, p. 381. Academic Press, New York, 1967.
29. L. Tiffany, Ph.D. Thesis, Univ. of Michigan, Ann Arbor, Michigan (1971).
30. J. Applequist, *J. Chem. Phys.* **45**, 3459 (1966).
31. G. Giacometti, *in* "Structural Chemistry and Molecular Biology" (A. Rich and N. Davidson, eds.), p. 67. Freeman, San Francisco, California, 1967.
32. Th. Ackerman and H. Ruterjans, *Ber. Bunsen. Phys. Chem.* **68**, 850 (1964).
33. M. Nagasawa and A. M. Holtzer, *J. Amer. Chem. Soc.* **86**, 538 (1964).
34. H. A. Scheraga, "Protein Structure," Chapter 6. Academic Press, New York, 1961.
35. T. N. Nekrasova, E. V. Anufrieva, A. M. Yelyashevich, and O. B. Ptotsyn, *Vysokomol. Soedin.* **7**, 913 (1965).
36. R. L. Snipp, W. G. Miller, and R. E. Nylund, *J. Amer. Chem. Soc.* **87**, 3547 (1965).
37. T. M. Birshtein and O. B. Ptitsyn, "Conformations of Macromolecules." Wiley (Interscience), New York, 1966.
38a. N. Go, P. N. Lewis, M. Go, and H. A. Scheraga, *Macromolecules* **4**, 692 (1971).
38b. J. T. Edsall, P. J. Flory, J. C. Kendrew, A. M. Liquori, G. Némethy, G. N. Ramachandran, and H. A. Scheraga, *Biopolymers* **4**, 121 (1966).
39. I. Z. Steinberg, W. F. Harrington, A. Berger, M. Sela, and E. Katchalski, *J. Amer. Chem. Soc.* **82**, 5263 (1960).
40. I. Z. Steinberg, A. Berger, and E. Katchalski, *Biochim. Biophys. Acta* **28**, 647 (1958).
41a. A. J. Hopfinger and A. G. Walton, *J. Macromol. Sci.-Phys.* **B3**(1), 195 (1969).
41b. L. Mandelkern, *in* "Poly-α-Amino Acids" (G. D. Fasman, ed.), p. 701. Dekker, New York, 1967.
42a. W. Traub and K. A. Piez, *Advan. Protein Chem.* **25**, 243 (1971).
42b. A. Rich and F. H. C. Crick, *J. Mol. Biol.* **3**, 483 (1961).
43. F. R. Brown III, A. J. Hopfinger, and E. R. Blout, *J. Mol. Biol.* **63**, 101 (1972).
44. A. Yonath and W. Traub, *J. Mol. Biol.* **43**, 461 (1969).
45. D. A. Brant, W. G. Miller, and P. J. Flory, *J. Mol. Biol.* **23**, 47 (1967).
46. Y. Kobayashi, R. Sakai, K. Kakiuchi, and T. Isemura, *Biopolymers* **9**, 415 (1970).
47. F. R. Brown III, A. DiCoroto, G. P. Lorenzi, and E. R. Blout, *J. Mol. Biol.* **63**, 85 (1972).
48. A. J. Hopfinger, unpublished results (1970).
49. D. M. Segal, *J. Mol. Biol.* **43**, 497 (1969).
50. J. C. Leyte and M. Mandel, *J. Polym. Sci. Part A* **2**, 1879 (1964).
51. A. M. Liquori, G. Barone, V. Crescenzi, F. Quadrifoglio, and V. Vitagliano, *J. Macromol. Chem.* **1**, 291 (1966).
52. M. Mandel and M. G. Stadhouder, *Makromol. Chem.* **80**, 141 (1964).
53. M. Mandel, J. C. Leyte, and M. G. Stadhouder, *J. Phys. Chem.* **71**, 603 (1963).
54. J. Semen, Ph.D. Thesis, Case Western Reserve Univ., Cleveland, Ohio (1972).
55. M. Nagasawa, T. Murose, and K. Kondo, *J. Phys. Chem.* **69**, 4005 (1965).
56. J. Semen, M.S. Thesis, Case Western Reserve Univ., Cleveland, Ohio (1969).
57. H. Tadokoro, Y. Chantani, H. Kusangi, and M. Yokoyama, *Macromolecules* **3**, 441 (1970); G. Allegra, E. Benedetti, and C. Pedone, *Macromolecules* **3**, 727 (1970).
58. G. Natta, P. Corradini, and P. Ganis, *Makromol. Chem.* **39**, 238 (1960).
59. H. Tadokoro, *J. Chem. Phys.* **33**, 1558 (1960).

60. J. L. Koenig and A. C. Angood, unpublished work (1965).
61. G. W. King, "Spectroscopy and Molecular Structure," Chapter 9. Holt, New York, 1965.
62. V. Crescenzi, *Advan. Polym. Sci.* **5**, 358 (1968).
63. J. B. Lando and W. W. Doll, *J. Macromol. Sci.-Phys.* **B2**(2), 205 (1968).
64. J. B. Lando, H. G. Olf, and A. Peterlin, *J. Polym. Sci. Part A1* **4**, 941 (1966).
65. W. W. Doll and J. B. Lando, *J. Macromol. Sci.-Phys.* **B4** (2), 309 (1970).
66. W. W. Doll, Ph.D. Thesis, Case Western Reserve Univ., Cleveland, Ohio (1969).
67. C. W. Wilson, *J. Polym. Sci. Part A1*, **1**, 1305 (1963).
68. H. R. Tadokoro, M. Hasegawa, M. Kobayashi, Y. Takahashi, and Y. Chantani, IUPAC Preprints, Boston, Massachusetts (1971).
69. H. Inoue and S. N. Timasheff, *Biochemistry* **7**, 2501 (1968).
70. J. A. Schellman, *C. R. Trav. Lab. Carlsberg Ser. Chim.* **29**, 223, 230 (1955).
71. P. L. Privalov, Study of Thermal Denaturing of Egg Albumin, *Biophysics U.S.S.R.* (English Transl.) **8**(3), 363 (1963).
72. G. Nemethy, I. Z. Steinberg, and H. A. Scheraga, *Biopolymers* **1**, 43 (1963).
73. K. D. Gibson and H. A. Scheraga, *Proc. Nat. Acad. Sci. U.S.* **5**, 420 (1967).
74. A. J. Hopfinger, *Macromolecules* **4**, 731 (1971).
75. A. J. Hopfinger, *Biopolymers* **10**, 1299 (1971).
76. P. J. Flory, "Statistical Mechanics of Chain Molecules." Wiley (Interscience), New York, 1969.
77. Y. Shindo, *Biopolymers* **10**, 1081 (1971).
78. E. Isaacon and H. B. Keller, "Analysis of Numerical Methods." Wiley, New York, 1966.
79. T. Ooi, R. A. Scott, G. Vanderkooi, and H. A. Scheraga, *J. Chem. Phys.* **46**, 4410 (1967).
80. R. E. Dickerson and I. Geis, "Structure and Action of Proteins." Harper, New York, 1969.
81. K. H. Forsythe and A. J. Hopfinger, *Macromolecules*, in press.
82. T. T. Herskovits and L. Mescanti, *J. Biol. Chem.* **240**, 639 (1965).

Chapter 5 | Applications of Conformational
Potential Energy Calculations

I. Absorption and Optical Rotation Spectroscopies

A. Introduction and Theory

Most spectral properties of a molecule depend in a more or less additive fashion on the chemical groups of which it is made. The relative geometries of the constituent groups alter such properties as a secondary effect. Optical rotation, on the other hand, is inextricably bound to the interactions among the groups and, therefore, to the molecular geometry. Apart from rare molecules which have intrinsically asymmetric chromophores, optical rotation has no primary source within the groups themselves but comes directly from their relative orientations. Since conformational calculations provide information concerning which relative orientations of the various groups are most probable, conformational analysis might be a means of predicting optical spectra.

The optical rotation properties of a molecule may be measured in terms of the optical rotatory dispersion (ORD) spectrum or the circular dichroism (CD) spectrum. These two types of curves are related by the well established Kronig–Kramers (KK) transform. Thus if we establish the theory responsible for one of these two curves we can predict the other spectrum by performing the transformation. We shall consider the theory of circular dichroism, since it has the simplifying property of vanishing except within an optically active absorption band.

Circular dichroism is defined as

$$\Delta\epsilon = \epsilon_l - \epsilon_r \qquad (5\text{-}1)$$

where ϵ_l and ϵ_r are the molecular extinction coefficients for left and right circularly polarized light, respectively. Conversion to the alternative measure of circular dichroism in terms of the molecular ellipticity is easily achieved by means of the formula

$$[M_\theta] = 2.303(4500/\pi)\,\Delta\epsilon^2 \qquad (5\text{-}2)$$

Perhaps the best way to introduce the molecular mechanisms which give rise to circular dichroism is to first describe the molecular origins of the rather familiar property of the absorption strength of a molecule, and then to show how the theory of circular dichroism parallels the absorption strength theory.

B. The Absorption Strength of a Molecule

The integrated intensity of an electronic absorption band is directly proportional to the square of the electric dipole moment μ for the transition

absorption strength $\propto |\langle 0|\mu|i\rangle|^2$ for the $0 \rightarrow i$th state transition

(5-3)

This is a quantity which plays the same role in quantum theory as the oscillating electric dipole in the classical theory of the absorption and emission of radiation. One principal difference is that the magnitude of the classical dipole depends on the strength of the field which induces it, whereas in quantum mechanics it is a fixed molecular quantity which depends on the wave functions of the ground and excited states. The magnitude of the transition dipole may be determined experimentally by the formula

$$D = \mu^2 \cong \frac{(2.303)\ 3hc}{8\pi^3\ N\lambda_{max}} \int \epsilon\ d\lambda$$

(5-4)

where λ_{max} is the wavelength at which the transition occurs, and h, c, N, and ϵ are respectively, Planck's constant, the velocity of light, Avagadro's number, and the molecular extinction coefficient of nonpolarized light. D is the dipole strength of the transition and λ is the wavelength. If accurate wave functions of the ith excited state ψ_i and the ground state ψ_0 are known, then

$$D = [\int \psi_i\ \mu\psi_0\ d\tau]^2$$

(5-5)

This is a quantitative expression for Eq. (5-3).

C. Origin of Circular Dichroism in a Molecule

Just as the observed absorption of a molecule may be said to arise from its dipole strength D, circular dichroism arises from an analogous quantity, the rotatory strength R, which can be experimentally calculated by

$$R \cong \frac{1}{4} \frac{(2.303)\ 3hc}{8\pi^3\ N\lambda_{max}} \int \Delta\epsilon\ d\lambda$$

(5-6)

In other words, the integrated intensity of the circular dichroism band is directly proportional to the rotatory strength.

The rotatory strength R is given quantum mechanically as

$$R = \text{Im}(\langle 0|\boldsymbol{\mu}|i\rangle \langle i|\mathbf{M}|0\rangle) \qquad (5\text{-}7)$$

for the $0 \rightarrow i$th state transition.

$\mathbf{M}$ is the magnetic dipole moment of the transition. We know that the electric dipole moment $\boldsymbol{\mu}$ arises from the polarizability of the atoms (or groups of atoms) in the molecule and the charge distribution within the molecule. How does the magnetic dipole moment arise? This is a most basic question for the principal objective of molecular theories of rotatory power is to demonstrate how the $\mathbf{M}$ arise in optically active transitions and how they couple with the electric transition moments to yield nonvanishing rotatory strengths. The magnetic dipole moment arises both in classical and quantum theories because of a net circulation of current about a point in space. Further, if the magnetic dipole moment $\mathbf{M}$ is known and the wave functions of the states 0 (ground) and i are known, then the magnetic transition moment $\mathbf{M_t}$ is given by

$$\mathbf{M_t} = \int \psi_i \mathbf{M}\psi_0 \, d\tau \qquad (5\text{-}8)$$

For many transitions it is possible to pick an origin of space such that the magnetic transition moment vanishes, $|\mathbf{M_t}| = 0$. This is the case because the molecular transitions may be regarded as linear charge displacements. A circulation of charge, the basic requirement for the existence of magnetic moments, clearly requires an intrinsic rotation of charge density. Whenever such a rotation of charge density is part of the electronic transition one can expect to observe optical rotation. In Fig. 5-1a are illustrations of linear charge displacement and rotation of charge density. From Fig. 5-1a it should be clear that optical rotation will occur whenever the electric moment of the molecule (or group) is rotated as well as translated during a transition. We can further distinguish between linear and rotational charge motions in the following way. The magnetic moment operator is

$$\mathbf{M} = \frac{\mathbf{r} \times e\mathbf{v}}{2c}$$

where $\mathbf{r}$ is the position vector of the electrons, $e\mathbf{v}$ is the current density, and c is the velocity of light. Using Fig. 5-1b as a reference it can be seen that $\mathbf{r} = \mathbf{R} + \mathbf{p}$ where $\mathbf{R}$ is the distance from the arbitrary origin to the optical center of the group in question (i.e., center of charge) and $\mathbf{p}$ defines the positions of the electrons relative to the optical center of the group. This geometry implies that

$$\mathbf{M} = \frac{\mathbf{r} \times e\mathbf{v}}{2c} = \frac{(\mathbf{R}+\mathbf{p}) \times e\mathbf{v}}{2c} = \frac{\mathbf{R} \times e\mathbf{v}}{2c} + \frac{\mathbf{p} \times e\mathbf{v}}{2c} \qquad (5\text{-}9)$$

or

$$\mathbf{M} = \frac{\mathbf{R} \times e\mathbf{v}}{2c} + \mathbf{m} \qquad \text{where} \qquad \mathbf{m} = \frac{\mathbf{p} \times e\mathbf{v}}{2c}$$

is defined as the local (intrinsic) magnetic moment operator.

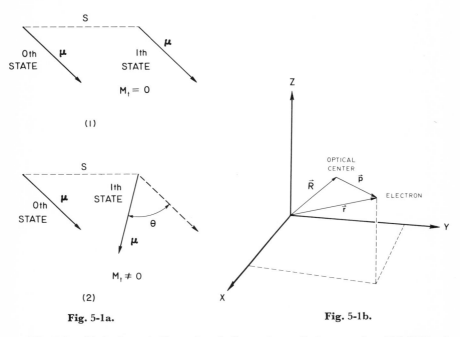

Fig. 5-1a.   Fig. 5-1b.

Fig. 5-1a. (1) A schematic illustration of a linear charge displacement for which $|\mathbf{M}_t| = 0$. (2) A schematic illustration of a linear *plus* a rotational charge displacement for which $|\mathbf{M}_t| \neq 0$. The transition is between states 0 and i and the $\boldsymbol{\mu}$ are the electric moments.

Fig. 5-1b. The geometry relating an arbitrary geometric center to the optical center of a monomer.

Let us look at the two limiting conditions: (a) the transition is purely electric, or (b) the transition is intrinsically magnetic;

$$\text{for case (a):} \quad \mathbf{M}_t \approx \frac{\mathbf{R} \times e\mathbf{v}}{2c}; \qquad \text{for case (b):} \quad \mathbf{M}_t \approx \mathbf{m} \qquad (5\text{-}10)$$

If there is only one group being considered, then the origin of the arbitrary frame and the origin of the optical center can be chosen to be identical and case (a) implies $|\mathbf{M}_t| = 0$. Thus, unless there is a local (intrinsic) magnetic moment for the group there will be no optical rotation from the group. This mechanism of molecular rotatory power is known as the *one-electron mechanism*.

The optical factor in this mechanism has virtually no geometric dependence since the relative orientations of the two transition moments are entirely determined by the electronic structure of the single chromophoric group. The perturbing effects on **m** by the rest of the group (or molecule) are the only possible conformational contributions.

The second mechanism which is responsible for optical rotation is the *dipole–coupling mechanism*. In this model the transition moments μ_i of two or more optically active groups couple and give rise to a net magnetic moment (or set of moments). In this mechanism the optical rotatory strength for the interaction is dependent upon the relative orientations and positions of the active groups. In other words the optical rotatory strength from this contribution is sensitive to changes in conformation.

The third mechanism, which is also sensitive to conformation, is the *electric–magnetic coupling mechanism*. Magnetic dipoles do not couple directly with electric dipoles. However, all magnetic transitions which are associated with the orbital motions of electrons produce electric moments as well. It is this interaction (coupling) of these electron electric moments with the electric dipoles of the optical active groups which makes the third contribution to the molecular rotatory strength.

Thus all three of these mechanisms make a contribution to the rotatory strength R_{ik} for the kth transition for the molecule in the ith conformation,

$$R_{ik} = R_{ik}(\text{one-electron}) + R_{ik}(\text{dipole-coupling}) + R_{ik}(\text{electric-magnetic})$$

(5-11)

D. A General Model for Absorption Strength and Rotatory Strength for an Isolated Optical Homopolymer

An optical homopolymer, abbreviated OPHOpolymer, is any macromolecule which may have monomer units of varying structure, but the optical transitions of all monomer units are identical. A polypeptide is an example of a OPHOpolymer. It may have different monomer units due to different possible residue sidechains, but the optical transitions of the amide (also imide) backbone group is the same regardless of side chain. Consider now any linear OPHOpolymer which is schematically shown in Fig. 5-2. Let there

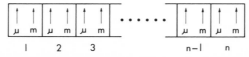

Fig. 5-2. A schematic illustration of a linear OPHOpolymer of n monomer units. The μ_i are the electric monomer moments and the m_i are the magnetic monomer moments.

be α (discrete) excited states associated with each of the n monomer units in the OPHOpolymer. Then $\mathbf{U}$, the electric moment column vector can be defined as

$$
\mathbf{U} = \begin{pmatrix}
\left.\begin{array}{l}
\mu_x(1,1)\hat{i} + \mu_y(1,1)\hat{j} + \mu_z(1,1)\hat{k} \\
\mu_x(1,2)\hat{i} + \mu_y(1,2)\hat{j} + \mu_z(1,2)\hat{k} \\
\vdots \\
\mu_x(1,n)\hat{i} + \mu_y(1,n)\hat{j} + \mu_z(1,n)\hat{k}
\end{array}\right\} \text{1st excited state transition} \\
\left.\begin{array}{l}
\mu_x(2,1)\hat{i} + \mu_y(2,1)\hat{j} + \mu_z(2,1)\hat{k} \\
\mu_x(2,2)\,\hat{i} + \mu_y(2,2)\hat{j} + \mu_z(2,2)k \\
\vdots \\
\mu_x(2,n)i + \mu_y(2,n)j + \mu_z(2,n)k
\end{array}\right\} \text{2nd excited state transition} \\
\vdots
\end{pmatrix} \qquad (5\text{-}12)
$$

In the notation given above for $\mu_k(i,j)$ the i refers to a specific excited state and j is a particular monomer. An analogous column vector $\mathbf{M}$ may be constructed for the set of magnetic moments.

We will assume that the $\boldsymbol{\mu}_i$ and the $\mathbf{m}_i$ are the electric and magnetic moments before the monomers interact. After the monomers interact within the OPHOpolymer for some specific conformation, there are new electric and magnetic moments $\mathbf{T}\boldsymbol{\mu}_i$ and $\mathbf{T}\mathbf{m}_i$. The $\mathbf{T}$ is the unitary matrix which diagonalizes the total interaction matrix $\mathbf{H}(i \rightarrow j)$ associated with the ith to the jth excited state transition which we denote by δ.

The $\mathbf{H}(i \rightarrow j)$ matrix is defined as

$$
\mathbf{H}(i \rightarrow j) = \left(\begin{array}{cccc|cccc}
\epsilon_i & & (\mathsf{H}^i) & & \chi & & (\mathsf{H}^{ij}) & \\
& \epsilon_i & & & & \chi & & \\
(\mathsf{H}^i) & & \ddots & & (\mathsf{H}^{ij}) & & \ddots & \\
& & & \epsilon_i & & & & \chi \\
\hline
\chi & & (\mathsf{H}^{ij}) & & \epsilon_j & & (\mathsf{H}^j) & \\
& \chi & & & & \epsilon_j & & \\
(\mathsf{H}^{ij}) & & \ddots & & (\mathsf{H}^j) & & \ddots & \\
& & & \chi & & & & \epsilon_j
\end{array}\right) \qquad (5\text{-}13)
$$

The elements of the $\mathbf{H}(i \rightarrow j)$ interaction matrix are:

(1) The ϵ_k is the transition energy associated with the kth excited state. This quantity is normally measured experimentally using model molecules.

(2) The (H^k) mix the n possible degenerate states of the monomers yielding all possible linear combinations. These are the exciton interactions,

$$[H^k]_{a,b} = \sum_{l,\lambda} (q_l{}^a(k) \, q_\lambda{}^b(k))/r_{al,b\lambda} \qquad (5\text{-}14)$$

where $q_l{}^a(k)$ is the partial charge on the lth atom in the ath monomer unit for the kth excited state. The symbol r denotes the interaction distance. The partial charges are usually calculated by some molecular orbital scheme.†

(3) The χ represent perturbations of the one-electron transition moment of one monomer caused by the average static field of the other monomers. For a strongly polar molecule in which the ionized charge distribution is known (should any exist), a set of partial charges $\{Q^k\}$ can be assigned to the atoms of each monomer (on the basis of bond moments) which reproduce the permanent moment:

$$\chi_a = \sum_{l,\lambda,b} (Q_l{}^a Q_\lambda{}^b)/r_{al,b\lambda} \qquad (5\text{-}15)$$

for the ground state only.

(4) The (H^{ij}) are interactions between charges in excited states i and j:

$$[H^{ij}]_{a,b} = \sum_{l,\lambda} (q_l{}^a(i) \, q_\lambda{}^b(j))/r_{al,b\lambda} \qquad (5\text{-}16)$$

E. The Optical Spectrum

The fundamental equations of absorption and rotation were shown before to be

$$D = D_0 \, \boldsymbol{\mu} \cdot \boldsymbol{\mu} \qquad \text{and} \qquad R = R_0 \, \text{Im}(\boldsymbol{\mu} \cdot \mathbf{m}) \qquad (5\text{-}17)$$

For some of the transitions the following will be true:

$$(\mathbf{T}\boldsymbol{\mu}) \cdot (\mathbf{T}\boldsymbol{\mu}) \leqslant \boldsymbol{\mu} \cdot \boldsymbol{\mu} \qquad \text{and/or} \qquad \text{Im}((\mathbf{T}\boldsymbol{\mu}) \cdot (\mathbf{T}\mathbf{m})) \leqslant \text{Im}(\boldsymbol{\mu} \cdot \mathbf{m})$$

When this inequality holds we say that the transition is *hypochromatic*. In this situation the strength of the transition for the interacting monomers is less than that for the isolated monomers. When

$$(\mathbf{T}\boldsymbol{\mu}) \cdot (\mathbf{T}\boldsymbol{\mu}) > \boldsymbol{\mu} \cdot \boldsymbol{\mu} \qquad \text{and/or} \qquad \text{Im}((\mathbf{T}\boldsymbol{\mu}) \cdot (\mathbf{T}\mathbf{m})) > \text{Im}(\boldsymbol{\mu} \cdot \mathbf{m})$$

we say that the transition is *hyperchromatic*. In this situation the strength of the transition of the interacting monomers is greater than that when the monomer units are isolated.

† See Streitwieser (1) for a discussion and review.

It is the existence of *hypo-* and *hyper*chromism which gives rise to a characteristic absorption spectrum. In the same way it is *hypo-* and *hyper*chromism which is responsible for a CD spectrum that is characteristic of the primary and secondary structures of a molecule.

As implied earlier, the electric moments are given by

$$\mu(i,j) = \sum_l q_l{}^j(i)\, \mathbf{R}_{jl} \tag{5-18}$$

for the ith excited state, where $\mathbf{R}_{jl}$ is the distance of the lth atom in the jth monomer from some fixed origin in j. Let us assume that the v $\rightarrow$ w transition is purely electric and its optical center is $\mathbf{p}$ from our fixed origin. Then for any excited state i,

$$\mathbf{m}(i,j) = i\pi\omega_{\text{v}\rightarrow\text{w}} \sum_l \mathbf{r}_{jl} \times (q_l{}^j(i)\, \mathbf{R}_{jl}) \tag{5-19}$$

where $\mathbf{r}_{jl}$ is the vector defining the positions of the lth atom in the jth monomer relative to the optical center of the v $\rightarrow$ w transition in the jth monomer. $\omega_{\text{v}\rightarrow\text{w}}$ is the frequency of the v $\rightarrow$ w transition. The expressions for $\mu(i,j)$ and $\mathbf{m}(i,j)$ enable us to compute $\mathbf{U}$ and $\mathbf{M}$.

Having $\mathbf{U}$, $\mathbf{M}$, and T allows us to proceed and to calculate TU and TM which in turn allows us to compute the absorption strength for the δ transition between any two excited states,

$$D = (\mathsf{TU}) \cdot (\mathsf{TU}) \tag{5-20}$$

$$R = \text{Im}\{[\mathsf{TU}] \cdot [\mathsf{TM}]\} \tag{5-21}$$

If we assume that the absorption bands and rotation bands are gaussian with known width parameters, then we can construct the absorption spectrum and CD spectrum of the particular conformation in question as is demonstrated in Figs. 5-3a and 5-3b. Pysh (2) has used the theory described in this section to predict the absorption and CD spectra of some homopolypeptides. The results of some of his calculations are shown in Figs. 5-3c and 5-3d. In constructing these figures Pysh adopted a gaussian envelope to describe the circular dichroism band. The particular envelope had the form

$$G(\nu) = G^0(\nu') \exp(-(\nu - \nu')^2/d^2) \tag{5-22}$$

where $G^0(\nu') = 0.811 \times 10^{42}(\nu'/d)\, R(\nu')$. Here $R(\nu')$ is the rotational strength in cgs units as a function of frequency ν, and d is the width parameter assigned the constant value of 2000 cm^{-1}. The choice of d for any transition must be found experimentally. The symbol ν' refers to the value of ν at which the transition occurs.

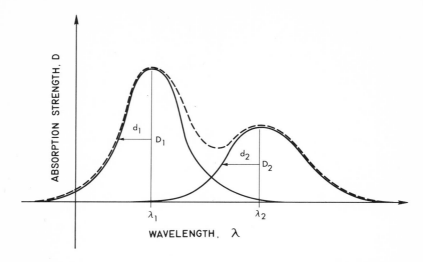

Fig. 5-3a. The predicted absorption spectrum of some molecule which has two transitions located at λ_1 and at λ_2. The actual absorption spectrum is indicated by the dotted line. This curve is the sum of the two gaussian curves centered about the wavelengths at which the transitions occur. The D's represent the absorption strengths as calculated by the theory. The d's represent the width parameters as determined by experiment.

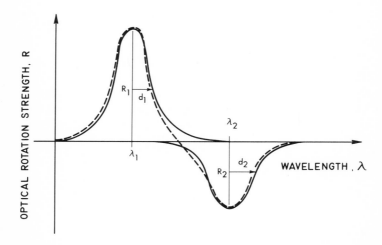

Fig. 5-3b. The predicted CD spectrum of some molecule which has two transitions located at λ_1 and at λ_2. The actual CD curve is indicated by the dotted line. This curve is the sum of the two gaussian curves. The R's represent the optical rotation strength as calculated by the theory and the d's are the width parameters as determined from experiment.

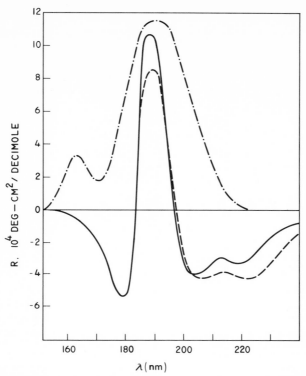

Fig. 5-3c. Three-state solution for the α-helix. —·—·— calculated oscillator strength in arbitrary units; ——— calculated circular dichroism; – – – circular dichroism observed with helical poly-L-alanine.

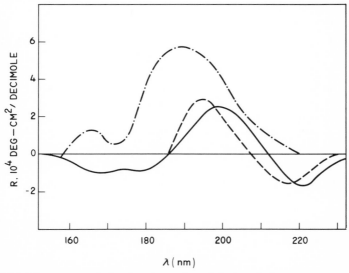

Fig. 5-3d. Three-state solution for the antiparallel pleated sheet. —·—·— calculated oscillator strength in arbitrary units; ——— calculated circular dichroism; – – – circular dichroism observed with β form of poly-L-lysine.

269

F. CD Spectrum of an Unordered Polypeptide Chain

In Chapter 4 we pointed out that poly(L-glutamic acid (PGA) in the charged form apparently is not in an unordered conformation as previously believed. Thus the CD/ORD spectrum of PGA in the charged form does not represent a disordered polypeptide chain, and the question arises as to what the CD/ORD spectra of a disordered chain actually is. Ronish and Krimm (3a) chose a dipeptide unit to represent a chain segment of an unordered polypeptide chain. This is consistent with the work of Flory, discussed in Chapter 4, who demonstrated that a dipeptide unit is a good model for an unordered polypeptide chain. They proceeded to compute the CD spectrum of this structural unit as a function of the backbone bond rotations (ϕ, ψ).

Their calculation is based on the monopole approximation to the far ultraviolet transitions of the peptide group, a method which seems to have at least a qualitative reliability. This calculation uses an SCF wave function for the peptide monomer ground state and the $n\pi^*$ and $\pi\pi^*$ transitions. The polypeptide wave function is constructed from this monomer wave function by configuration interactions.

The $n\pi^*$ and $\pi\pi^*$ transitions were both included in the secular determinant instead of treating the $\pi\pi^*$ by an exciton calculation and then adding in the $n\pi^*$ as a perturbation. By including both transitions, the calculation becomes more difficult numerically, but it is conceptually simpler and seems to lead to more reliable results.

The secular determinant of the polypeptide wave function was estimated by a monopole approximation where point charges were placed on or near atoms of the chain. The magnitudes and positions of the monopoles were estimated from the SCF wave function for the monomer. Monopoles were chosen which optimized the agreement between observed and calculated CD spectra of a known polypeptide chain structure, viz., the right-handed α-helix.

In order to test the proposal that the CD of a polypeptide chain with an arbitrary sequence of ϕ, ψ angles can be adequately represented by a sum of the corresponding dipeptide CD spectra, calculations were made on oligopeptides containing 28 residues. Then, using the same set of 27 ϕ, ψ angles, the average CD based on dipeptide spectra were determined. In all cases the dipeptide sum spectrum was a reasonable approximation to that of the oligopeptide.

The calculations of the dipeptide CD spectra were done at $10°$ intervals in ϕ and ψ for the entire range of ϕ, ψ angles. It is of interest to note that the resulting spectra fall into a relatively small number of types, thus permitting a plot of a dipeptide conformational CD map. These spectra have been classified according to the following scheme, which is illustrated by specific examples in Fig. 5-4a.

Type 1:   A net negative CD at wavelengths greater than 215 nm followed by a positive CD with a maximum at a wavelength greater than 187 nm.

Type 2:   Similar to type 1 except that the positive maximum occurs at wavelengths shorter than 187 nm.

Type 3:   The CD is essentially zero at all wavelengths.

Type 4:   A positive CD at wavelengths greater than 215 nm followed by a negative CD with a minimum at a wavelength shorter than 185 nm.

Type 5:   Similar to type 4 except that the minimum occurs between 185 and 198 nm.

Type 6:   Similar to type 4 except that the minimum occurs at wavelengths greater than 198 nm.

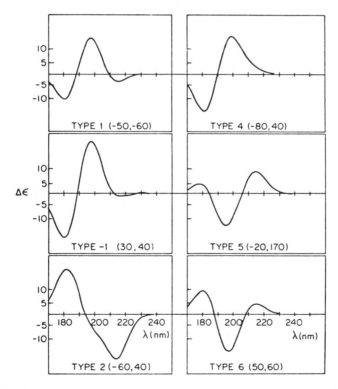

Fig. 5-4a. Examples of types of dipeptide CD spectra classified in the text. The numbers in parentheses are the (ϕ, ψ) values used for the examples. From Ronish and Krimm (3a).

A conformational CD map based on this classification is given in Fig. 5-4b. Some ϕ, ψ points do not quite fit into the above classification in that small contributions are present in the 230–250 nm region. These have been designated with a minus sign in Fig. 5-4b, indicating that there is a small band in this region which is positive for types 1 and 2 and negative for types 4, 5, and 6.

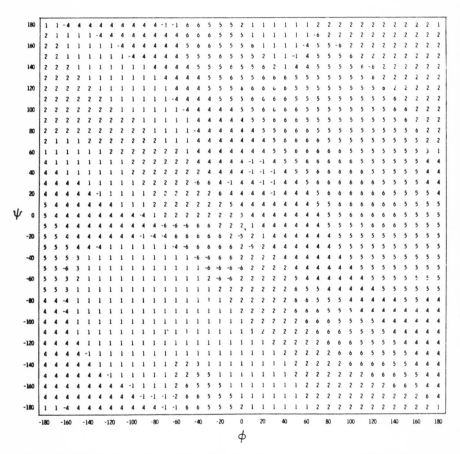

Fig. 5-4b. Dipeptide conformational CD map. Numbers at each ϕ, ψ represent spectral types, and are described in text. From Ronish and Krimm (3a). The new convention for (ϕ, ψ), given by the IUPAC-IUP Commission (3b) is used here.

The CD spectrum of the dipeptide unit was calculated by Boltzmann averaging the six types of spectra given in the CD conformational map using probabilities computed from (ϕ, ψ) conformational energy calculations of the dipeptide unit. Thus the resultant spectrum is dependent upon the types of energy interactions considered in the calculations.

The calculated CD spectrum of an unordered polypeptide chain shown in Fig. 5-5 disagrees with previous calculations in not exhibiting any long wavelength positive band. On the other hand, this result is in agreement with detailed experimental studies characterizing this spectrum.† The calculated negative band, at about 213 nm, is at a somewhat longer wavelength than that observed. It must be remembered however, that this is directly related to the location of the $n\pi^*$ transition, about which there may be some uncertainty for such structures.

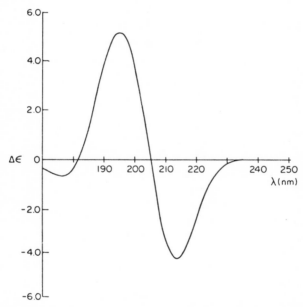

Fig. 5-5. Calculated CD spectrum of an unordered polypeptide chain. From Ronish and Krimm (3a).

It is perhaps worth emphasizing that there is probably no single unordered polypeptide chain structure. Different nonregular systems probably sample the ϕ, ψ space in different ways, depending on such factors as interaction with solvent, side-chain steric and electrostatic interactions, etc. This model of the unordered chain is based on a dipeptide energy map in which such factors are minimal. An examination of the CD spectra of a wide variety of unordered polypeptides indicates that the spectrum of Fig. 5-5 is more closely approached the greater is the tendency toward randomization of order in the system. In

† See Ronish and Krimm (3a) for a complete set of references.

this sense, the above calculated spectrum could be considered as that of a truly random chain, which is approached more or less closely by real polypeptide systems.

G. Prediction of the Optical Spectra of Molecules in Solution

How might the theory developed in the previous section be used to determine the optical spectrum of a particular molecular species which composes the solute portion of some solution? One answer to this question is conceivable if the following two assumptions are valid for the solution under consideration:

(1) The solvent molecules do not have a characteristic optical spectrum in the same spectral region as the solute molecules.

(2) The solute molecules do not interact.

The first assumption makes certain that the observed spectra are due only to solutes. The second assumption insists that the individual solute molecule–solute molecule interactions are zero *or* the sum of these interactions make a contribution to each of the various optical spectra which averages to zero. Molecular aggregation in solution will thus be treated as a solid state calculation (i.e., as a crystal structure calculation).

Consider now the spectrum (say CD) of some molecule which is composed of n electronic transitions. For any particular conformation k of the molecule it is possible to determine the rotational strength $R(i,k)$ (where i refers to a particular transition) for each of the n transitions. But it is also possible, using conformational potential functions, to compute the *solvent-dependent* conformational energy $E(k)$ for conformation k. Hence it is possible to compute an approximate *solvent-dependent* partition function Q for the molecule.

$$Q = \sum_{l=1}^{m} \exp(-E(l)\,\beta) \qquad (5\text{-}23)$$

where m is the number of sample conformations used in the estimate of Q and $\beta = 1/RT$ where T is the temperature of the solution. Once we have Q, we can proceed and calculate the *average* rotational strength $\langle R(i)\rangle$ for the ith transition from the expression

$$\langle R(i)\rangle = Q^{-1} \sum_{l=1}^{m} R(i,l)\,\exp(-E(l)\,\beta) \qquad (5\text{-}24)$$

Once we have the set of $\{\langle R(i)\rangle\}$ for a particular molecule in a specific solution we construct the *solvent-dependent* CD spectrum of the molecule. An analogous procedure may be used to determine the *solvent-dependent* absorption spectrum.

In the scheme of Ronish and Krimm (3a), discussed in the previous section, the effect of solvent on the CD spectrum of an unordered polypeptide chain can be accounted for in the following way: (1) compute the solvent-dependent conformational energy (ϕ, ψ) map for the dipeptide; (2) estimate the approximate partition function for the dipeptide from the (ϕ, ψ) map; (3) compute the corresponding conformational probability (ϕ, ψ) map using the computed value of the partition function; (4) weigh each of the types of CD spectra in the CD conformational (ϕ, ψ) map by the appropriate probabilities; and (5) sum up each of the weighted CD spectra to obtain the resultant spectrum.

II. Epitaxial Crystallization of Macromolecules†

A. General Theory of Epitaxial Crystallization of Macromolecules

Whenever a fluid exists atop a substrate there is an interaction between the substrate and the fluid. On the molecular level one is interested in the interactions of the molecules of the fluid with the atoms of the substrate. Thus atoms in the substrate which may be quite a distance from the actual interface between fluid and substrate may have interactions with the fluid molecules which are very important in characterizing the gross properties of the system. Most often one is interested in being able to determine if the interfacial interaction is such that the fluid molecules will "stick" to the substrate surface, and if they do, being able to characterize the mode and distribution of the growth of fluid-molecule "packets" on the surface. That is, one is interested in the *nucleation kinetics* of the fluid molecules at the interface. For small molecules, those which may be treated as simple geometric shapes and do not have the capacity to assume different geometric shapes (conformations), the nucleation kinetics may be treated rather well by straightforward statistical mechanics. The Langmuir adsorption isotherm theory (4) is a classic example of a statistical mechanical treatment of this surface problem. Water condensation on glass is an example of a simple nucleation kinetic problem which can be treated by Langmuir theory.

However, when the substrate is composed of a lattice of charges and the fluid is multicomponent in composition (i.e., there are two or more chemical or structural species such as a polymer in solution), then simple statistical mechanical theories cannot be used to describe the interfacial phenomenon.

† The author expresses his thanks to E. Baer and S. Wellinghoff who allowed him to discuss their experimental studies of the epitaxial crystallization of polyethylene. K. A. Mauritz carried out the theoretical calculations on epitaxial crystallization and the author appreciates his help.

The deposition of macromolecules on an ionic surface in crystalline clusters which is called *epitaxial crystallization* of macromolecules is an example of a nucleation process which is not easily characterized by statistical mechanical theories. The complex relationship between interfacial energy and the geometry of the system make it very difficult to compute the partition function of the system.

The following observations are noted from the limited experimental studies (5–9) of the epitaxial crystallization of macromolecules:

(1) There is a high specificity between the interface and the macromolecule. Only certain macromolecules will "stick" to certain surfaces and form microcrystals; that is, undergo epitaxial crystallization.

(2) Macromolecules which deposit epitaxially on a surface do so in very precise ways. Most macromolecules will lie down in one preferred orientation for a particular substrate surface. It appears in some cases that macromolecules which do not undergo epitaxial crystallization for a particular substrate cannot lie down in the preferred orientation dictated by the surface because of high interfacial interactions.

(3) The crystal growth of macromolecules on a surface is dependent upon the substrate to the extent that the preferred orientations of laying down on the substrate for the initial layer of macromolecules dictates a definite geometric structure. Further crystallization of the macromolecule must conform to this geometry.

(4) The rate of epitaxial crystal growth and the extent of crystal growth are dependent upon solvent and the temperature. Apparently the competition between macromolecule–solvent, macromolecule–substrate, macromolecule–macromolecule, solvent–solvent, and solvent–substrate interactions play an important role in the epitaxy phenomenon. Temperature seems to be a critical factor via the solvent molecules. By speeding up or slowing down the motion of solvent molecules, by raising or lowering temperatures, appears to affect both the rate and amount of epitaxial growth.

Theoretical conformational analysis is being applied to the epitaxy phenomenon in order to explain the experimental findings in molecular terms (10). Epitaxial crystallization of macromolecules can be thought of as a conformational and/or configurational calculation in which the macromolecules are subjected to an external force field which is generated by the substrate. Van der Waals, electrostatic, dipole, polymer–solvent, hydrogen bond, and torsional energies can be included in the computation of the total interfacial energy. Preliminary results of some calculations which have been carried out and will be presented below suggests that by applying theoretical conformational analysis to the phenomenon of epitaxial crystallization it might be possible, by scanning the interfacial energy as a function of the spatial variables of the system, to predict:

(1) whether or not epitaxy crystallization will occur for a particular substrate and macromolecule;

(2) what the orientation of a macromolecule on a substrate surface will be as a function of conformation;

(3) the crystal packing of the macromolecules on the surface.

B. The Epitaxial Crystallization of Polyethylene

The most complete investigation of the epitaxial crystallization of a macromolecule is polyethylene on alkali halide substrates (5, 6). The findings of this research are summarized below, and may serve as a guide for further research in this area.

1. Experimental

The epitaxial crystallization of polyethylene on alkali halide surfaces has been studied as a function of concentration, molecular weight, and degree of supercooling. Crystallization can be achieved by isothermal cleavage in solution. This procedure eliminates the effects of air, water and temperature gradients. Regardless of the crystallization conditions used, electron diffraction indicates the presence of a monoclinic phase in the interfacial areas of polyethylene that has been epitaxially deposited on NaCl. In the thinnest deposits (<50 Å), only a monoclinic phase with its (010) plane parallel to the substrate (001) plane and its c axis parallel to the alkali halide $\langle 110 \rangle$ directions is found. As the polyethylene layer grows thicker, growth of the usual orthorhombic form becomes more favorable with respect to the monoclinic, ultimately resulting in a complete transformation into the orthorhombic form with its (110) growth plane parallel to the substrate surface and the c axis of its unit cell parallel to the $\langle 110 \rangle$ directions of the alkali halide surface at thicknesses of about 500 Å. A schematic illustration of the modes of epitaxial crystal growth of polyethylene on NaCl are shown in Fig. 5-6.

Annealing of the samples on the alkali halide substrate has been carried out in air in order to determine the relative thermodynamic stability of the two packings of polyethylene which is skewed. Complete transformation to the orthorhombic packing is seen, as has been observed previously with the compressively stressed bulk material (11), indicating the instability of the monoclinic form relative to the orthorhombic form with heat treatment. However, high temperatures (130°C) are required for the complete transformation. A strong interaction between the interfacial polyethylene layers and the substrate surface is implied.

Preliminary experiments have also been carried out using KCl (10% change in lattice dimension from NaCl) and RbBr (20% change) as substrates. Electron diffraction indicates the presence of both the orthorhombic and monoclinic forms of packing with the c axes in the $\langle 110 \rangle$ direction for polyethylene crystallized on a KCl substrate. In some areas of the surface, however, it is possible to obtain only orthorhombic reflections. Polymers crystallized on RbBr are found only in the orthorhombic state. This implies that lattice matching has little to do with orientation of the initial nucleus, but does exert an influence on the type of packing at the substrate polymer interface once a certain degree of mismatch has been reached.

In the case of polyethylene crystallized epitaxially on NaCl, the first polymer layer to deposit consists of crystallites (of fold period of approximately

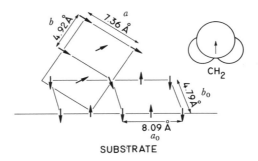

SUBSTRATE

Fig. 5-6. The epitaxial crystallization of polyethylene on sodium chloride crystals. The view is along the c axis of the polyethylene chains which, in turn, is the (110) or ($\bar{1}$10) directions on the NaCl lattice. The lattice constants, (a_0, b_0), refer to the monoclinic phase nearest the substrate surface, and the (a, b), refer to the lattice constants of the more common orthorhombic mode of packing. There is a transformation from the monoclinic to the orthorhombic packing as the distance from the substrate surface increases.

150 Å) that have grown rapidly in the plane of the substrate surface as compared to growth perpendicular to it. This indicates preferred polymer–solvent interactions. These structures are organized into domains containing crystals of similar dimensions, oriented in either the $\langle 110 \rangle$ or $\langle \bar{1}10 \rangle$ directions of the substrate. The domains are bounded by areas of little growth oriented along steps in the $\langle 100 \rangle$ directions of the substrate. Also, crystallites which have grown in the direction of the initial layer, but whose growth rates parallel and perpendicular to the surface are similar, are randomly nucleated on the contact layer. Provided growth conditions are favorable, these structures can grow into an interlocking membrane that consists mainly of the orthorhombic form. In RbBr epitaxy, which gives only orthorhombic reflections, the thicker structures are found to be nucleated directly on the surface to the exclusion

of the initial domained contact layer. This evidence appears to show that there is a sharp transition in packing mode with layer thickness in NaCl epitaxy, but none in RbBr epitaxy where interaction with the surface does not seem to favor an initial monoclinic state.

When crystal height (measured perpendicular to the surface) is plotted against time a limiting growth height can be deduced. This final thickness is a direct function of polymer concentration in that the higher the polymer concentration the greater the thickness. This phenomenon is thought to occur as a result of viscosity effects in the immediate region of the growth face slowing the nucleation of high molecular weight polymer. The lowest molecular weight crystallizes most slowly at all temperatures because of the low thermodynamic stability of its growth nucleus.

As has been observed previously with polydisperse polymer, the degree of supercooling exerts a noticeable effect on growth rate. Increased supercooling resulted in increased growth for all conditions.

2. Theoretical Calculations

Theoretical conformational analysis has been used to determine the molecular geometry of the epitaxial crystallization of polyethylene. The goal of this work was to explain the particular orientation and packing of polyethylene molecules on an alkali halide substrate in terms of pairwise atomic interactions. Nine monomer units of polyethylene were assigned the planar zigzag conformation and were allowed to interact with a NaCl substrate. The geometry of the system is shown in Fig. 5-7. The planar zigzag conformation polyethylene was chosen because the experimental crystal packing modes presented above could only be rationalized in terms of the zigzag conformation. The pairwise energy contributions to the total interaction energy consisted of:

(1) *Partial charge–charge electrostatic interactions between atoms in the polymer and ions in the substrate respectively.* The partial charges on the atoms of the polymer were calculated using the *Del Re* technique (12) which gave the saturated contributions to the partial charges and the *reproduction of bond moments* technique† which yielded the π contributions to the partial charges. All pairwise interactions were based upon point charge distributions.

(2) *Dipole–charge interactions between atoms in the polymer and ions in the substrate.* The atoms in the polymer were assumed to be polarizable [using a set of coefficients of polarizability $\{\alpha\}$ given by Pitzer (14)] while the ions in the substrate were considered nonpolarizable. This made it possible to decouple the dipole interaction matrix between ions and atoms, and to calculate the total induced dipolar energy using scalar operations.

† See Poland and Scheraga (13) for description of the method.

(3) *Steric interactions*. Since the van der Waals part of the attractive interaction in ionic crystals makes a relatively small contribution to the cohesive
energy ($\simeq$ 1 to 2%) (15) *and* because there are no reliable nonbonded
potential functions for the interactions between ions and atoms† it was necessary to use the "hard sphere" model to simulate steric interaction. Polymer–
substrate interaction energies were calculated only for those situations in
which the distance between each ion and atom was equal to or greater than
the respective sum of the ionic and van der Waals radii of the ion and atom.

(4) *The medium of the crystal* was treated as being electrically homogeneous
by assuming a constant dielectric within the crystal. Solvent enters into the

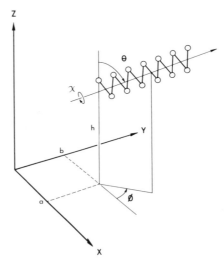

Fig. 5-7. The macromolecule-planar substrate geometry for a linear "zigzag" chain of
polyethylene and an alkali halide substrate. The parameters $\{a, b, h, \theta, \phi, \chi\}$ are defined.

energy calculations only through a specific dielectric constant (representative
of the bulk solvent) which assumes the medium in which the polymer exists
(the solution) is also electrically homogeneous.

From Fig. 5-7 it can be seen that the determination of the preferred
orientation of the polyethylene chain on the substrate surface requires one to
locate the global minimum in total interaction energy for a function of six
variables $\{a, b, h, \theta, \phi, \chi\}$. This global minimum (which is degenerate) has been
found and an energy contour map of h versus ϕ, the two most critical variables,
is shown in Fig. 5-8 with the remaining four variables optimized at each point

† Appendix I gives a set of *approximate* nonbonded potential functions for the interactions
between ions and atoms involved in covalent bonds.

on the map. This map is for one single polyethylene chain. The results of the calculations indicate that:

(1) The present computer model used to simulate epitaxial crystallization is realistic since the global minimum corresponds to an orientation of the polymer in the $\langle 110 \rangle$ or $\langle \bar{1}10 \rangle$ direction on the substrate surface which is in agreement with experiment.

(2) The interaction between the substrate and the single polyethylene chain is either acting in unison with other types of molecular interactions

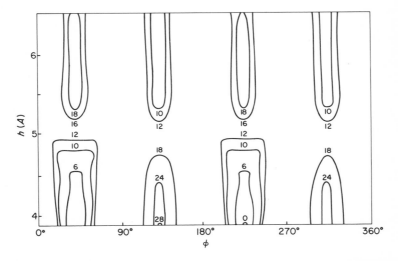

Fig. 5-8. An h versus ϕ energy contour map for five monomer units of a single polyethylene chain in the linear "zigzag" conformation interacting with a planar NaCl substrate. The energy contours are plotted relative to the minima ($h \simeq 3.8$ Å, $\phi = 45°$, $135°$, $225°$, $315°$) which are assigned a value of zero energy. The energy contours are in kcal/mole/5 CH_2 units. The parameters $\{a, b, \theta, \chi\}$ have been optimized for each point on the map.

and/or is the dominant force in dictating the molecular geometry. If one, or both, of these explanations were not correct, then one would not expect a single polyethylene chain to orient in the direction observed for polymer crystal growth *and* to also assume a value of χ which corresponds to a rotation about the chain axis, that is necessary to generate the crystal packings that are found experimentally. Figure 5-9 is an illustration of the molecular geometry of a single polyethylene chain on a planar NaCl surface when the total interaction energy is that of the global minimum.

(3) The partial charge–charge interactions between atoms in the chain and

ions in the substrate are more sensitive to changes in molecular geometry than are the induced dipolar interactions. Thus, at least for polyethylene on NaCl, dipolar forces are not as important in controlling epitaxial crystallization as had been previously suggested (8).

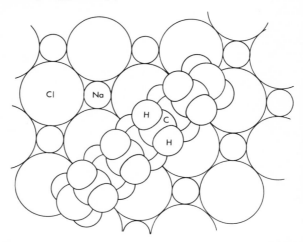

Fig. 5-9. A space filling model representation of the orientation of a single polyethylene chain on a planar NaCl substrate. This orientation (110) corresponds to a minimum in total interfacial energy.

III. NMR and Conformational Analysis

A. Structural Usefulness of NMR

Simple accurate rules for determining the structures of macromolecules would be very helpful in understanding their chemistry. X-Ray diffraction provides the necessary information needed to establish such rules for crystalline materials but is not applicable to solutes which crystallize poorly or for determining the conformation of molecules in solution. Whatever spectroscopic means is chosen to study the solution-state structures of macromolecules it must give information not only on the overall size and shape of the macromolecule but also on local structural features. Nuclear magnetic resonance, NMR, which is, in principle, capable of revealing something of the spatial arrangement of the immediate neighbors of any magnetic nucleus in the molecule, seems to be the most promising spectroscopic technique presently available.

There are two major drawbacks to applying NMR to structural investigations. First of all, overlapping resonances often make it impossible to obtain the complete NMR spectrum of the molecule. Thus, resonance assignments are often ambiguous and structural identification is meaningless. Technological advances in the resolution and sensitivity of spectrometers as well as schemes of deuterating sample species has and will continue to make it easier to obtain resolved spectra of ever-increasing complex molecules. However, even if the complete NMR spectrum of a molecule can be resolved and assigned there is no assurance that a structural determination is possible. This is a consequence of the second major drawback to using NMR for structural studies. There is a lack of understanding of the relationship between spectra and molecular structure. In an analogy to X-ray crystallography, the equivalent relationship to the Fourier transformation which allows a structural determination to be made from a diffraction pattern has not been available in NMR spectroscopy.

This is not to say that NMR has not proved valuable in determining certain aspects of molecular structure. NMR spectroscopy can be very helpful in the elucidation of the primary structures of macromolecules. For example, the distribution of isotactic, syndiotactic, and atactic polystyrene resulting from a particular scheme of synthesis has been determined by NMR. NMR is also very useful for measuring the percent of structural and/or chemical defects in polymers. Studies along this line have been carried out for poly(vinylidene fluoride) (16, 17). These studies make direct use of the fact that the environment of a magnetic nucleus is extremely sensitive to the nearby atomic species. This environmental sensitivity has been used on innumerable occasions to determine what atomic species are "close to one another" in all types of chemical compounds. These types of investigations make up the majority of experimental investigations being carried out using NMR spectroscopy. Still, structural studies leading to precise secondary and tertiary geometry have not been possible in the majority of molecules, especially those of high molecular weight. We will spend the remainder of this section discussing some new schemes which, hopefully, will allow NMR spectroscopy to be used to determine precise molecular structure.

In the early 1960's both experimental and theoretical investigations indicated that the vicinal coupling constants observed in high resolution NMR spectroscopy could be used to determine molecular structure. Specifically, the $J_{HH'}$ vicinal coupling constant in ethanic, ethylenic, and related systems, $HC\overset{\curvearrowright}{\underset{\theta}{\smile}}H'C'$, was found to vary in ordered fashion with the dihedral angle θ. LCAO–MO calculations were carried out in order to determine the functional relationship between θ and $J_{HH'}$ (18–20). Karplus proposed his now famous relationship

$$J_{HH'} = A + B\cos\theta + C\cos 2\theta \qquad (5\text{-}25)$$

TABLE 5-1

Estimated and Observed Vinyl–Allylic Proton Spin Couplings[a]

Structure	θ	^3J(est)[b]	^4J(est)[b]	^3J	^4J
(R = H or CN)	0	6.6	1.3	6.6	1.3
	0	6.6	1.3	6.1	1.6
Norbornenes	20	6.1	0.8	2.8–3.0	0.55–0.7
Norbornadiene	20	6.1	0.8	2.7	0.95
Cyclopentenone	60	3.6	−1.6	2.8	±2.2
4-Bromocyclopent-2-enone	60	3.6	−1.6	3.0	±1.3
	40 and 80	3.9	−1.4	2.1	±1.8
Cyclopentene	40 and 80	3.9	−1.4	2.1 (3.7)	−1.4
Cyclohexenone	40 and 80	3.9	−1.4	3.7	±1.5
	40 and 80	3.9	−1.4	2.6	±1.3
	40 and 80	3.9	−1.4	3.5	±1.8
Cyclohexene	40 and 80	3.9	−1.4	3.1 (3.6) (3.6–4.1)	−1.4
	80	—	−2.4	—	±2.3
	80	—	−2.4	—	±2.1

TABLE 5-1 (continued)

Structure	θ	$^3J^{(est)\,b}$	$^4J^{(est)\,b}$	^3J	^4J
	80	—	−2.4	—	±2.1
	80	2.8	—	1.8–2.5	—
	80	2.8	—	2.5	...
	40	5.0	−0.3	5.0	None obsd. (<1 Hz)
Cycloheptene	15 and 135	6.7	−0.2	5.7 (5.6)	−1.0
1-tert-Butylcycloheptene	15 and 135	6.7	−0.2	6.8	≤0.6 ($W_H = 1.2$)
cis-Cyclooctene	25 and 145	7.3	−0.1	7.8 (7.8)	−0.7
1-Methylcyclooctene	25 and 145	7.3	—	7.8	...
	25 and 145	7.3	−0.1	7.9	≤0.7 ($W_H = 1.4$)
cis-Cyclononene	35 and 155	7.7	−0.2	8.2	−0.8
cis-Cyclodecene	60 and 180 0 and 120	6.7	−0.6	6.8	−0.8
	160	—	−0.3	—	±0.8 (R = H) ±0.7 (R = Me)
Propene	60, 60, and 180	6.3	−1.1	6.4	−1.3
3-Iodo-1-propene	60 and 180	7.6	−0.8	7.7	−0.4
1,1-Dimethyl-3,3-di-tert-butyl-1-propene	180	11.6	—	11.4	—
3,3-Di-tert-butyl-1-propene	180	11.6	0.0	10.6	−0.1

[a] See Garbisch (22) for the set of references used to construct this table. Reprinted from *J. Amer. Chem. Soc.* **86**, 5561 (1964). Copyright 1964 by the American Chemical Society. Reprinted by permission of the copyright owner.

[b] The values of $^3J^{(est)}$ and $^4J^{(est)}$ were computed using the following relationships:

$$^3J^{(est)} \cong \begin{cases} 6.6\cos^2\phi + 2.6\sin^2\phi & (0° \leqslant \phi \leqslant 90°), \\ 11.6\cos^2\phi + 2.6\sin^2\phi & (180° > \phi > 90°), \end{cases}$$

$$^4J^{(est)} \cong \begin{cases} 1.3\cos^2\phi - 2.6\sin^2\phi & (0° \leqslant \phi \leqslant 90°) \\ -2.6\sin^2\phi & (180° \geqslant \phi \geqslant 90°) \end{cases}$$

could be used to approximate the complex expression found from the quantum mechanical calculations. For a $C(sp^3)$—$C(sp^3)$ bond having a bond length of 1.543 Å and a bond energy of 9 eV, the constants are

$$A = 4.22, \qquad B = -0.5, \qquad C = 4.5$$

Theoretical considerations suggested that serious errors could result if relationships similar to the Karplus relationship given by Eq. (5-25) were assumed for systems in which one or more of the chemical species bonded to either or both of the carbons was different from hydrogen. However, Bothner-By (21) has proposed Karplus-type functions for alkylethylenes and Garbisch (22) has found coupling relationships for vinyl–alkyl compounds. Table 5-1 lists some of the vicinal coupling constants. Very recently Bystrov et al. (23) proposed a relationship between the dihedral angle and the vicinal coupling constant $^3J_{NC}$ for HC–NH systems. This relationship is based upon combined infrared and NMR studies of N-acyl dialanyl methyl esters, with variable degrees of methylation of the amide NH groups. The important implication of this last proposed relationship between $^3J_{NC}$ and the dihedral angle, at least to protein chemists, is that the rotations about N—C^α bonds, the ϕ rotations, along with the ψ rotations are responsible for the backbone conformations of peptides and proteins. Thus the Bystrov–Karplus relationship would make it possible to use NMR spectroscopy to determine the backbone N—C^α conformations in peptides and, perhaps, even proteins.

Unfortunately, all of the relationships between the magnitudes of coupling constants J and the values of the dihedral angles θ are periodic in θ. Thus the determination of the magnitude of a coupling constant will not uniquely define θ, but yield a few, normally no more than four, values of possible θ. For a molecule possessing n internal bond rotations where coupling constant data suggests m possible values for each of the internal bond rotations results in n^m trial conformations. For a tripeptide $n = 6$ and, let us assume, $m = 3$ for each internal bond rotation there would be $6^3 = 216$ trial backbone conformations for the ϕ rotations. Clearly, macromolecules composed of ordered monomer sequences, i.e., homopolymers with respect to some simple repeating sequence, may be easier to treat by this analysis since the repetitious primary structure may allow simplifying assumptions, such as the equivalence condition, to be invoked. Polymers composed of heterosequences of monomer units, such as proteins, would be extremely difficult to handle by analysis of coupling constants.

Gibbons et al. (24) have suggested that conformational potential energy calculations might be used to choose the correct value of internal bond rotation from those values found to be consistent with coupling constant data. A simple demonstration of how this might be carried out can be shown with the aid of Fig. 5-10. In this figure we have plotted the proposed Bystrov–

Karplus relationship in such a way that the θ axis is parallel to the ϕ axis of a (ϕ,ψ) conformational potential energy map of an L-alanine amino acid residue. Suppose now the coupling constant for the HN—C$^\alpha$H group was found to have a value of $^3J_{NC} = 3.0$. The Bystrov–Karplus relationship

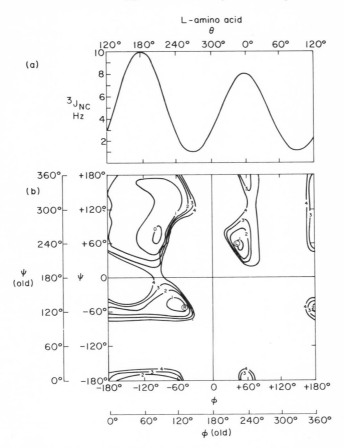

Fig. 5-10. (a) A plot of the proposed Bystrov–Karplus relationship of ϕ versus $^3J_{N-C}$ for an L-amino acid residue. (b) A conformational map of an L-alanine amino acid residue. From Gibbons *et al.* (24).

would yield four values for ϕ: 0°, 120°, 180°, 300°. From the conformational potential energy map one could conclude that $\phi = 120°$ is the most probable value for this rotation, and further, the ψ rotation about the C$^\alpha$—C′ bond would have one of two values, 120° or 300°. In practice the determination of the values of the dihedral angles will not be so simple. It would seem that consideration of isolated monomer units is not sufficient to accurately characterize the conformational energy surface of the monomer when in a polymer. Thus the elucidation of the dihedral angles will probably require that conformational potential energy calculations be carried out for groups of monomer units.

B. An Example of an NMR Vicinal Coupling Constant Structural Calculation

Torchia carried out vicinal coupling constant calculations to determine the poly-L-proline II ring conformations (25). The geometry of the pyrrolidine ring of proline is shown in Fig. 5-11.

In order to calculate the spectrum of this complex spin system, seven chemical shifts, three geminal couplings, and ten vicinal couplings are required as input data for a spectral simulation program. Four of the ring

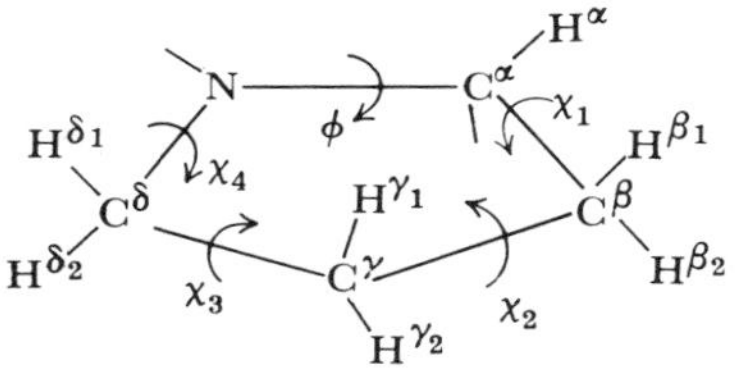

Fig. 5-11. The geometry of the pyrrolidine ring. From Torchia (25). Reprinted from *Macromolecules* **4**, 440 (1971). Copyright 1971 by the American Chemical Society. Reprinted by permission of the copyright owner.

protons, H^{α}, the two H^{δ}'s and an H^{β}, give distinct resonances at about 5.2, 6.2, 6.4, and 7.7τ, respectively. The second H^{β} at 8.1τ partially overlaps the approximately equivalent H^{γ} protons located at 8.0τ. Initially the three geminal couplings were taken as −12.0 Hz on the basis of published data for five- and six-membered rings, and the two $J_{\alpha\beta}$ vicinal couplings were estimated to be 8.0 and 5.0 Hz from the fine structure of the α-proton resonance. Initially all $J_{\beta\gamma}$ and $J_{\gamma\delta}$ were taken to be 6.5 Hz. These initial values of chemical shifts and coupling constants were varied and the ranges of vicinal couplings which gave satisfactory simulations were determined. These are summarized in Table 5-2. The predicted and observed NMR spectra of the proline ring are shown in Fig. 5-12. The chemical shifts and couplings used to generate the satisfactory simulation, Fig. 5-12, are summarized in the chart at the top of the figure. While the experimental and calculated spectra are almost identical, it is emphasized that equally good simulations were obtained using γ-proton vicinal couplings anywhere within the ranges given in Table 5-2.

TABLE 5-2

Vicinal Couplings Which Give Satisfactory Simulations of Poly-L-proline II Spectrum[a]

Vicinal coupling	Mean value (Hz)	Acceptable range (Hz)
$\left\{\begin{matrix} J_{\alpha\beta1} \\ J_{\alpha\beta2} \end{matrix}\right\}$ or $\left\{\begin{matrix} J_{\alpha\beta2} \\ J_{\alpha\beta1} \end{matrix}\right\}$	8.5	8.0 to 9.0
	5.5	5.0 to 6.0
$J_{\beta1\gamma1} + J_{\beta1\gamma2}$	13.0	12.0 to 14.0
$J_{\beta2\gamma1} + J_{\beta2\gamma2}$	13.0	12.0 to 14.0
$J_{\gamma1\delta1} + J_{\gamma1\delta2}$	13.0	11.5 to 14.5
$J_{\gamma2\delta1} + J_{\gamma2\delta2}$	13.0	11.5 to 14.5
$J_{\beta1\gamma1} - J_{\beta1\gamma2}$	0.0	−5.0 to 5.0
$J_{\beta2\gamma1} - J_{\beta2\gamma2}$	0.0	−5.0 to 5.0
$J_{\gamma1\delta1} - J_{\gamma1\delta2}$	0.0	−5.0 to 5.0
$J_{\gamma2\delta1} - J_{\gamma2\delta2}$	0.0	−5.0 to 5.0

[a] From Torchia (25). Reprinted from *Macromolecules* **4**, 440 (1971). Copyright 1971 by the American Chemical Society. Reprinted by permission of the copyright owner.

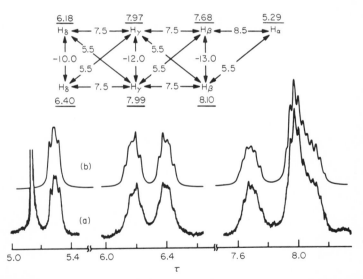

Fig. 5-12. Comparison of 220 MHz spectra of proline ring protons of poly-L-proline II: (a) experimental spectrum in D_2O, HDO resonance at 5.15 τ, (b) computer-simulated spectrum. At the top is a schematic summary of the coupling constants (Hz) and chemical shifts (underlined) used in the simulation. Line width assumed in simulation, 5.0 Hz. Chemical shifts are in τ units downfield from DSS (sodium 2,2-dimethyl-2-silapentane-5-sulfonate). From Torchia (25). Reprinted from *Macromolecules* **4**, 440 (1971). Copyright 1971 by the American Chemical Society. Reprinted by permission of the copyright owner.

The vicinal couplings in Table 5-2 can be related to the dihedral angles, χ_d, between the proline ring protons by a Karplus equation

$$J_{H,H} = A \cos^2 \chi_d + B \quad (0° \leqslant \chi_d \leqslant 90°) \tag{5-26}$$

$$J_{H,H} = A' \cos^2 \chi_d + B \quad (90° \leqslant \chi_d \leqslant 180°) \tag{5-27}$$

Experimental work indicates that a set of coefficients which applies to proline ring couplings is

$$A = 8.5 \quad Hz, \qquad A' = 10.5 \quad Hz, \qquad B = 1.4 \quad Hz$$

These coefficients result from an analysis of cyclo(tri-L-prolyl) ring couplings (26) and agree closely with coefficients proposed from an analysis of ring vicinal coupling data for six membered rings.†

While the vicinal couplings obtained from the simulation of the measured spectrum do not specify a unique ring conformation, they do limit the types of acceptable conformations. For instance, the planar conformation ($\phi = 120°$, $\chi_i = 0°$, $i = 1,2,3,4$) is excluded since using Eqs. (5-26) and (5-27) one calculates $J_{\alpha\beta1}/J_{\alpha\beta2} = 2.5$, a result which deviates by over 65% from the experimental values of 1.5 or 0.7 obtained from Table 5-2. Individual nonplanar conformations, in which the ring is markedly puckered (at a single atom) or twisted (with atoms i and $i + 2$ on opposite sides of the plane of the three other ring atoms) are also prohibited. Consider for example a conformation puckered at C^γ ($\phi = 120°$, $-\chi_1 = \chi_4 = 15°$, $\chi_2 = -\chi_3 = 25°$). From Eqs. (5-26) and (5-27) and the χ_d corresponding to this conformation one finds

$$J_{\gamma^1\beta^2} + J_{\gamma^2\beta^1} = J_{\gamma^1\delta^1} + J_{\gamma^2\delta^1} = 10 \text{ Hz}$$

and

$$J_{\gamma^1\beta^2} + J_{\gamma^1\beta^2} = J_{\gamma^1\delta^2} + J_{\gamma^2\delta^2} = 17 \text{ Hz}$$

which are well outside the limits given in Table 5-2. In similar fashion, one can exclude every individual puckered or twisted conformation in which ϕ or χ_i, $i = 1, 2, 3, 4$ deviates by more than about 15° from the planar ring values of these angles. This result is of interest, since it shows that the ring conformations proposed for poly-L-proline in the solid state (28, 29) (C^β puckered, with $\chi_1 = -\chi_2 \sim 25°$) and for cyclo(tri-L-prolyl) in solution ($\chi_1 \sim 30°$) are incompatible with the couplings in Table 5-2.

The only (single) conformations which are compatible with the NMR data have C^α slightly above (exo) or C^β slightly below (endo) the plane of the remaining four atoms, (i.e., $100° \lesssim \phi \lesssim 120°$, $\chi_1 \sim -10°$, $-10° \lesssim \chi_2, \chi_3, \chi_4 \lesssim 10°$).

Individual conformations which have large puckerings cannot account for

† See Kopple and Marr (27) for references.

the NMR data. However, if such conformations are in rapid equilibrium, so that their J values average, the resultant couplings are in good agreement with those in Table 5-2. It is proposed that in solution the proline rings of poly-L-proline II rapidly interconvert, via the planar conformation, between two equally populated C^γ puckered conformations.

In a similar investigation (30) of poly(hydroxy-L-proline), which differs from poly-L-proline in that a hydroxyl group is bonded to the C^γ of the pyrrolidine ring in place of a proton, Torchia has found that the ring maintains a single nonplanar conformation in aqueous solution. This puckered conformation, having C^γ *exo*, is described by the angles $\phi \sim 120°$, $\chi_1 \sim \chi_4 \sim -25°$, $\chi_2 \sim \chi_3 \sim 45°$. A possible explanation for the single ring conformation may be that a $C=O\cdots H-O$ intrachain hydrogen bond is formed between the hydroxyl group and an adjacent backbone carbonyl group.

In summary, it should be stressed that the main emphasis in this section is to point out relationships between a proton magnetic resonance spectrum and the structure of the molecule. The Karplus–Bystrov relationship requires more rigorous theoretical and experimental verification. Conformational potential energy calculations, as we have already noted, contain many approximations. In spite of these obstacles, we feel that the types of structural studies outlined here will be carried out and, in turn, stimulate the refinement of some of the proposed concepts.

C. Carbon-13 Magnetic Resonance Spectroscopy

Nuclear magnetic resonance spectroscopy can be carried out only for nuclei which contain magnetic moments. In naturally occurring molecules the predominant carbon isotope is ^{12}C which has no spin and does not exhibit magnetic resonance. The ^{13}C isotope has a spin of $\frac{1}{2}$ but a natural abundance of only 1.1%. Nevertheless the ^{13}C spectra of simple molecules, steroids (31) and some amino acids (32) have been recorded in enriched samples and at natural abundance. The synthesis of ^{13}C isotopic molecules, although costly, does provide a means to study the ^{13}C magnetic resonance spectra of such molecules. Gibbons *et al.* (33) have recorded the ^{13}C NMR spectrum of gramicidin S-A, a cyclic decapeptide. The structure of gramicidin S-A is shown in Fig. 5-13. Figure 5-14 shows the ^{13}C spectrum of gramicidin S-A in two solvents, methanol and DMSO. The assignment of peaks in the spectrum to individual carbon atoms is based on a comparison of peak positions with the ^{13}C chemical shifts found for the N-acetyl amino acid methyl esters (34). Individual peaks in the monomer spectra were assigned by comparison with the spectra of the amino acids, both as observed and as calculated from

Grant's rules (35). Agreement between observed and predicted chemical shifts for amino acids is excellent, as shown in Table 5-3.

The spectrum extends over 160 ppm of applied field relative to $^{13}CS_2$ as an external standard. The complicating feature of ^{13}C–proton coupling was eliminated by noise decoupling and ^{13}C–^{13}C direct couplings, $J_{CC'}$, are not seen, for the chance that an individual ^{13}C atom will have another ^{13}C

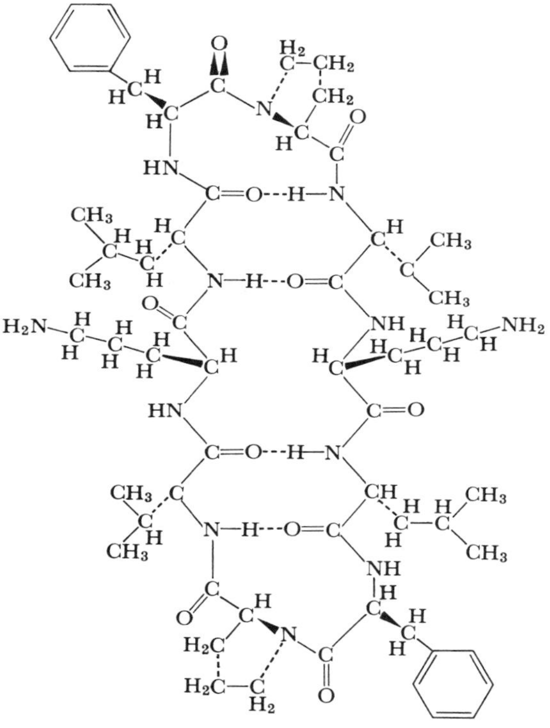

Fig. 5-13. Schematic drawing of the three-dimensional structure of gramicidin S-A, cyclo(-Phe-Pro-Val-Orn-Leu-)$_2$, proposed from proton NMR studies. All ω angles are 0° except for Leu = 10°. The ϕ angles of Val, Orn, Leu = 30°, Phe = 150°. The ψ angles of Val, Orn, Leu = 0°, Phe and Pro = 130°. From Gibbons *et al.* (33).

atom adjacent to it is approximately 1 in 10^4. Each carbon thus appears as a singlet. Later, we will discuss the possible use of direct J_{AB} coupling constants in ^{13}C-enriched molecules to deduce conformation.

The gramicidin S-A spectra are readily divided into well-separated regions: (a) carbonyl carbons (19–24 ppm), (b) aromatic-heteroaromatic carbons (50–70 ppm), (c) C^{α} carbons (130–140 ppm), (d) C^{β} carbons (150–165 ppm), and (e) C^{γ} and methyl carbons (165–180 ppm).

The symmetry of gramicidin S-A is definitely C_2 from the ^{13}C NMR spectra. The five pairs of residues yield only five carbonyl, five C^α, and five C^β chemical shifts. There is some interest in the two methyl group ^{13}C lines of

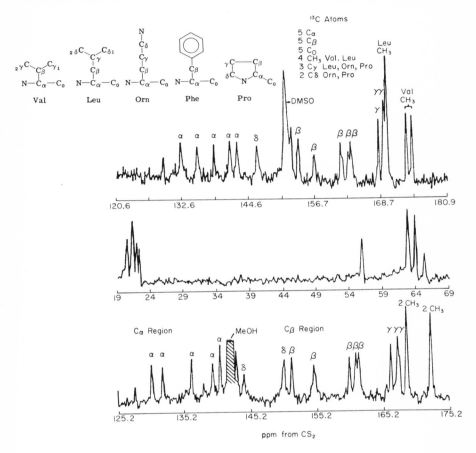

Fig. 5-14. The natural abundance ^{13}C NMR spectrum of gramicidin S-A. (a) The high field part of the spectrum in DMSO; (b) the low field part of the spectrum in DMSO; (c) the high field part in methanol. All chemical shifts are in ppm upfield from $^{13}CS_2$ as an external standard. From Gibbons *et al.* (33).

valine. These groups are not magnetically equivalent, so two lines are to be expected. However, the chemical shift difference between them is observed to vary from amino acid to derivative to gramicidin. This might imply that the valine side chain has different average rotational conformations in the three cases.

Let us now consider the relative heights and intensities of the various ^{13}C lines in the NMR spectra. There should be a general enhancement of all ^{13}C resonances on noise decoupling of protons because of the nuclear Overhauser effect, and the enhancement factor of 2.988 is theoretically possible wherever the ^{13}C relaxation mechanism is dominated by proton–^{13}C dipole–dipole

TABLE 5-3

Peak Positions and Assignments[a]

Carbon	Amino acid pred[b]	Amino acid obs[c]	Deriva- tives[d]	Gramicidin S-A in DMSO	Gramicidin S-A in methanol
Pro α	125.5	132.6	134.0	132.6	130.3
Val α	127.3	132.6	134.7	135.7	131.9
Orn α	132.7	—	138.5	138.6	136.4
Phe α	—	136.9	138.6	141.4	139.6
Leu α	135.8	139.5	142.1	142.8	140.7
Pro δ	148.3	147.5	144.8	146.1	144.3
Orn δ	153.0	—	152.0	Solvent	150.4
Leu β	153.2	153.0	151.0	154.1	151.5
Phe β	—	156.5	155.3	156.8	154.9
Orn β	164.7	—	162.0	161.3	160.2
Val β	167.4	163.8	162.1	162.8	161.2
Pro β	166.6	164.3	163.0	163.4	161.6
Pro γ	169.5	169.6	167.6	168.5	166.4
Leu γ	170.3	168.6	168.0	169.5	167.5
Orn γ	168.2	—	168.0	169.5	167.5
Leu CH$_2$	170.7	170.8	169.0	170.0	168.9
Leu CH$_3$	170.7	171.9	171.0	170.0	168.9
Val CH$_2$	176.5	174.9	173.0	173.6	172.5
Val CH$_3$	176.5	176.1	173.7	174.5	172.5

[a] From Gibbons *et al.* (33).

[b] From parameters in Horsley *et al.* (34) and Grant and Paul (35). Phenylalanine predictions are not possible from these parameters.

[c] From Horsley *et al.* (34). Ornithine was not reported. All positions are chemical shifts in ppm upfield from $^{13}CS_2$.

[d] *N*-acetyl amino acid methyl esters in DMSO.

coupling. Large deviations from this figure should only be found when the ^{13}C nucleus is not directly bonded to a proton or when the ^{13}C–H separation greatly exceeds that of a normal carbon–hydrogen bond. The intensity relationships will help identify chemical shifts and, when properly identified, hopefully yield information on relaxation times, transition probabilities, and the molecular environment of the various carbon atoms in the molecule.

Although most effort in ^{13}C NMR spectroscopy is concerned with obtaining and assigning chemical shifts, this method may possess the intrinsic capability of elucidating information on molecular conformation. For any bond A-B, where A and B both possess nuclear spins, the direct coupling constant J_{AB} may have a periodic dependence on θ in the same way the vicinal coupling constants in proton NMR depend upon θ. This periodic dependence may be very small and it may prove impossible to use the J_{AB}–θ relationships in structural studies. Nevertheless, in the hope that such a J_{AB} versus θ dependence will prove practical we report here the derivation of Pople and Santry (36) who developed a general theory of direct coupling constants. They found it convenient to introduce a *reduced coupling* constant K_{AB} which is independent of the nuclear moments of A and B and depends only on the electronic environment. This will be defined as

$$K_{AB} = (2\pi/\hbar\gamma_A \gamma_B) J_{AB} \tag{5-28}$$

where γ_A, γ_B are the nuclear magnetogyric ratios. It should be noted that K_{AB} and J_{AB} will have different signs if one of γ_A and γ_B is negative. Such would be the case for ^{15}N and ^{17}O.

K_{AB} is the mean value of a *reduced coupling tensor* $(K_{AB})_{\alpha\beta}$ which is such that the energy perturbations caused by fixed nuclear magnetic moments $\mu_{A\alpha}$ and $\mu_{B\beta}$ includes a cross term

$$E_{AB} = (K_{AB})_{\alpha\beta}\, \mu_{A\alpha}\mu_{B\beta} \tag{5-29}$$

using the summation convention for the tensor suffixes α and β. These fixed nuclear moments must be treated as if they produce a dipolar magnetic field:

$$F_{\alpha}^{(dip)} = (3r_\alpha r_\beta - \delta_{\alpha\beta}^2)\, r^{-5}\, \mu_\beta \tag{5-30}$$

together with a "contact field":

$$F_{\alpha}^{(cont)} = (8\pi/3)\, \delta(\mathbf{r})\, \mu_\alpha \tag{5-31}$$

the latter acting only on the electron spins.

This gives rise to three contributions to the total coupling constant K,

(1) orbital effects of $F_{\alpha}^{(dip)}$,
(2) the action of $F_{\alpha}^{(dip)}$ on electron spins,
(3) the action of $F_{\alpha}^{(con)}$ on electron spins.

Ramsey showed that there were no cross terms between these effects in E_{AB} so that each reduced coupling constant may be written in terms of three separate contributions,

$$K_{AB} = K_{AB}^{(1)} + K_{AB}^{(2)} + K_{AB}^{(3)} \tag{5-32}$$

We shall first consider the contact contribution $K_{AB}^{(3)}$, since it turns out to make the dominant contribution to the value of K_{AB} in most systems.

$$K_{AB}^{(3)} = (64\pi^2/9) \, \beta^2 (^3\!\varDelta E)^{-1} \, (S_A| \, \delta(\mathbf{r}_A)| \, S_A) \, (S_B| \, \delta(\mathbf{r}_B)| \, S_B) \, P_{S_A S_B}^2 \quad (5\text{-}33)$$

where β is the Bohr magneton, $^3\!\varDelta E$ the mean excitation energy approximation for $^3\!\varDelta E_{i\rightarrow j}$, $(S_A|\delta(\mathbf{r}_A)|S_A)$ the one center integral of s orbitals on the A nuclear center, and $P_{S_A S_B}$ the bond order associated with the S_A and S_B orbitals. It is defined as

$$P_{\mu\nu} = 2 \sum_{i=1}^{N} C_{i\mu} C_{i\nu} \quad (5\text{-}34)$$

in which N is the number of fully occupied molecular orbitals and the C_i are the atomic orbital wave function coefficients. Values for the one-electron integrals are given in Table 5-4. $K_{AB}^{(3)}$ is independent of θ since the s orbitals are

TABLE 5-4

One Electron Integrals[a]

Atom	$(s|\delta(r)|s)$	Atom	$(s|\delta(r)|s)$
H	0.5500	N	4.770
B	1.408	O	7.638
C	2.767	F	11.966

[a] See Pople and Santry (36). Atomic orbital constants in atomic units.

constant with respect to the C–C bond. Hence, a varying functional dependence of $\gamma_{CC'}$ on θ must come from the less significant contributions to K_{AB}, $K_{AB}^{(1)}$ and $K_{AB}^{(2)}$. $K_{AB}^{(2)}$ is given by

$$
\begin{aligned}
K_{AB}^{(2)} = (4/25) \, \beta^2 \, \langle r^{-3}\rangle_A \, \langle r^{-3}\rangle_B \, (^3\!\varDelta E)^{-1} \, \{ & 2(P_{X_A X_B}^2 + P_{Y_A Y_B}^2 + P_{Z_A Z_B}^2) \\
& + 3(P_{X_A X_B} P_{Y_A Y_B} + P_{Y_A Y_B} P_{Z_A Z_B} + P_{Z_A Z_B} P_{X_A X_B}) \\
& - (P_{X_A Y_B}^2 + P_{Y_A X_B}^2 + P_{Y_A Z_B}^2 + P_{Z_A Y_B}^2 + P_{Z_A X_B}^2 + P_{X_A Z_B}^2) \\
& + 3(P_{X_A Y_B} P_{Y_A X_B} + P_{Y_A Z_B} P_{Z_A Y_B} + P_{Z_A X_B} P_{X_A Z_B}) \}
\end{aligned}
\quad (5\text{-}35)
$$

where $\langle r^{-3}\rangle_A$ is the mean value of r^{-3} for the 2p orbitals of atom A. The set of bond orders for $K_{AB}^{(2)}$ are those for 2p orbitals. Since p orbitals are directed relative to a A–B bond, $K_{AB}^{(2)}$ will vary with θ. This is the case because the various cross terms such as $P_{X_A X_B}$ will vary in magnitude depending upon the relative orientations of the p orbitals on A and B. The relative orientations of the p orbitals, in turn, are dependent upon θ. Unfortunately all of the cross terms are small or zero so that we will have to be able to distinguish between very small changes in J_{AB} in order to determine values of θ.

The θ dependence of $K_{AB}^{(1)}$ is also almost constant. Further $K_{AB}^{(1)} = 0$ unless there is multiple bonding between A and B. The value of $K_{AB}^{(1)}$ is given by

$$K_{AB}^{(1)} = (8/3)\,\beta^2 \langle r^{-3} \rangle_A \langle r^{-3} \rangle_B (\varDelta E)^{-1}$$
$$\times \{ P_{X_A X_B} P_{Y_A Y_B} + P_{Y_A Y_B} P_{Z_A Z_B} + P_{Z_A Z_B} P_{X_A X_B}$$
$$- P_{X_A Y_B} P_{Y_A X_B} - P_{Y_A Z_B} P_{Z_A Y_B} - P_{Z_A X_B} P_{X_A Z_B} \} \qquad (5\text{-}36)$$

Obviously both $K_{AB}^{(2)}$ and $K_{AB}^{(1)}$ will be zero if A and/or B is hydrogen. At this point there is no way to know whether or not the rather slightly periodic dependence of J_{AB} upon θ will be able to be used in conformational elucidation in ^{13}C NMR spectroscopy. Nevertheless the possible importance of this relationship merits its mention as a future area of research in conformational analysis.

IV. Conformational Fluctuations in Macromolecules

A. Introduction

A macromolecule in solution is subject to fluctuation forces which arise from the motion of the solvent molecules, other macromolecules, and the macromolecule itself. If the macromolecule is in a conformational state which corresponds to a shallow and broad conformational energy minimum, then the conformational fluctuations will be relatively large and the probability of an instantaneous conformational transition due to random fluctuation forces may also be large. On the other hand, if the conformational state of the macromolecule corresponds to a deep and sharp conformational energy minimum, then a conformational transition is very improbable and the size of the conformational fluctuations is small. In the former case we say the entropy of the conformational state is high and the relaxation time is long. In the latter case we say the entropy of the conformational state is low and relaxation time is short. Thus, a measure of the conformational fluctuations of a macromolecule yields information about both the time-dependent and time-independent properties of the macromolecule.

By studying the conformational fluctuations of macromolecules in solution it is possible to build up time-dependent profiles of the behavior of the macromolecules under varying conditions. One obvious application of such a time-dependent knowledge is the elucidation of the dynamic, in contrast to static, function of biological macromolecules.

In the following section a method is presented for the computer simulation of conformational fluctuations in macromolecules. This method assumes the validity of theoretical conformational analysis and adopts the potential energy expressions discussed in Chapter II in the derivation of motion equations.

B. Theory of Random Conformational Fluctuations

The Langevin equation of motion (37) for a particle undergoing Brownian motion in the presence of an external force field is

$$m\frac{d\mathbf{v}}{dt} = \mathsf{F} + \mathsf{U}(t) + a\mathbf{v} \tag{5-37}$$

where F is the external force field, $\mathsf{U}(t)$ is the force due to the Brownian motion (we assume the fluctuations are nonordered), and $a\mathbf{v}$ represents the dynamic friction which the particle encounters. Consider now a general linear chain of the type shown in Fig. 5-15 and discussed in Chapter 1. A Langevin equation can be used to describe the change in the value of each of the internal bond rotations as a function of time. The equation takes the general form

$$I_l \ddot{\boldsymbol{\chi}}_l = -\mathbf{f}(l)\,\dot{\boldsymbol{\chi}}_l + \sum_{i,j}^{m} \mathsf{T}(i,j) + \mathsf{U}(t) \tag{5-38}$$

where I_l is the moment of inertia about the lth axis (i.e., bond about which χ_l takes place), and $\mathbf{f}(l)$ is the rotational friction coefficient tensor for the lth internal bond rotation angle $\boldsymbol{\chi}_l$. The sum over the tensor $\mathsf{T}(i,j)$ represents the resultant torque on the lth internal bond rotation angle $\boldsymbol{\chi}_l$, due to bonded and nonbonded interactions for a macromolecule having m atoms. The tensor $\mathsf{U}(t)$ represents the force due to Brownian motion, that is, random collisions by the solvent molecules. In Eq. (5-38) the term on the left side is the inertia. For very viscous solvent interactions the inertia is approximately zero. For these conditions

$$\mathbf{f}(l)\,\dot{\boldsymbol{\chi}}_l = \sum_{i,j}^{m} \mathsf{T}(i,j) + \mathsf{U}(t) \tag{5-39}$$

If we further make the assumption that the rotational friction coefficient tensor is independent of position and orientation of the lth axis, then

$$\mathbf{f}(l) \Longleftrightarrow f \quad \text{(a constant scalar)}$$

Then expressing Eq. (5-39) in difference form, we obtain

$$\boldsymbol{\chi}_l(t+\tau) - \boldsymbol{\chi}_l(t) = f^{-1}\tau \sum_{i,j}^{m} \mathsf{T}(i,j) + f^{-1}\tau \mathsf{U}(t) \tag{5-40}$$

The value of f can usually be approximated by the Stokes equation (38)

$$f = 8\pi R_0^3 \eta_0 \qquad (5\text{-}41)$$

where R_0 is the hydrodynamic radius of the ith monomer unit and η_0 is the solvent viscosity. R_0 is usually taken to be the radius of a sphere whose volume is equal to the sum of the van der Waals sphere volumes of the atoms composing the monomer unit, i.e.,

$$R_0 = \left[\sum_k r_{ik}^3 \right]^{1/3} \qquad (5\text{-}42)$$

where r_{ik} is the van der Waals radius of the kth atom in the ith monomer unit. The determination of $U(t)$ is usually not possible and we are forced to adopt a simulation procedure to approximate $U(t)$. In general a random number

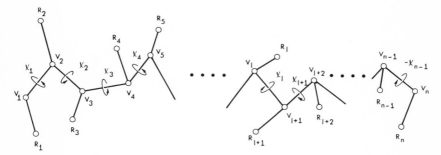

Fig. 5-15. The geometry of the general linear chain.

generation scheme may be used to compute the elements of $U(t)$ *provided* reasonable bounds can be placed on the magnitudes of the elements. The following reasoning can be used to estimate an upper limit ζ_l to be used in the random number generator. From the Einstein definition of the two-dimensional rotational diffusion coefficient D_{χ_l} (39), one finds for, say, the ith monomer unit that

$$D_{\chi_l} = \langle \Delta \chi_l^2 \rangle / 4\tau = KT/f \qquad (5\text{-}43)$$

where T is the absolute temperature and

$$\langle \Delta \chi_l^2 \rangle = \int_{-\zeta}^{\zeta} \chi_l^2 \, p(\chi_l) \, d\chi_l / \int_{-\zeta}^{\zeta} p(\chi_l) \, d\chi_l \qquad (5\text{-}44)$$

where $p(\chi_l)$ is the probability distribution function of the χ_l internal bond rotation angle. For a set of uniform random numbers, $p(\chi_l) = $ constant or

$$\langle \Delta \chi_l^2 \rangle = \zeta_l^2 / 3 \qquad (5\text{-}45)$$

Then from the definition of the two-dimensional rotational diffusion co-efficient

$$\zeta_l = [12(\tau/f) KT]^{1/2} \tag{5-46}$$

Thus for any given temperature and solvent we need only specify the value of τ which fixes the time scale in the simulation calculations.

The torque about a fixed axis defined by atoms u and v with the corresponding rotation angle χ_l is given by

$$\mathbf{T}_{ij} = \mathbf{n}_{uv} \cdot (\mathbf{r}_i \times \mathbf{F}_{ij}) \tag{5-47}$$

where i indicates, as shown in Fig. 5-16, the atoms rigidly attached to the rotating axis, and j refers to the atoms exerting an external force on the atoms

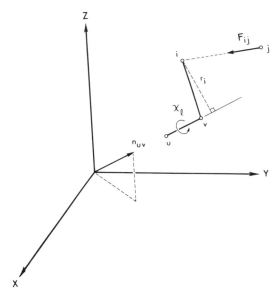

Fig. 5-16. Geometry used to calculate the torque, $\mathbf{T}_{ij}$, for an internal bond rotation in a molecule.

rigidly attached to the rotating axis. The n_{uv} is the vector in the direction of the fixed axis which is located in a right-handed cartesian frame. The vector r_i is the perpendicular distance of the ith atom to a projection along the fixed axis. Explicitly carrying out the indicated vector operations lead to

$$T_{ij} = \hat{n}_{uv}^{(x)}[Y_i F_{ij}(x) - Z_i F_{ij}(y)] + \hat{n}_{uv}^{(y)}[Z_i F_{ij}(x) - X_i F_{ij}(z)]$$
$$+ \hat{n}_{uv}^{(z)}[X_i F_{ij}(y) - Y_i F_{ij}(x)] \tag{5-48}$$

where the $n_{uv}(\delta)$ are the orthogonal components of $\mathbf{n}$, $F_{ij}(\delta)$ are the orthogonal components of F_{ij} and (X_i, Y_i, Z_i) are the coordinates of atom i. The elements of T_{ij} are computed using the conformational potential functions discussed in Chapter 2. In other words $\mathbf{F}_{ij}$ can be assumed to be given by†

$$\mathbf{F}_{ij} = \mathbf{F}_{ij}(NB) + \mathbf{F}_{ij}(ES) + \mathbf{F}_{ij}(tor)$$
$$+ \mathbf{F}_{ij}(HB) + \mathbf{F}_{ij}(p\text{-}s) + \mathbf{F}_{ij}(vib) \qquad (5\text{-}49)$$

In principle, at least, the terms on the right in (5-49) can be the result of considering as many macromolecules as desired. However, the size of the associated configurational hyperspace is usually so large as to make multi-macromolecular assemblies unreasonable to simulate. Thus the various $\mathbf{F}_{ij}$ in Eq. (5-49) are the result of a single-isolated intrachain macromolecular interaction. Obviously the neglect of any one or more of the terms on the right-hand side of Eq. (5-49) will probably simplify the calculations, but at the same time, impose additional restrictions on the interpretation of results of the calculations. For example, neglect of $\mathbf{F}_{ij}(vib)$ imposes the constraint of a rigid-nonflexing valence bond geometry.

Calculations done to date have actually included only the first three terms in Eq. (5-49) due to the complexity of such computations. Further, these three terms, especially $\mathbf{F}_{ij}(NB)$, make the largest contribution, by far, to $\mathbf{F}_{ij}$.

If the nonbonded potential energy is given by a Lennard-Jones 6-12 function,

$$P_{ij} = \frac{B_{ij}}{r_{ij}^{12}} - \frac{A_{ij}}{r_{ij}^{6}} \qquad (5\text{-}50)$$

then

$$\mathbf{F}_{ij}(NB) = \frac{\partial P_{ij}}{\partial r_{ij}} \frac{\mathbf{r}_{ij}}{|\mathbf{r}_{ij}|} = \left[\frac{-12B_{ij}}{r_{ij}^{13}} + \frac{6A_{ij}}{r_{ij}^{7}} \right] \frac{\mathbf{r}_{ij}}{|\mathbf{r}_{ij}|} \qquad (5\text{-}51)$$

with components

$$\mathbf{F}_{ij}^{(\delta)}(NB) = \frac{\partial P_{ij}}{\partial r_{ij}} \frac{\partial r_{ij}}{\partial \delta_j} \qquad \text{where} \quad \delta \text{ is } x, y, \text{ or } z$$

$$= \mathbf{F}_{ij}(NB) \frac{(\delta_i - \delta)}{r_{ij}}$$

where

$$r_{ij} = [(x_i - x_j)^2 + (y_i - y_j)^2 + (z_i - z_j)^2]^{1/2} \qquad (5\text{-}52)$$

---

† NB, nonbonded; ES, electrostatic; tor, torsional; HB, hydrogen bond; p-s, polymer–solvent; vib, vibrational.

In the same way the coulombic electrostatic force is given by

$$\mathbf{F}_{ij}(\text{ES}) = \frac{\partial D_{ij}}{\partial r_{ij}} \frac{\mathbf{r}_{ij}}{|\mathbf{r}_{ij}|} = \left[\frac{-kQ_i Q_j}{\epsilon r_{ij}^2} \right] \frac{\mathbf{r}_{ij}}{|\mathbf{r}_{ij}|} \tag{5-53}$$

with components

$$F_{ij}^{(\delta)}(\text{ES}) = \frac{\partial D_{ij}}{\partial r_{ij}} \frac{\partial r_{ij}}{\partial \delta_j} = F_{ij}(\text{ES}) \frac{(\delta_i - \delta_j)}{r_{ij}}$$

What we ultimately must achieve is a set of expressions relating the elements of T_{ij} to the set of internal bond rotations $\{\boldsymbol{\chi}_l\}$. For a single isolated macromolecule each T_{ij} has the following functional relationship to $\{\boldsymbol{\chi}_l\}$,

$$T_{ij} = G[H_{ij}(r_{ij}(X_i\{\boldsymbol{\chi}_l\}, Y_i\{\boldsymbol{\chi}_l\}, Z_i\{\boldsymbol{\chi}_l\}, X_j\{\boldsymbol{\chi}_l\}, Y_j\{\boldsymbol{\chi}_l\}, Z_j\{\boldsymbol{\chi}_l\})))] \tag{5-54}$$

where H_{ij} is the sum of the nonbonded, electrostatic, and torsional $\Phi(\boldsymbol{\chi}_l)$ potential energies between atoms i and j,

$$H_{ij} = P_{ij} + D_{ij} + \Phi(\boldsymbol{\chi}_l) \tag{5-55}$$

Upon repeated application of the chain rule we obtain

$$T_{ij} = F_{ij}(\chi_l) = \frac{H_{ij}}{\partial \chi_l}$$

$$= \frac{H_{ij}}{r_{ij}} \left[\frac{\partial r_{ij}}{\partial X_i} \frac{\partial X_i}{\partial \chi_l} + \frac{\partial r_{ij}}{\partial Y_i} \frac{\partial Y_i}{\partial \chi_l} + \frac{\partial r_{ij}}{\partial Z_i} \frac{\partial Z_i}{\partial \chi_l} + \frac{\partial r_{ij}}{\partial X_j} \frac{\partial X_j}{\partial \chi_l} \right.$$

$$\left. + \frac{\partial r_{ij}}{\partial Y_j} \frac{\partial Y_j}{\partial \chi_l} + \frac{\partial r_{ij}}{\partial Z_j} \frac{\partial Z_j}{\partial \chi_l} \right] + F_{ij}(\chi_l)(\text{tor}) \tag{5-56}$$

The only problem which now remains is the derivation on an expression for the $\partial \delta_m / \partial \chi_l$ where δ_m can be X_m, Y_m, Z_m. These derivatives can be computed by differentiating Eqs. (1-62) and/or (1-63) of Chapter 1 with respect to χ_l.

Backbone atoms

$$\frac{\partial}{\partial \chi_l} \mathbf{v}_m = \mathbf{v}_0^0 + \sum_{\mu=1}^{m} \left\{ \frac{\partial}{\partial \chi_l} \prod_{\eta=1}^{\mu} \mathsf{T}_{\chi_\eta} (\mathbf{v}_\mu^0 - \mathbf{v}_{\mu-1}^0) \right. \tag{5-57}$$

Side chain atoms

$$\frac{\partial}{\partial \chi_l} \mathbf{V}_{m,n} = \mathbf{v}_0^0 + \sum_{\mu=1}^{m-1} \left\{ \frac{\partial}{\partial \chi_l} \prod_{\eta=1}^{\mu} \mathsf{T}_{\chi_l} (\mathbf{v}_\mu^0 - \mathbf{v}_{\mu-1}^0) \right\} + \frac{\partial}{\partial \chi_l} \prod_{\eta=1}^{m} \mathsf{T}_{\chi_\eta} (\mathbf{V}_{m,n}^0 - \mathbf{v}_{m-1}^0)$$

$$\tag{5-58}$$

The contributions to $F_{ij}(\chi_l)$ (tor) can be computed directly from the potential energy expression $\Phi(\boldsymbol{\chi}_l)$,

$$\Phi(\boldsymbol{\chi}_l) = h(\boldsymbol{\chi}_l) \, g(\boldsymbol{\chi}_l) \tag{5-59}$$

where $h(\chi_l)$ refers to the barrier height and $g(\chi_l)$ the shape of the rotational barrier. Then

$$\mathbf{F}(\chi_l)(\text{tor}) = \frac{\partial \Phi(\chi_l)}{\partial \chi_l} \cdot \frac{\boldsymbol{\chi}_l}{|\boldsymbol{\chi}_l|} \tag{5-60}$$

Since the derivatives $\partial \delta_m / \partial \chi_l$ are explicit trigonometric functions of the $\{\boldsymbol{\chi}_l\}$, as can be inferred from Eqs. (5-57) and (5-58), it follows that Eq. (5-40) represents one of a set of equations (one for each $\boldsymbol{\chi}_l$) involving some or all of the $\boldsymbol{\chi}_l$. This is a set of nonlinear equations which must be solved simultaneously each time a new set of random $\{\zeta_l\}$ is chosen. The sequential solution of this set of nonlinear equations allows one to follow the time-dependent conformational path of a macromolecule as a function of the set of random fluctuations. Such sets of nonlinear equations can often be solved using either the Euler expansion method or the generalized Newton technique (40).

C. Computations on Diglycine

Simon (41a) has carried out a series of fluctuation calculations for two planar peptide units joined at the common C^α carbon atom and having a proton for a side chain—diglycine. This molecular structure is shown in Fig. 5-17. Parameters suggested by Scott and Scheraga (42) were used in the conformational potential energy calculations. The viscosity parameter η_0 was set equal to that of water at 20°C and R_0 was taken to be 2 Å. Each computational iteration in a calculation corresponds to a specific value of τ which is known when the same value of $|\zeta|$ is assigned to both rotational angles ϕ and ψ. Thus by keeping track of $|\zeta|$ and the number of computation cycles involved in a given simulation calculation it was possible to study time-dependent fluctuations in diglycine. $|\zeta|$, as defined earlier, is a measure of the absolute value of the maximum allowed value of angular fluctuation. Motion from the right-handed α-helical conformation as a function of $|\zeta|$ is summarized in Table 5-5. Only when the maximum value of $|\zeta|$ exceeded $\pm 10°$ did a transition (that is, a "jump" out of the right-handed α-helical region) occur. On the other hand, motion away from the planar peptide conformation having $\phi = 0°$ and $\psi = 0°$ took place for very small values of $|\zeta|$. Table 5-6 describes the motion of diglycine when initially placed in the $\phi = 0°$, $\psi = 0°$ planar conformation. Figure 5-18 is an energy contour map of diglycine with a typical fluctuation path superimposed.

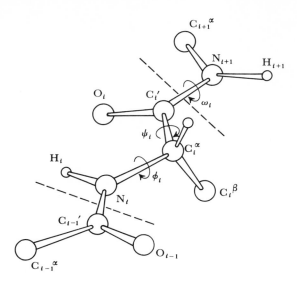

Fig. 5-17. Perspective drawing of a section of a polypeptide chain representing two peptide units. The limits of a residue are indicated by dashed lines. The recommended notation for the backbone atoms and for bond rotations is shown. From Edsall *et al.* (41b).

For diglycine, the "jump" from the right-handed α-helix was often followed by a "jump" into the left-handed α-helical conformation. This occurred more frequently than "jumps" from the right-handed α-helix to "jumps" into the global minima (degenerate) located at $\phi = 80°$, $\psi = 240°$, and $\phi = 280°$, $\psi = 120°$. Still less frequent were "jumps" from the left-handed α-helical region to the right-handed α-helical region.

The relaxation time $\langle t \rangle$, here defined as the statistical average time required

TABLE 5-5

Motion out of the Right-Handed α-Helical Region for Diglycine[a]

Initial coordinates				Standard deviation	Relaxation time	Standard deviation
ϕ	ψ	$\max|\zeta|$	$\langle \tau \rangle$ cycles	S_τ	$\langle t \rangle$ (sec)	S_t (sec)
120°	120°	2°		—		—
		5°		—		—
		10°		—		—
		15°	34	7	7.65×10^{-8}	1.48×10^{-8}
		20°	17	8	6.80×10^{-8}	3.20×10^{-8}

[a] Temperature fixed at 20°C, solvent is water.

TABLE 5-6

Motion from the $(0,°0°)$ Conformation for Diglycine[a]

Initial coordinates				Standard deviation	Relaxation time	Standard deviation
ϕ	ψ	$\max\|\zeta\|$	$\langle\tau\rangle$	S_τ	$\langle t\rangle$ (sec)	S_t (sec)
$0°$	$0°$	$2°$	62	25	2.48×10^{-9}	1.00×10^{-9}
		$5°$	48	15	1.20×10^{-8}	3.75×10^{-8}
		$7°$	10	2	4.90×10^{-9}	0.98×10^{-9}
		$15°$	5	1	1.23×10^{-8}	2.25×10^{-9}

[a] Temperature fixed at 20°C, solvent is water.

for the molecule to pass from one stable conformational state to a different stable conformational state for all processes is on the same order of magnitude as the relaxation time observed for the helix–coil transitions in some polypeptides. For example, Schwarz (43) has deduced the relaxation time for the helix–coil transition for polypeptides and lists a range of values from 10^{-8} to

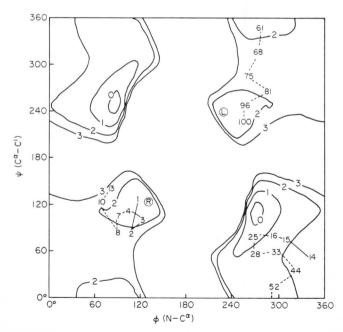

Fig. 5-18. Computer run for diglycine. This is an example of motion from the right-handed α-helical region to the vicinity of the global minimum and then to the left-handed α-helical region. $\max\|\zeta\| = 15°$. From Simon (41a).

10^{-5} sec. Lastly, one would intuitively expect that the relaxation time should decrease with increased size of the average fluctuation. This is not the case when the temperature of the system is held constant. Under these conditions the time increase associated with increasing the size of a fluctuation compensates for the decreasing number of fluctuations needed to pass from a given conformational region.

D. Application of the Computer Simulation of Conformational Fluctuations

1. Introduction

The limiting condition which dictates the size of the molecule which can be handled in the computer simulation of conformational fluctuations is the number of allowed internal bond rotations. Every allowed internal bond rotation contributes an equation to the set of nonlinear equations of motion. Therefore, very large molecules, such as proteins, are not practically suited to be studied by this method since there are hundreds of internal bond rotations in these macromolecules. Thus, for example, the role of conformational fluctuations in the catalytic activity of biological enzymes should not be studied by this technique.

However, for ordered macromolecules, or simple copolymers, it is possible to describe the conformational properties of the entire macromolecule by a representative short sequence of the macromolecule. That is, the conformational properties of these macromolecules are locally uniform. For these cases the number of internal bond rotations should be sufficiently small (less than 50) so that the computer simulation of conformational fluctuations can be carried out.

2. Conformational Transitions

If the ordered macromolecule, or simple copolymer, under consideration is known to undergo a conformational transition, then the computer simulation of conformational fluctuations provides an ideal means of studying the time-dependent nature of the transition. Such calculations should, for the first time, shed light on the conformational nature of the macromolecule during the transition. Questions such as

(1) Is the transition cooperative?
(2) How is the final state reached?
(3) What is the velocity of the transition?
(4) Are there any stable intermediate states?

may be answered by following the time-dependent conformational path of the macromolecule during the simulation calculations. Transitions resulting from nonrandom fluctuations (up to now we have only considered random fluctuations) may be simulated by picking the elements of $U(t)$ in such a way so as to describe the ordered nature of the fluctuations. If, for example, the elements of each row of $U(t)$ were chosen so that the sum of the squares of each row was constant and equal to all other rows, then the fluctuational force would be constant at all times. The transition, if it were to occur, would be driven by a constant force. $U(t)$ thus controls the external nature of the forces which generate the conformational fluctuations.

The helix–coil transition of a polypeptide with ionizable side chains might be studied by repeatedly performing computer-simulation conformational-fluctuation calculations on the polypeptide, always initially in the right-handed α-helical conformation, as its side chains progressively are ionized in, say, a random manner. The temperature-dependent helix–coil transition of a polypeptide can be studied by repeatedly performing computer-simulation conformational-fluctuation calculations on the polypeptide, always initially in the right-handed α-helical conformation, for increasing size in the van der Waals radii of the polymer atoms. As discussed in the section of the helix–coil transition, raising the temperature can, stereochemically, be thought the equivalent of enlarging the van der Waals radius of an atom. Poly-L-glutamic acid, poly-L-lysine, polyglycine, and poly-L-alanine and their copolymers are ideal candidates to be studied.

3. Conformational Properties of "Small" Macromolecules

There is a vast number of molecules of biological importance whose sizes are less than that of macromolecules, but which still have the capacity to assume any of a large number of conformations. Some hormone related peptides, neural molecules, and many biologically active peptide fragments such as glutathione, bradykinin, and angiotensin, some of whose primary structures are listed in Table 5-7, are examples of these types of molecules. In the vast majority of cases the experimental elucidation of the conformation(s) of these molecules in solution appears to be very remote. These molecules are not large enough to be structurally characterized by classical methods such as light-scattering or viscosity measurements. Such molecules do not possess regions of extended order so that most of the spectroscopic methods are of limited value. Solid state X-ray crystallography can be used to elucidate the solid state conformation of some of these molecules, but, due to packing considerations, the solid state conformation is very probably different from the solution conformation.

High resolution NMR and/or CD/ORD in *conjunction* with conformational energy calculations offers some hope in the elucidation of the solution conformation of these molecules, especially the simple neural molecules. However, while the number of conformational degrees of freedom (internal bond rotations) is small enough to allow the conformational energy minimization, one can never be certain as to how many other stable energy minima exist and how easy it is to pass between stable conformations. For a fixed set of conditions, a series of conformational fluctuation calculations should project a time-dependent profile of the solution-state structure of the molecule in all of its preferred conformational hyperspace. The repeated solution of the equations of conformational motion tells us how the molecule wishes to behave rather than forcing us to search for its pattern of behavior. Such

TABLE 5-7

Primary Structure of Some "Mini" Macromolecules[a]

A. Glutathione	H-γ-Glu-Cys-Gly-OH
B. Angiotensin	H-Asp-Arg-Val-Tyr-Ile-His-Pro-Phe-OH
	(Val)
C. Bradykinin	H-Arg-Pro-Gly-Phe-Ser-Pro-Phe-Arg-OH
	(Gly)
D. Phenethylamine	$\langle\bigcirc\rangle$—CH_2—CH_2—NH_2
E. Amphetamine	$\langle\bigcirc\rangle$—CH_2—$\underset{\underset{CH_3}{\vert}}{CH}$—$NH_2$
F. Decapeptide having growth hormone releasing activity	H-Val-His-Leu-Ser-Ala-Glu-Glu-Lys-Glu-Ala-OH

[a] The residues in parentheses may be substituted without loss of activity.

computer simulated conformational-fluctuation calculations on these "small" macromolecules may well provide the first means of elucidating the conformational properties of such molecules in solution. In turn, this information may provide a foundation for understanding the structure-function relationship in some of these molecules.

V. Configurational Properties of Chain-Molecule Fluids

A series of calculations which are extensions to those required to carry out the fluctuation computations for macromolecules are needed in simulation studies of chain-molecule fluids. Chain-molecule fluids are collections of linear oligomeric chains which are allowed to interact. In the macromolecular fluctuation calculations the time-dependent Langevin equation of motion is solved or repeatedly approximated by Monte Carlo statistics. For the chain-molecule fluids the associated time-dependent set of Newtonian equations must be solved or repeatedly approximated by Monte Carlo statistics.

Both the equilibrium and nonequilibrium properties of chain fluids are not well understood in terms of the configurational degrees of freedom of the fluid. The fundamental phenomenon of how a collection of flexible chains will interact and effect the thermodynamic properties of a fluid which they compose has not been clearly elucidated. The reason for this is due to the experimental difficulties of relating the macroscopic properties of the fluid to the molecular structure of the fluid. Theoretically, there has not been a universally reliable means of partitioning the configurational states of the chain-fluid molecules so that accurate values of all associated thermodynamic quantities can be predicted from the partition function.

It is the cell model which was first generalized to chain-molecule fluids by the Belgian school (44). One important assumption in making the transition from "monomer" to polymer fluid relates to the dynamics of flexible chains with many volume dependent degrees of freedom (see below). Taken all together the assumptions are such that the characteristic parameters p^*, V^*, and T^* could actually become functions of the variables of state and therefore not represent true reducing parameters.

The predictions of the cell theory have been extensively tested by Nanda *et al.* (45) by constructing from experimental data master curves for the various projections of the reduced p-V-T surface and of such pressure quantities as the internal pressure and the cohesive energy density (46, 47). These master curves are representative of both oligomer liquids and amorphous high polymers above the glass temperature. The results of these studies indicate that the cell theory predicts the equation of state very well, but the predicted values of other thermodynamic variables, such as the internal pressure, are found to differ considerably with experimental findings.

An improved theory was developed by Simha and Somcynsky (48) which is based on the introduction of vacancies in the quasi lattice. The resulting minimization of the complete configurational free energy yields several gratifying results derived from the dependence of the number of holes on V and T. The resulting equation of state is in very good agreement with the empirical reduced isotherms and isobars obtained previously as well as recent measurements up to 2 kbar for polystyrene and poly(orthomethyl styrene). The nature of the hole model and the extent of the agreement invite applications to the thermodynamic aspects of the glass transition process.

Reverting to simple liquids, the second of the theoretical approaches, namely computer experimentation, has been pursued by two groups of investigators. One of these proceeds by the Monte Carlo techniques in which the statistical mechanical averaging is replaced by an average over a Markoff chain (49). The other solves the classical equations of motion for a many-body system and obtains therefrom the pressure equation. Moreover, nonequilibrium properties can be studied as well.

As a reminder, we indicate briefly the generalization of the computer techniques to the system of molecules with rotational and internal conformational degrees of freedom. In the Monte Carlo procedure all molecules are initially assumed to be in some configurational state K_0 with internal energy U_0. On a random basis, one of these is permitted to change one or more of its configurational coordinates. The maximum allowable change depends on the temperature. Let the new configurational state K_1 have an energy U_1. If $U_1 < U_0$, K_1 represents the new initial state. If $U_1 > U_0$, the computer randomly selects a quantity ϵ between zero and unity. If $\epsilon < \exp[-(U_1 - U_0)/KT]$, the system remains in K_1, returning otherwise to K_0. These iterations are to be continued until some criterion for iterative convergence is satisfied.

In the dynamic method all molecules are assumed to be at rest in the state K_0, U_0. To a given molecule there are now imparted one or more translational, external, or internal rotational velocity components. Their (realistic) magnitude depends on temperature. After some time interval Δt, the system is examined, the equations of motion are solved, and a second set of velocity quantities is obtained. The process is continued and in this manner a sequence of iterations is carried out. Convergence results when $|U_n - U_{n-1}| <$ some prechosen value.

Except for the molecular dynamics approach, our discussion has emphasized the thermodynamic equilibrium in one-component systems. The methods to be developed may subsequently prove useful for the simulation of a glass transition and the thermodynamic analysis of multicomponent systems.

References

1. A. Streitwieser, Jr., "Molecular Orbital Theory." Wiley, New York, 1962.
2. E. S. Pysh, *J. Chem. Phys.* **52**, 4723 (1969).
3a. E. W. Ronish and S. Krimm, *Biopolymers* **11**, 1919 (1972).
3b. IUPAC-IUB Commission on Biochemical Nomenclature, *Biochemistry* **9**, 3471 (1970).
4. T. L. Hill, "Introduction to Statistical Thermodynamics." Addison-Wesley, Reading, Massachusetts, 1962.
5. J. Willems, *Discuss. Faraday Soc.* **25**, 111 (1957).
6. E. W. Fisher, *Kolloid-Z.* **159**, 108 (1958).
7. J. A. Koutsky, A. G. Walton, and E. Baer, *J. Polym. Sci. Part A* **45**, 611 (1966).

8. S. H. Carr, A. G. Walton, and E. Baer, *Biopolymers* **6**, 469 (1968).
9. S. H. Carr, Ph.D. Thesis, Case Western Reserve Univ. (1970).
10. K. A. Mauritz, E. Baer, and A. J. Hopfinger, *J. Polym. Sci.-Phys.* (in press).
11. T. Seto, T. Hora, and K. Tanaka, *Jap. J. Appl. Phys.* **7**, 31 (1968).
12. G. Del Re, B. Pullman, and T. Yonezawa, *Biochim. Biophys. Acta* **75**, 153 (1963).
13. D. Poland and H. A. Scheraga, *Biochemistry* **6**, 3791 (1967).
14. K. S. Pitzer, *in* "Advances in Chemical Physics" (I. Prigogine, ed.) Vol. II, p. 59. Wiley (Interscience), New York, 1959.
15. C. Kittel, "Introduction to Solid State Physics." Wiley, New York, 1968.
16. C. W. Wilson, *J. Polym. Sci. Part A* **1**, 1305 (1963).
17. C. W. Wilson and E. R. Santee, Jr., *J. Polym. Sci. Part C* **8**, 97 (1965).
18. H. M. McConnell, *J. Mol. Spectrosc.* **1**, 11 (1957).
19. M. J. S. Dewar and R. C. Fahey, *J. Amer. Chem. Soc.* **85**, 2704 (1963).
20. M. Karplus, *J. Amer. Chem. Soc.* **85**, 2870 (1963).
21. A. A. Bothner-By, C. Naar-Colin, and H. Gunther, *J. Amer. Chem. Soc.* **84**, 2748 (1962).
22. E. W. Garbisch, Jr., *J. Amer. Chem. Soc.* **86**, 5561 (1964).
23. V. F. Bystrov, S. L. Portnova, V. I. Tsetlin, V. T. Ivanov, and Yu. A. Ovchinnikov, *Tetrahedron* **25**, 493 (1969).
24. W. A. Gibbons, G. Nemethy, A. Stern, and L. C. Craig, *Proc. Nat. Acad. Sci. U.S.* **67**, 239 (1970).
25. D. A. Torchia, *Macromolecules* **4**, 440 (1971).
26. C. M. Deber, D. A. Torchia, and E. R. Blout, *J. Amer. Chem. Soc.* **93**, 4893 (1971).
27. K. D. Kopple and D. H. Marr, *J. Amer. Chem. Soc.* **89**, 6193 (1967).
28. V. Sasisekharan, *Acta Crystallogr.* **12**, 897 (1959).
29. Y. C. Leung and R. E. Marsh, *Acta Crystallogr.* **11**, 17 (1958).
30. D. A. Torchia, unpublished results (1972).
31. H. J. Reich, M. Jautelat, M. T. Messe, F. J. Weigert, and J. D. Roberts, *J. Amer. Chem. Soc.* **91**, 7445 (1969).
32. W. J. Horsley and H. Sternlicht, *J. Amer. Chem. Soc.* **90**, 3738 (1968).
33. W. A. Gibbons, J. A. Sogn, A. Stern, L. C. Craig, and L. F. Johnson, *Nature (London)* **227**, 840 (1970).
34. W. J. Horsley, H. Sternlicht, and J. Cohen, *J. Am. Chem. Soc.* **92**, 680 (1970).
35. D. M. Grant and E. G. Paul, *J. Amer. Chem. Soc.* **86**, 2984 (1964).
36. J. A. Pople and D. P. Santry, *Mol. Phys.* **11**, 1 (1964).
37. G. E. Uhlenbeck and L. S. Ornstein, *Phys. Rev.* **36**, 823 (1930).
38. H. Lamb, "Hydrodynamics," p. 589. Dover, New York, 1945.
39. A. Einstein, *Elektrochemistry* **14**, 235 (1908).
40. E. Isaacson and H. B. Keller, "Analysis of Numerical Methods." Wiley, New York, 1966.
41a. E. M. Simon, *Biopolymers* **10**, 973 (1971).
41b. J. T. Edsall, P. J. Flory, J. C. Kendrew, A. M. Liquori, G. Némethy, G. N. Ramachandran, and H. A. Scheraga, *Biopolymers* **4**, 121 (1966).
42. R. A. Scott and H. A. Scheraga, *J. Chem. Phys.* **45**, 2091 (1966).
43. G. Schwarz, *J. Mol. Biol.* **11**, 64 (1965).
44. I. Prigogine, N. Trappeniers, and V. Mathot, *Discuss. Faraday Soc.* **15**, 93 (1953); I. Prigogine, "The Molecular Theory of Solutions." North-Holland Publ., Amsterdam, 1957.
45. V. S. Nanda, R. Simha, and T. Somcynsky, *J. Polym. Sci. Part C* **12**, 277 (1966).
46. A. Quach and R. Simha, *J. Appl. Phys.* **42**, 4592 (1971).
47. A. Quach, Ph.D. Thesis, Case Western Reserve Univ. Cleveland, Ohio (1971).
48. R. Simha and T. Somcynsky, *Macromolecules* **2**, 342 (1969).
49. W. Wood, F. Parker, and J. Jacobson, *Nuovo Cimento Suppl.* **9**, 133 (1958).

Appendix I | Empirical Nonbonded Potential Functions between Ions and Atoms in Molecules

In Chapter 2 we discussed how potential energy functions which represent the dispersion attractive forces and the steric-overlap repulsive forces between two chemically saturated and spherically symmetrical molecules can be modified to describe the interaction between pairs of noncovalently bonded atoms in a molecule or between molecules. This approximation forms part of the basis for conformational energy calculations.

It is found that a Lennard-Jones 6-12 potential function P_{ij}, of the form

$$P_{ij} = \frac{B_{ij}}{R_{ij}^{12}} - \frac{A_{ij}}{R_{ij}^{6}} \tag{I-1}$$

can be used to represent this pairwise interaction. The A_{ij} are calculated using a modified Slater–Kirkwood equation

$$A_{ij(a-b)} = \frac{\frac{3}{2}e(\hbar/m^{1/2})\,\alpha_a\,\alpha_b}{(\alpha_a/N_{\text{eff}}^{(a)})^{1/2} + (\alpha_b/N_{\text{eff}}^{(b)})^{1/2}} \tag{I-2a}$$

which is discussed in Chapter 2. The $B_{ij(a-b)}$ are determined by insisting that minimum in potential energy occur at the contact distance, R_0, for the pairwise interaction

$$B_{ij(a-b)} = \tfrac{1}{2}A_{ij(a-b)}\,R_0^{6} \tag{I-2b}$$

It is becoming increasingly important to consider the influence of ions on macromolecular conformation. The interaction of ions with the charged side chains of lysine, glutamic, aspartic, and ornithine peptide residues can seriously modify chain conformation (1).

In Chapter 5 we discussed epitaxial crystallization. The basic pairwise interactions which dictate the properties of this phenomenon are between ions in the lattice substrate and atoms in the polymer. While such interactions are predominantly electrostatic, there is nevertheless a meaningful contribution to the total energy from the attractive dispersion forces, and, when the repulsive forces begin to dominate, the shape of the potential curve will have serious consequences upon the tightness of molecular packing.

The question to which we must now address ourselves is how to represent these ion–atom interactions. To answer this question we assume that the lattice energy of any ionic crystal can be represented as the sum of all pairwise interactions between the ions. The pairwise interactions, in turn, are composed of two terms, nonbonded and electrostatic:

$$E_{\mathrm{T}} = \sum_{\substack{i=1 \\ i \neq j}}^{N} \sum_{\substack{j=1 \\ i \neq j}}^{N} \left(\frac{-A_{ij}}{R_{ij}^6} + \frac{B_{ij}}{R_{ij}^{12}} \right) + \sum_{\substack{i=1 \\ i \neq j}}^{N} \sum_{\substack{j=1 \\ i \neq j}}^{N} \frac{Q_i Q_j}{R_{ij}} \qquad \text{(I-3)}$$

Then the total interaction energy $E_{\mathrm{T}}(i)$, associated with an ion i immersed in the crystal is

$$E_{\mathrm{T}}(i) = \frac{1}{2} \left[\sum_{j=1}^{N'} \left(\frac{-A_{ij}}{R_{ij}^6} + \frac{B_{ij}}{R_{ij}^{12}} \right) + \sum_{j=1}^{N'} \frac{Q_i Q_j}{R_{ij}} \right] \qquad \text{(I-4)}$$

where N' indexes some ion located sufficiently far from ion i so that the energy contribution from this and all other more distant ions from i is negligible. The value of N' can be determined by computing the size of the energy contribution from electrostatic interactions as a function of distance. Satisfactory convergence of the electrostatic energy guarantees convergence of the nonbonded terms which depend on higher powers of $1/r$. The values of E_{T} and then $E_{\mathrm{T}}(i)$ may be determined by melting the crystal at constant volume. Equation (I-4) contains six unknowns for an ionic crystal composed of two ionic species a and b; $A_{ij(a-a)}$, $A_{ij(a-b)}$, $A_{ij(b-b)}$, $B_{ij(a-a)}$, $B_{ij(a-b)}$, and $B_{ij(b-b)}$. In Chapter 3 we showed that the laws of static equilibrium must be valid for atoms in a unit cell and for the unit cells themselves in any crystal.

By using the equations resulting from applying the laws of static equilibrium and Eq. (I-3), for the value of N' determined from Eq. (I-4), we have 13 equations in six unknowns which may be estimated by a least squares method.

Having values for $A_{ij(a-a)}$ and $A_{ij(b-b)}$ allows us to determine expressions for α_a and α_b in terms of $N_{\mathrm{eff}}^{(a)}$ and $N_{\mathrm{eff}}^{(b)}$, respectively, using Eq. (I-2a):

$$\alpha_a = \left(\frac{4A_{ij(a-a)}^2}{N_{\mathrm{eff}}^{(a)} e_{\frac{3}{2}} (\hbar/m^{1/2})^2} \right)^{1/3} \qquad \text{(I-5a)}$$

$$\alpha_b = \left(\frac{4A_{ij(b-b)}^2}{N_{\mathrm{eff}}^{(b)} e_{\frac{3}{2}} (\hbar/m^{1/2})^2} \right)^{1/3} \qquad \text{(I-5b)}$$

Substituting these expressions into Eq. (I-2a) used in the computation of $A_{ij(a-b)}$ yields an equation of the form

$$A_{ij(a-b)} = f(N_{\mathrm{eff}}^{(a)}, N_{\mathrm{eff}}^{(b)}) \qquad \text{(I-6)}$$

This results in one equation in two unknowns. Fortunately, we have a very good idea of the values of $N_{\mathrm{eff}}^{(a)}$, $N_{\mathrm{eff}}^{(b)}$ needed to balance Eq. (I-6). The N_{eff} of an ion should very nearly equal the number of electrons surrounding the ion. For example in a Na^+-Cl^- lattice, $N_{eff}^{Na^+} = 10$ and $N_{eff}^{Cl^-} = 18$. Using such reliable initial values of $N_{eff}^{(a)}$ and $N_{eff}^{(b)}$ one can quickly compute, using a trial-and-error scheme, the values of $N_{eff}^{(a)}$ and $N_{eff}^{(b)}$ which balance Eq. (I-6). Once $N_{eff}^{(a)}$ and $N_{eff}^{(b)}$ have been determined, Eqs. (I-5a) and (I-5b) provide values for α_a and α_b, respectively. At this point we are in a position to compute the $A_{ij(a-b)}$ coefficients for the "6-12" potential used to describe the non-bonded interactions between ions and atoms involved in covalent bonds.

To calculate the $B_{ij(a-b)}$ for the ion–atom nonbonded interaction requires that we have R_0; see Eq. (I-2b). The simplest means of obtaining R_0 is to take the sum of the van der Waals radius of the atom and the ionic radius of the ion. However, there may be several serious errors in this method. Indeed, the major factor which controls the size of the ionic radius is the electrostatic interaction between ionic species. The small (at least compared to the ionic charge) residual charge on an atom involved in covalent bonds diminishes the contribution of the electrostatic term to the characteristic value of R_0. Therefore, a more reasonable means of determining R_0 is to consider an average value $\langle R_0 \rangle$ for each specific type of interaction $(a-b)$ determined from the crystal structures of complexes containing ions. The monomers and dimers of many peptides crystallized with various ions provides one source to compute $\langle R_0 \rangle$.†

In Table I-1 we report a set of $\langle R_0 \rangle$, A_{ij}, and B_{ij} for a number of ion–atom nonbonded interactions described by a Lennard-Jones "6-12" potential function. Remember that the B_{ij} were computed on the basis of minimizing both the nonbonded and the electrostatic energies at $\langle R_0 \rangle$. The characteristic residual charges assigned to the atoms in these calculations were $Q_C = +.060$ eu, $Q_H = +0.050$ eu, $Q_N = -0.200$ eu, and $Q_O = -0.300$ eu. When the residual charge on an atom in an actual conformational energy calculation differs from that assigned in these calculations, we must rely upon the electrostatic term to make the appropriate energy correction. The most common example of an atom having a significantly different residual charge from that used in these calculations is a hydrogen atom involved in a hydrogen bond; i.e., $Q_H \cong +0.250$ eu in this case. Obviously, these potential functions have been determined using many approximations and assumptions. Therefore, we can, at best, only be cautiously optimistic that such functions reasonably describe ion–atom interactions. Surprisingly, a comparison of the values of corresponding parameters in Tables 2-2, 2-4, and I-1 for F, Cl and F^-, Cl^- indicates that the Lennard-Jones "6-12" potential functions of the ions are rather similar to those of the atoms. The electrostatic term is primarily responsible for the different interaction properties of the atoms and the corresponding ions.

---

† See Lakshiminarayanan *et al.* (2) for a list of amino acid crystals.

TABLE I-1

Ion–Atom Interaction Parameters $\langle R_0 \rangle$, $A_{ij(a-b)}$, and $B_{ij(a-b)}$ for a
Lennard-Jones "6-12" Potential Function[a]

Interaction (a–b)	$\langle R_0 \rangle$ (Å)	$A_{ij(a-b)}$ (kcal-A^6/mole)	$B_{ij(a-b)} \times 10^{-4}$ (kcal-A^{12}/mole)
Na$^+$–H	2.35	18.9	0.1512
Na$^+$–C	2.75	95.1	2.210
Na$^+$–N	2.25	93.2	0.6831
Na$^+$–O	2.20	94.3	0.6145
K$^+$–H	2.65	73.8	1.180
K$^+$–C	3.05	362.5	15.68
K$^+$–N	2.60	365.5	5.031
K$^+$–O	2.50	357.7	4.580
Rb$^+$–H	2.75	116.9	2.518
Rb$^+$–C	3.20	567.8	37.35
Rb$^+$–N	2.70	557.4	13.65
Rb$^+$–O	2.70	561.8	12.10
F$^-$–H	2.45	63.5	0.5685
F$^-$–C	2.90	305.8	9.080
F$^-$–N	2.75	298.3	5.430
F$^-$–O	2.75	303.4	5.550
Cl$^-$–H	2.95	163.7	5.325
Cl$^-$–C	3.40	809.5	69.65
Cl$^-$–N	3.20	793.7	51.86
Cl$^-$–O	3.25	797.8	53.44
Br$^-$–H	3.15	209.8	13.35
Br$^-$–C	3.60	1040.0	112.5
Br$^-$–N	3.45	1026.5	79.55
Br$^-$–O	3.50	1029.9	87.31

[a] The carbon has an sp^3 orbital configuration, N is in an sp^2 configuration, and O is in an sp^2 state.

References

1. W. A. Hiltner, A. J. Hopfinger, and A. G. Walton, *J. Amer. Chem. Soc.* **94,** 4324 (1972).
2. A. V. Lakshiminarayanan, V. Sasisekharan, and G. N. Ramachandran, *in* "Conformation of Biopolymers" (G. N. Ramadrandran, ed.), p. 61. Academic Press, New York, 1967.

Appendix II | Detailed Balancing Approach to Equilibrium Properties of Linear Chains [†]

I. Introduction

In recent years there has been much theoretical interest in certain one-dimensional models for linear chain systems. Most of these models are variants of one form or another of the one-dimensional Ising model and they have been used to treat a wide range of systems such as ligand binding to a polymer chain and helix–coil transitions in proteins or DNA. The most common approach to treating these models has been through the standard partition function technique. This method is exemplified by the work of Zimm and Bragg (1) and an extensive review has been given by Poland and Scheraga (2). It is our purpose here to point out an alternative technique for handling these problems which has many advantageous features. The approach is to apply directly the principle of detailed balancing to the elementary reaction processes which describe the model. It turns out that this technique provides all the information one usually seeks, plus a number of other quantities not directly available from a partition function. This approach arose originally from a kinetic theory of the Ising model. In Section II we will treat the case of the infinite homogeneous chain in some detail, to illustrate the main features of this approach. Section III briefly deals with the finite homogeneous chain and the features peculiar to it. Finally Section IV is concerned with the extremely difficult, but important problem of disordered copolymer chains.

II. Infinite Homogeneous Chain

This section will be an account of the equilibrium properties of the general kinetic equations developed by Silberberg and Simha (3) and discussed further by Rabinowitz *et al.* (4). The basic model is an infinite chain consisting of individual sites, each of which can be in of two possible states labelled "0"

† The author gratefully acknowledges the contribution of this tailored manuscript by Robert H. Lacombe and Robert Simha of the Division of Macromolecular Science, Case Western Reserve University.

and "1" according to much used convention

$$\cdots 10111101010001011\cdots$$

Considering only nearest neighbor interactions, each site can undergo transitions from a "0" to a "1" state and vice versa according to the following reaction scheme:

$$\cdots 000 \cdots \underset{k_3'}{\overset{k_1}{\rightleftharpoons}} \cdots 010 \cdots$$

$$\cdots 001 \cdots \underset{k_2'}{\overset{k_2}{\rightleftharpoons}} \cdots 011 \cdots \qquad\qquad \text{(II-1)}$$

$$\cdots 101 \cdots \underset{k_1'}{\overset{k_3}{\rightleftharpoons}} \cdots 111 \cdots$$

One defines configurational probabilities to describe the state of the chain as follows:

Singlets	$n(0)$ = total fraction of sites in state "0"
	$n(1)$ = total fraction of sites in state "1"
Normalization	$n(0) + n(1) = 1$
Doublets	$n(00)$ = total fraction of adjacent sites in the configuration "00"
	$n(01)$ = total fraction of adjacent sites in the configuration "01"

Triplet and higher configurational probabilities are defined in a similar fashion.

At equilibrium Eqs. (II-1) and detailed balancing imply the following set of equations involving triplets:

$$k_1 n(000) = k_3' n(010), \qquad k_2 n(001) = k_2' n(011), \qquad k_3 n(101) = k_1' n(111)$$

$$\text{(II-2)}$$

At this stage one introduces a set of nearest neighbor conditional probabilities defined as

$$P_{00} = n(00)/n(0), \qquad P_{01} = n(01)/n(0)$$
$$P_{10} = n(10)/n(1), \qquad P_{11} = n(11)/n(1) \qquad\qquad \text{(II-3)}$$

with the normalization

$$P_{00} + P_{01} = 1, \qquad P_{10} + P_{11} = 1$$

The nearest neighbor property implies the relations

$$P_{00} = n(00)/n(0) = n(100)/n(10) = n(000)/n(00)$$
$$P_{10} = n(10)/n(1) = n(010)/n(01) = n(110)/n(11), \qquad \text{etc.} \qquad \text{(II-4)}$$

Using the conditional probabilities and the nearest neighbor property, Eqs. (II-2) can be written as

$$P_{00}^2 = K_0{}^2 P_{01} P_{10}, \qquad K_0{}^2 = k_3{}'/k_1 \qquad \text{(II-5a)}$$

$$P_{00} = K_2 P_{11}, \qquad K_2 = k_2{}'/k_2 \qquad \text{(II-5b)}$$

$$P_{10} P_{01} = K_1^{-2} P_{11}^2, \qquad K_1^{-2} = k_1{}'/k_3 \qquad \text{(II-5c)}$$

Equations (II-5a) and (II-5c) can be combined to give

$$P_{00} = (K_0/K_1)\, P_{11}$$

This result is consistent with Eq. (II-5b), provided that

$$K_2 = K_0/K_1 \qquad \text{(II-6)}$$

Equations (II-5) are thus not independent and we therefore delete Eq. (II-5b) to obtain

$$\begin{aligned} P_{00}^2 &= K_0{}^2 P_{01} P_{10} \\ P_{10} P_{01} &= K_1^{-2} P_{11}^2 \end{aligned} \qquad \text{(II-7)}$$

Equations (II-7) plus the normalization conditions give two equations in the unknowns P_{01} and P_{10}. The solution is

$$P_{01} = \frac{1}{1 + K_0 S}, \qquad P_{10} = \frac{1}{1 + (K_1/S)} \qquad \text{(II-8)}$$

with

$$S = (K_0 - K_1)/2 + [(K_0 - K_1)^2/4 + 1]^{1/2}$$

The singlet probabilities can be found directly by using the symmetry relation for doublets:

$$n(01) = n(10)$$

From Eqs. (II-3),

$$\frac{n(0)}{n(1)} = \frac{P_{10}}{P_{01}}, \qquad n(0) = \frac{P_{10}/P_{01}}{1 + (P_{10}/P_{01})}$$

and from Eq. (II-8),

$$n(0) = S^2/(1 + S^2) \qquad \text{(II-9)}$$

Equations (II-8) and (II-9) now contain the entire equilibrium solution, since any desired configurational probability from doublets upwards can be found by means of Eqs. (II-4). Weight and number average sequence lengths are also directly obtained, viz.:

$$\langle 0 \rangle = \sum_{j=1}^{\infty} jn(10, 1) = \sum_{j=1}^{\infty} jn(1) P_{10} P_{00}^{j-1} P_{01} = 1 + K_0 S$$

$$\langle 0 \rangle_w = \sum_{j=1}^{\infty} j^2 n(10, 1) = \sum_{j=1}^{\infty} j^2 n(1) P_{10} P_{00}^{j-1} P_{01} = 1 + 2K_0 S$$

One feature to note is that by this method one must first obtain all the conditional probabilities P_{ij} before calculating singlets or any of the higher order probabilities. This turns out to be a distinct advantage in situations where a knowledge of doublet and higher order variables is required. We further note that moments of the sequence distribution beyond the first are not readily available by means of partition function techniques.

III. Finite Homogeneous Chains

Two new but related features appear here, namely the effect of the chain ends and the dependence of all probabilities on position in the chain (5).
    Position dependent singlets are defined as follows:

$$n^i(0) = \text{probability that site } i \text{ is a "0"}$$

$$n^i(1) = \text{probability that site } i \text{ is a "1"}$$

For normalization

$$n^i(0) + n^i(1) = 1/N \qquad N = \text{total number of chain units}$$

Likewise for the conditional probabilities

$$P_{00}^i = \frac{n^i(00)}{n^i(0)} = \text{probability that given site } i \text{ is in state "0",}$$
$$\text{site } i + 1 \text{ is in state "0"}$$

Analogous definitions hold for P_{01}^i, P_{10}^i, and P_{11}^i.
    A set of separate reactions must now be introduced for the chain ends, viz:

$$00 \cdots \underset{k_5{}'}{\overset{k_4}{\rightleftarrows}} 10 \cdots, \qquad 01 \cdots \underset{k_4{}'}{\overset{k_5}{\rightleftarrows}} 11 \cdots \qquad \text{(II-10)}$$

With these modifications the counterparts to Eqs. (II-5) are

$$k_4 n^1(0) P_{00}^1 = k_5{}' n^1(1) P_{10}^1 \qquad \text{(II-11a)}$$

$$k_5 n^1(0) P_{01}^1 = k_4{}' n^1(1) P_{11}^1 \qquad \text{(II-11b)}$$

$$k_1 P_{00}^i P_{00}^{i+1} = k_3{}' P_{01}^i P_{10}^{i+1} \qquad \text{(II-11c)}$$

$$k_2 P_{00}^i P_{01}^{i+1} = k_2' P_{01}^i P_{11}^{i+1} \tag{II-11d}$$

$$k_2 P_{10}^i P_{00}^{i+1} = k_2' P_{11}^i P_{10}^{i+1} \tag{II-11e}$$

$$k_3 P_{10}^i P_{01}^{i+1} = k_1' P_{11}^i P_{11}^{i+1} \tag{II-11f}$$

$$k_4 P_{00}^{N-1} = k_5' P_{01}^{N-1} \tag{II-11g}$$

$$k_5 P_{10}^{N-1} = k_4' P_{11}^{N-1} \tag{II-11h}$$

Equations (II-11a), (II-11b), and (II-11g), (II-11h) arise from the left- and right-hand chain ends, respectively. In analogy to Eq. (II-6) we have the conditions

$$(k_2/k_2')^2 = (k_3/k_1')\,(k_1/k_3') \tag{II-12a}$$

$$(k_2/k_2')\,(k_1'/k_3) = (k_4/k_5')\,(k_4'/k_5) \tag{II-12b}$$

Equation (II-12a) is precisely the condition Eq. (II-6) obtained for the infinite chain. Equation (II-12b) arises from the presence of chain ends. Equations (II-11c) and (II-11d) can be combined to give a set of finite difference equations for the conditional probabilities, viz:

$$P_{00}^{i+1} = a/b - b^{-1}(P_{00}^i)^{-1}$$

$$P_{10}^i = \frac{(k_2'/k_2)\,P_{00}^i}{(k_3'/k_1) - (k_3'/k_1 - k_2'/k_2)\,(P_{00}^i)} \tag{II-13}$$

with

$$a = (1 + (k_2/k_2')), \qquad b = (k_2/k_2') - (k_1/k_3')$$

Equation (II-11g) provides the boundary condition at the right-hand chain end, namely

$$P_{00}^{N-1} = 1/(1 + (k_4/k_5')) \tag{II-14}$$

Equations (II-13)–(II-14) are readily solved for the complete set of conditional probabilities. The details have been given elsewhere (5) and the results are

$$P_{00}^i = \frac{\alpha_1^{N-i-1}(P_{00}^{N-1}\alpha_2 - 1) - \alpha_2^{N-i-1}(P_{00}^{N-1}\alpha_1 - 1)}{\alpha_1^{N-i}(P_{00}^{N-1}\alpha_2 - 1) - \alpha_2^{N-i}(P_{00}^{N-1}\alpha_1 - 1)}$$

$$P_{10}^i = (\alpha_1 - 1)\,(1 - \alpha_2) \tag{II-15}$$

$$\times \frac{\alpha_1^{N-i-1}(P_{00}^{N-1}\alpha_2 - 1) - \alpha_2^{N-i-1}(P_{00}^{N-1}\alpha_1 - 1)}{\alpha_1^{N-i}(\alpha_1 - 1)\,(P_{00}^{N-1}\alpha_2 - 1) - \alpha_2^{N-i}(\alpha_2 - 1)(P_{00}^{N-1}\alpha_1 - 1)}$$

with

$$\alpha_{1,2} = \tfrac{1}{2}\{a \pm (a^2 - 4b)^{1/2}\} \qquad 1 \leqslant i \leqslant N - 1$$

The singlets are calculated from the following conservation relation:

$$n^i(0) = n^{i-1}(00) + n^{i-1}(10)$$

or

$$n^i(0) = (P_{10}^{i-1}/N) + (P_{00}^{i-1} - P_{10}^{i-1}) n^{i-1}(0) \qquad 2 \leqslant i \leqslant N \qquad \text{(II-16)}$$

Equation (II-11a) provides a boundary condition at the left-hand chain end:

$$Nn^1(0) = \frac{k_5' P_{10}^1}{k_4 P_{00}^1 + k_5' P_{10}^1} \qquad \text{(II-17)}$$

Equations (II-15)–(II-17) can now be solved for the complete set of singlet probabilities. The result is

$$n^i(0) = \frac{(\alpha_1 - 1)(1 - \alpha_2) [\alpha_1^{i-1}(P_{00}^{N-1} \alpha_2 - 1) - \alpha_2^{i-1}(P_{00}^{N-1} \alpha_1 - 1)]}{N(\alpha_1 - \alpha_2) [\alpha_1^{N-1}(\alpha_1 - 1)(P_{00}^{N-1} \alpha_2 - 1)^2 + \alpha_2^{N-1}(1 - \alpha_2)(P_{00}^{N-1} \alpha_1 - 1)^2]}$$
$$\times [\alpha_1^{N-i}(P_{00}^{N-1} \alpha_2 - 1) - \alpha_2^{N-i}(P_{00}^{N-1} \alpha_1 - 1)] \qquad \text{(II-18)}$$

Finally the variable of most physical significance is the sum of all the $n^i(0)$:

$$n(0) = \sum_{i=1}^{N} n^i(0) = \frac{1 - \alpha_2}{\alpha_1 - \alpha_2} \frac{\alpha_1^{N-1} + K^2 \alpha_2^{N-1} - 2(K/N)(\alpha_1^N - \alpha_2^N)/(\alpha_1 - \alpha_2)}{\alpha_1^{N-1} + [(1 - \alpha_2)/(\alpha_1 - 1)] K^2 \alpha_2^{N-1}}$$
$$\text{(II-19)}$$

with

$$K = (P_{00}^{N-1} \alpha_1 - 1)/(P_{00}^{N-1} \alpha_2 - 1)$$

Equations (II-15), (II-18), and (II-19) now specify the complete solution for the finite chain up to singlet level. Higher order joint probabilities can be obtained in the usual fashion, i.e., for doublets,

$$n^i(00) = n^i(0) P_{00}^i$$

$$n(00) = \sum_{i=1}^{N-1} n^i(00)$$

Likewise triplet and higher configurational probabilities can be derived. Thus, finite homogeneous chains offer more computational work, but no essential difficulty when approached by the detailed balancing method.

IV. Copolymer Chains

The detailed balancing technique has recently been applied to binary copolymer chains (units A and B) of arbitrary chain length (6). The elementary reaction processes are generalized in the obvious way:

$$
\begin{array}{ccccccc}
\begin{array}{c}000\\ \cdots \text{XYZ}\cdots\end{array} & \xrightleftharpoons[k_1'(XYZ)]{k_1(XYZ)} & \begin{array}{c}010\\ \cdots \text{XYZ}\cdots\end{array}, & & \begin{array}{c}100\\ \cdots \text{XYZ}\cdots\end{array} & \xrightleftharpoons[k_3'(XYZ)]{k_3(XYZ)} & \begin{array}{c}110\\ \cdots \text{XYZ}\cdots\end{array}\\[2em]
\begin{array}{c}001\\ \cdots \text{XYZ}\cdots\end{array} & \xrightleftharpoons[k_2'(XYZ)]{k_2(XYZ)} & \begin{array}{c}011\\ \cdots \text{XYZ}\cdots\end{array}, & & \begin{array}{c}101\\ \cdots \text{XYZ}\cdots\end{array} & \xrightleftharpoons[k_4'(XYZ)]{k_4(XYZ)} & \begin{array}{c}111\\ \cdots \text{XYZ}\cdots\end{array}
\end{array}
$$

$$(\text{II-20})$$

X, Y, and Z may be either A or B, thus the sequence XYZ represents eight separate A,B configurations, and Eqs. (II-20) represent thirty-two separate reactions. All probabilities now depend not only on site position with respect to the chain end, but also on the particular arrangements of the A,B units in the chain.

Equations (II-13), (II-14), (II-16), and (II-17) now generalize to the following:

$$
P_{XY}^{i-1}(00) = \frac{K_1(XYZ)\, K_2(XYZ)}{K_1(XYZ)\,[1 + K_2(XYZ)] - [K_1(XYZ) - K_2(XYZ)]\, P_{YZ}^i(00)}
$$

$$(\text{II-21a})$$

$$
P_{YZ}^i(10) = \frac{K_2(XYZ)\, P_{YZ}^i(00)}{K_1(XYZ) - [K_1(XYZ) - K_2(XYZ)]\, P_{YZ}^i(00)}
$$

$$(\text{II-21b})$$

$$
n_Y^i(0) = P_{XY}^{i-1}(10)/N + [P_{XY}^{i-1}(00) - P_{XY}^{i-1}(10)]\, n_X^{i-1}(0)
$$

$$(\text{II-21c})$$

with the boundary conditions

$$
P_{XY}^{N-1}(00) = \frac{K_E(XY)}{K_E(XY)+1}
$$

$$(\text{II-21d})$$

$$
n_X^1(0) = \frac{1}{N} \frac{K_L(XY)\, P_{XY}^1(10)}{P_{XY}^1(00) + K_L(XY)\, P_{XY}^1(10)}
$$

$$(\text{II-21e})$$

and the definitions

$$
K_1(XYZ) = k_1'(XYZ)/k_1(XYZ), \qquad K_2(XYZ) = k_2'(XYZ)/k_2(XYZ)
$$

Equations (II-20) imply a set of thirty-two equilibrium constants. However, conditions similar to Eq. (II-6) apply and only sixteen are independent. Thus the final form of the equations involves only the quantities $K_1(XYZ)$ and $K_2(XYZ)$. The parameters $K_E(XY)$ and $K_L(XY)$ are equilibrium constants governing transitions at the right and left chain ends, respectively.

The $P_{XY}^i(00)$ are now the conditional probabilities that given site i is an X and in state "0," then site $i + 1$ is a Y and in state "0," and similarly for $P_{YZ}^i(10)$. The term $n_Y^i(0)$ is the singlet probability that site i is a Y and in state "0." Doublet and higher configurational probabilities are defined by the obvious generalization of these definitions. For a statistical copolymer, Eqs. (II-21) would be exceedingly difficult to deal with by analytical techniques, since the parameters $K_1(XYZ)$ and $K_2(XYZ)$ will vary in a complicated fashion. However, chains exhibiting some simple periodicity can be treated in an elementary fashion and accurate, efficient numerical techniques can be readily applied to the case of an arbitrary sequence distribution.

To further illustrate the calculational techniques, we consider a special case of the general model which illustrates many of the new features to be expected in copolymer chains. We thus write

$$K_1(XYZ) = \begin{cases} KK_A & \text{if Y is an A} \\ KK_B & \text{if Y is a B} \end{cases}, \qquad K_2(XYZ) = \begin{cases} K_A & \text{if Y is an A} \\ K_B & \text{if Y is a B} \end{cases}$$

$$(II\text{-}22)$$

Even using the simplified model of Eqs. (II-22), Eqs. (II-19) are still highly intractable to analytical solution for general A,B sequences. Consider, however, the simple example of a strictly alternating infinite chain. Equations (II-21) with Eqs. (II-22) now reduce to

$$P_{BA}(00) = \frac{KK_A}{K(1 + K_A) - (K - 1)\,P_{AB}(00)}$$

$$(II\text{-}23)$$

$$P_{AB}(00) = \frac{KK_B}{K(1 + K_B) - (K - 1)\,P_{BA}(00)}$$

$$n_A(0) = \tfrac{1}{2}P_{BA}(10) + [P_{BA}(00) - P_{BA}(10)]\,n_B(0)$$

$$n_B(0) = \tfrac{1}{2}P_{AB}(10) + [P_{AB}(00) - P_{AB}(10)]\,n_A(0)$$

The solution is

$$P_{AB}(00) = \{S_1 - [S_1{}^2 - 4K(K - 1)(1 + K_A)(1 + K_B)]^{1/2}\}/[2(1 + K_B)(K - 1)]$$

$$P_{AB}(10) = P_{AB}(00)/[K - (K - 1)\,P_{AB}(00)] \qquad (II\text{-}24)$$

$$n_A(0)/n_A(1) = [S_2 + R - 2(K_B + K)]/[2(K_B + K) - S_2 + R]$$

with

$$S_1 = K + K_A + K_B(2K + KK_A - 1), \qquad S_2 = K + K_A + K_B + KK_A K_B$$

$$R = [(S_2)^2 - 4K_A K_B(K - 1)^2]^{1/2}$$

The results for $P_{BA}(00)$, $P_{BA}(10)$, and $n_B(0)/n_B(1)$ are obtained by interchange of A and B everywhere in Eqs. (II-24). We thus have the complete solution for singlets, doublets, and higher-order runs for the infinite alternating copolymer chain. In general Eqs. (II-21) are readily treated for chains exhibiting a simple periodicity.

Another class of problems involves disordered chains where there is no discernible chain periodicity whatever. Simple solutions of Eqs. (II-21) no longer exist. However, recent work (6) has shown that these equations can be evaluated numerically by a Monte Carlo technique. For a long statistical copolymer, a suitable random number generator may be employed to generate a specific chain with the same probability of occurrence as the one in question. Thus the values of $K_1(\text{XYZ})$ and $K_2(\text{XYZ})$ for every chain site are now known. To proceed, Eqs. (II-21a), (II-21b), and (II-21d) are iterated to calculate all the conditional probabilities $P_{\text{XY}}^i(00)$ and $P_{\text{XY}}^i(10)$ which are then saved. Equations (II-21c) and (II-21e) can then be iterated to calculate all the singlets. However, one is usually interested in the quantities $n(0) = \sum_{i=1}^{N} n_{\text{X}}^i(0)$ and $n(00) = \sum_{i=1}^{N-1} n_{\text{X}}^i(0) P_{\text{XY}}^i(00)$, etc., where X and Y represent A or B, depending on location. These may be formed as the iteration proceeds, thus making it unnecessary to save any of the individual singlets.

The solutions $n(0)$, $n(00)$, etc., turn out for large chains to be very close to the ensemble average (for all chains having the same probability of occurrence), as obtained by Lehman (7) who solves numerically a functional equation for the logarithm of the partition function. For chains of 10,000 units, the standard deviation from the ensemble average ranges from 1 to 6% depending on the degree of cooperation in the chain. For larger chains, the deviations decrease roughly as $1/N^{1/2}$.

V. Summary

The detailed balancing technique has been shown to be a powerful and elegant method for dealing with the equilibrium of linear chain systems. The method is especially advantageous when information on doublet and higher-order configurations is desirable. As an example of this, recent studies (6, 8) have treated synthetic RNA chains using this model. These chains are known to exhibit a slight departure from randomness in their sequential structure. It was shown that only the doublet variables exhibited any significant change in going from a random to a slightly nonrandom copolymer. Thus any attempt to use the transition behavior of the chain to gain information about sequential chain structure will have to consider at least the doublet variables.

Another application lies in the helix–coil transition of short polypeptide chains where the sequence of chain units is known. Conformational analysis techniques could possibly be used to estimate the values of the equilibrium parameters $K_1(\text{XYZ})$ and $K_2(\text{XYZ})$ as a function of temperature. Equations (II-21) would then give full information on the nature of the helix–coil transition including the transition temperature, transition width for singlets, and weight and number averages of helical and coiled sections. Finally, even though we have considered only two-state chains, both homogeneous and copolymeric, it is clear that the theory can be generalized to include chains with three or more states for each site and heteropolymer compositions with three or more individual subunits.

We note finally that the equilibrium solutions for the singlets and doublets represent the boundary conditions for the kinetic equations of our model (4).

References

1. B. H. Zimm and J. K. Bragg, *J. Chem. Phys.* **31**, 526 (1959).
2. D. Poland and H. Scheraga, "Theory of Helix–Coil Transitions in Biopolymers." Academic Press, New York, 1970.
3. A. Silberberg and R. Simha, *Biopolymers* **6**, 479 (1968).
4. P. Rabinowitz, A. Silberberg, R. Simha, and E. Loftus, *in* "Stochastic Processes in Chemical Physics" (K. E. Schuler, ed.), p. 281. Wiley (Interscience), New York, 1969.
5. R. Simha and R. H. Lacombe, *J. Chem. Phys.* **55**, 2936 (1971).
6. R. H. Lacombe and R. Simha, *J. Chem. Phys.* **58**, 1043 (1973).
7. G. Lehman, *in* "Statistical Mechanics" (T. A. Bak, ed.). Benjamin, New York, 1967.
8. R. Simha and J. M. Zimmerman, *J. Polymer Sci.* **42**, 309 (1960); *J. Theoret. Biol.* **2**, 87 (1962).

Author Index

Numbers in parentheses are reference numbers and indicate that an author's work is referred to although his name is not cited in the text. Numbers in italics show the page on which the complete reference is listed.

Subject Index

A

Absorption band, 260–269
Absorption strength, 261
Acetic acid, 79, 185, 217, 242
N-acetyl-N'-methyl-alaninamide, 143, 147–148
N-acetyl-N'-methyl-glycylamide, 143
N-acetyl amino acid methyl esters, 291
N-acyl dialanyl methyl esters, 286
Alanyl residue, 79, 143–147, 287
Alkali halides, 277
Amorphous state, 18
Amphetamine, 139–142, 308
Angiotensin, 307–308
Atactic, 216, 219–221
Atomic orbital, 52, 95
Atomic polarizability, 45–46, 262, 279

B

Backbone of a polymer chain, 25
Barrier height to rotation, 119–122
Benzene crystals, 158
Benzyl alcohol, 185
Bond
    hydrogen, 241
    hydrophobic, 64
    imide, 183–187
    valence, 112, 115
        angle, 112
    virtual, 7–8
Bond angle, 2–4, 98
Bond length, 2–4, 98
Bond moments, 49–50, 54
Bond order, 296–297
Bohr magneton, 296
Bradykinin, 307–308

Branch atoms, 25
Brownian motion, 298
Buckingham potential function, 48, 157
n-Butanol, 185
Bystrov–Karplus relationship, 286

C

Calcium ions, 173
Cell theory, 309–310
Cellulose, 22
Chain folding, 15–18, 83
Chain-molecule fluid, 308–310
Charge
    partial, 151
    residual, 50–55
    π, 51–55
    σ, 51–53
Charge density, 262
Charge distribution, 53
Chemical shift, 281
Chloroethanol, 185
Chlorocruorin, 24
α-Chymotrypsin, 147, 238–257
Circular dichroism (CD), 260–274
    of collagen triple-helix models, 191–194
    dipeptide map, 177
    of the α-helix, 172–175
    of polypeptides, 78
Circularly polarized light, 260
Coefficient
    molecular extinction, 260
    of polarizability, 46, 56
    polymer–solvent exposure, 195–197
    rotational friction, 298
    second virial, 45
    two-dimensional rotational diffusion, 299–300

Molecular Biology

An International Series of Monographs and Textbooks

Editors

BERNARD HORECKER

Department of Molecular Biology
Albert Einstein College of Medicine
Yeshiva University
Bronx, New York

NATHAN O. KAPLAN

Department of Chemistry
University of California
At San Diego
La Jolla, California

JULIUS MARMUR

Department of Biochemistry
Albert Einstein College of Medicine
Yeshiva University
Bronx, New York

HAROLD A. SCHERAGA

Department of Chemistry
Cornell University
Ithaca, New York

WALTER W. WAINIO. The Mammalian Mitochondrial Respiratory Chain. 1970

LAWRENCE I. ROTHFIELD (Editor). Structure and Function of Biological Membranes. 1971

ALAN G. WALTON AND JOHN BLACKWELL. Biopolymers. 1973

WALTER LOVENBERG (Editor). Iron-Sulfur Proteins. Volume I, Biological Properties — 1973. Volume II, Molecular Properties — 1973

A. J. HOPFINGER. Conformational Properties of Macromolecules. 1973

R. D. B. FRASER AND T. P. MACRAE. Conformation in Fibrous Proteins. 1973

In preparation

OSAMU HAYAISHI (Editor). Molecular Mechanisms of Oxygen Activation